INNOVATIONS IN COMPETITIVE MANUFACTURING

INNOVATIONS IN COMPETITIVE MANUFACTURING

Edited by
PAUL M. SWAMIDASS

Professor of Operations Management
College of Business

And

Associate Director
Thomas Walter Center for Technology Management
Tiger Drive, Room 104
Auburn University, AL 36849-5358

KLUWER ACADEMIC PUBLISHERS
BOSTON / DORDRECHT / LONDON

Distributors for North, Central and South America:
Kluwer Academic Publishers
101 Philip Drive
Assinippi Park
Norwell, Massachusetts 02061 USA
Telephone (781) 871-6600
Fax (781) 871-6528
E-Mail <kluwer@wkap.com>

Distributors for all other countries:
Kluwer Academic Publishers Group
Distribution Centre
Post Office Box 322
3300 AH Dordrecht, THE NETHERLANDS
Telephone 31 78 6392 392
Fax 31 78 6546 474
E-Mail <orderdept@wkap.nl>

 Electronic Services <http://www.wkap.nl>

Library of Congress Cataloging-in-Publication

Innovations in competitive manufacturing/ edited by Paul M. Swamidass.
   p. cm.
  Includes bibliographical references and index.
  ISBN 0-7923-7896-2
  1. Production management. 2. Manufactures--Technological
innovations--Management. 3. Industrial management. 4. Competition.
I. Swamidass, Paul. M.

TS155 .I536 2000
658.5--dc21

                                    00-055105

Printed on acid-free paper.

Printed in the United States of America

CONTENTS

ALPHABETICAL LIST OF AUTHORS

Black, JT, *Auburn University, Alabama, USA*

Bloom, Matthew, *University of Notre Dame, Indiana, USA*

Burnham, John M., *Tennessee Technological University, Tennessee, USA*

Chinnaiah, Pratap, *Northeastern University, Massachusetts, USA*

Cooper, M. Bixby, *Michigan State University, Michigan, USA*

Darlow, Neil, *Cranfield University, UK*

De Meyer, Arnoud, *INSEAD, France*

Duray, Rebecca, *University of Colorado-Colorado Springs, Colorado, USA*

Edland, Timothy W., *Morgan State University, Maryland, USA*

Evans, James, *University of Cincinnati, Ohio, USA*

Fawcett, Stanley, *Brigham Young University, Utah, USA*

Goo, Gee-In, *Morgan State University, Maryland, USA*

Hadjinicola, George, *University of Cypress, Cypress*

Kamarthi, Sagar V., *Northeastern University, Massachusetts, USA*

Kamauff, Jr., John W., *University of Virginia, Virginia, USA*

Klassen, Robert, *University of Western Ontario, Ontario, Canada*

Kumar, Ravi, *University of Southern California, California, USA*

LaForge, R. Lawrence, *Clemson University, South Carolina, USA*

Lambert, David, *Tennessee Technological University, Tennessee, USA*

Landel, Robert D., *University of Virginia, Virginia, USA*

Leavy, Brian, *Dublin City University, Ireland*

Loch, Christoph, *INSEAD, France*

McCreery, John, *North Carolina State University—Raleigh, North Carolina, USA*

McKone, Kathleen, *University of Minnesota, Minnesota, USA*

Natarajan, R., *Tennessee Technological University, Tennessee, USA*

Parker, Rodney P., *University of Michigan, Michigan, USA*

Raju, P.K., *Auburn University, Alabama, USA*

Ramanan, Ramachandran, *University of Notre Dame, Indiana, USA*

Richards, Larry, *University of Virginia, Virginia, USA*

Sankar, Chetan S., *Auburn University, Alabama, USA*

Smith, Jr., Charles W., *Tennessee Technological University, Tennessee, USA*

Smunt, Timothy, *Wake Forest University, North Carolina, USA*

Sridharan, V., *Clemson University, South Carolina, USA*

Stecke, Kathryn, *University of Michigan, Michigan, USA*

Swamidass, Paul M., *Auburn University, Alabama, USA*

Swink, Morgan, *Michigan State University, Michigan, USA*

Terwiesch, Christian, *University of Pennsylvania, Pennsylvania, USA*

Umble, Elisabeth, *Texas A&M University, Texas, USA*

Weiss, Elliot, *University of Virginia, Virginia, USA*

Welch, John A., *Tennessee Technological University, Tennessee, USA*

White, Gregory, *Southern Illinois University-Carbondale, Illinois, USA*

Wilson, Dale A., *Tennessee Technological University, Tennessee, USA*

PREFACE

In the last fifty years, U.S. manufacturing has gone through a cycle: from strength in the fifties and sixties, to weakness in the seventies and eighties, and back again to strength in the nineties. There has never been a time in history when manufacturing management saw so many advances in a short period of two decades. The roaring competition for markets has forced U.S. manufacturers to become more competitive than ever.

During the fifties and sixties, manufacturing was treated as a technical detail within U.S. manufacturing firms while technical personnel without a strategic or competitive perspective managed the function. In the postwar economy, it worked very well because competition for U.S. products was negligible. The resurgence of European and Japanese manufacturing by the end of sixties posed unprecedented competition for U.S. manufacturers. To make matters worse, in the early seventies, under the belief that the U.S. was a postindustrial society, investment in manufacturing began to be neglected.

Competitive manufacturing in the U.S. was made possible by the progress made in a number of areas. For example, progress in competitive manufacturing is attributable to advances in the strategic use of manufacturing, cellular manufacturing, lean manufacturing, focused manufacturing, flexible automation, total quality management, supply chain management, design for manufacturing, mass customization, improved costing, and so on.

With the arrival of the first wave of high quality, competitive Japanese products in the U.S. in the seventies, the lack of strategic thinking in U.S. manufacturing became more and more evident. It was followed by a period, when U.S. products progressively lost their market shares to more competitive Japanese products.

Pressured by competition, U.S. manufacturers began the journey to competitive manufacturing in the late seventies; their efforts have brought revolutionary changes in U.S. manufacturing. The progress made by U.S. manufacturers in the area of competitive manufacturing has transformed them into a very competitive bunch and the envy of the world. A survey[1] of over 1,000 U.S. manufacturers in the discrete products manufacturing industries (SIC 34-38) reveals that, between 1993 and 1997, U.S. manufacturers improved their average return on investment from about 13 percent to 16.85 percent. The average inventory turns improved from 6 to 9.7 between 1990 and 1997. During 1990 and 1997, the marked increase in the use of hard technologies such as local area networks (LAN) and soft technologies such as manufacturing cells was noticeable.

Manufacturing managers whom I frequently meet in the U.S. are under constant pressure to think and act more competitively than ever before. For these managers and executives, who have either direct or indirect responsibility for competitive manufacturing, this book should serve as a good overview of the progress in manufacturing thinking and practices in the last two decades. This book is different from the books in the market devoted to single topics such as total quality management, just-in-time manufacturing, activity-based costing, concurrent engineering, and so on. This book covers all topics in this genre in 34 different articles written by various experts. Thus, this book should be attractive to manufacturing as well as non-manufacturing managers who would like to read a single book to get an overview of the progress in manufacturing over the last two decades, and to understand what it means to their own company and their function.

Almost any introductory text on operations management would touch all the topics covered in this book. However, introductory texts cannot do full justice to every topic. In this book, each article addresses one topic in-depth. The treatment of the topic is such that the articles are interesting and easy to read while presenting essential material. The more serious reader is provided with a list of references at the end of each article. This book is the product of over 40 experts. Competitive manufacturing covers so many topics, it is impossible for one writer to cover all aspects of it in sufficient depth. A single writer cannot duplicate the collective expertise of over 40 writers.

While practicing managers and manufacturing professionals would find this book relevant, thousands of MBA and EMBA students, who take courses in operations management and manufacturing strategy, should find this book to be a valuable supplement to their textbooks. The references at the end of each chapter should be of value to students enrolled in MBA and EMBA programs.

The book is arranged in 13 different chapters, each covering a major subject within manufacturing management (see the Table of Contents). Each chapter consists of

---

[1]Swamidass, Paul M. *Technology on the Factory Floor III: Technology Use and Training in the United States*. Washington, D.C.: The Manufacturing Institute of the National Association of Manufacturers, 1998; and Swamidass, Paul M. *Technology on the Factory Floor*. Washington, D.C.: The National Association of Manufacturers, 1992.

one or more articles. Each article is led by an abstract, which calls attention to the manufacturing and other organizations discussed or mentioned in the article. The many examples from industry should enable the reader to appreciate and comprehend the concepts presented in the article. The articles address topics that are proven to be useful to the industry. While the tone of the articles is informational, they also discuss implementation issues concerning new ideas and manufacturing practices.

Four in-depth descriptions of real-life cases provide illustration of key principles. In-depth illustrative cases deal with (1) the implementation of cell manufacturing at Duriron; (2) Integrated Product Development at Westinghouse; and (3) the development of an international supply partner by Black & Decker, and (4) predictive maintenance at Della steam plant.

The order in which the chapters are arranged and the order of articles within a chapter are not based on any particular logic. Therefore, the reader may begin this book anywhere after reading the introductory article "Innovations in Competitive Manufacturing: From JIT to E-Business."

I enjoyed working with several dozen experts contributing to this book. As always, it has been a pleasure to work with Gary Folven, the Kluwer editor and various professional staff at Kluwer.

Paul M. Swamidass
Auburn, AL, USA
May 2000

I. INTRODUCTION

1. INNOVATIONS IN COMPETITIVE MANUFACTURING: FROM JIT TO E-BUSINESS[1]

PAUL M. SWAMIDASS

Auburn University, Auburn, Alabama, USA

ABSTRACT

This chapter provides an overview of the book. In the seventies, there was a tendency to dismiss the value of manufacturing to the US economy. US manufacturing suffered from a lack of new investment and new ideas to propel it to new heights. Consequently, US manufacturing firms suffered competitively. However, around 1980, once manufacturers realized the magnitude of their uncompetitiveness, they made an historic turnaround in less than twenty years that has many lessons for the future. There are lessons for those who are working hard to catch up with the leaders in their industry. Towards the end of the chapter, the trend towards *e-business* is addressed. This new trend will enhance and strengthen the gains of the last twenty years. The remaining chapters of this book address in detail the various topics covered in this introductory chapter.

UNPRECEDENTED PRODUCTIVITY IMPROVEMENT

There is a revolution going on in manufacturing productivity in the USA. While the entire non-farm sector productivity rose by 2.8 percent and 3.0 in 1998 and 1999 respectively, the manufacturing sector productivity growth was even more remarkable at 4.8 percent and 6.0 percent during the same period (Reuters, March 7, 2000). In the

---

[1] Adapted from the address to the International Seminar on Manufacturing Technology Beyond 2000, Bangalore, India, November 18, 1999.

fifties and sixties, annual productivity growth was over 2 percent. The cost of oil and inflation are blamed for the low one percent annual productivity growth in the seventies.

Unprecedented productivity growth in the nineties is causing us to rethink fundamental assumptions about the US economy. For decades it was considered that annual GDP growth rates in excess of 2.5 percent or unemployment below 6 percent are inflationary. In the nineties, unemployment level has been running around a 30-year low of 4.1 percent, and GDP growth rate has been steadily in excess of 3 percent; manufacturing GDP between 1992 and 1997 grew at a remarkable 5 percent annual rate. The US economy has shown that with high productivity growth, the economy is able to sustain higher economic growth rate and low unemployment without inflation.

Federal Reserve Bank (US) Chairman Alan Greenspan, in explaining the impressive growth in US economy in the nineties remarked that "A key factor behind this impressive performance has been the remarkable acceleration in labor productivity" (Reuters, March 22, 2000). Further, speaking to the National Technology Forum on April 7, 2000 (Reuters, april 7, 2000) he is quoted as saying that "rising productivity was helping curb inflation in the U.S. economy and there is a good chance that technology can keep productivity rates growing." Low inflation and productivity growth create the envirable "virtuous cycle" in the economy.

While U.S. manufacturers are still the most productive (output per employee) in the world, they have also become more effective in terms of customer satisfaction, quality, delivery and new product introduction.

REGAINING MANUFACTURING COMPETITIVENESS: THE CASE OF USA

In the seventies, the uncompetitiveness of US manufacturing was exposed. The twenty years from 1980 to 1999 represent a unique turnaround in the history of manufacturing in the United States. During this period, there was an escalation of competitive pressures on US manufacturers. The competitive pressures manifested in several ways:

1. Higher US wage rates relative to many new Asian exporters.
2. The free access of US domestic markets to exporters from other countries, not vice versa.
3. Rapidly changing product technology in computer and electronics related products required frequent new product introduction; some requiring new product introduction every 9 to 12 months.
4. The rising expectations of customers concerning product quality.
5. The rising expectations of customers concerning customer service.

MANUFACTURING "TECHNOLOGIES" TO THE RESCUE

In response to the pressures, manufacturers in the U.S. turned to extensive use of inventions and/or imitations of exceptionally sound manufacturing technologies. Manufacturing technologies are classified as hard and soft technologies. Hard technologies are hardware and software intensive. The examples are CAD, CAM, CNC machines, CIM, FMS, automated inspection, robots, LAN, WAN, and so on.

Soft technologies are manufacturing and production know-how, techniques and procedures. Examples are: bar codes, concurrent engineering, JIT manufacturing, manufacturing cells, MRP, SQC, simulation and modeling, TQM, TPM, and so on. Soft technologies are not necessarily hardware/software dependent.

Many of these newer manufacturing technologies are now an integral part of manufacturing in the United States. As a direct result of the use of these hard and soft technologies, U.S. manufacturers report many improvements.

In a study sponsored by the National Association of Manufacturers (USA) and the National Science Foundation (USA), I investigated the use of manufacturing technologies in the U.S. Out of 1025 manufacturing plants participating in the study, more than two thirds of the respondents reported that the benefits from manufacturing technology use included: decrease in cycle time, decrease in manufacturing costs, increase in product lines, and increase in return on investment (ROI). They reported that the increase in product line during 1994–1997 was 25 percent, increase in ROI was 22 percent, increase in market share was 21 percent, decrease in manufacturing costs were 11 percent and the decrease in cycle time was 16 percent (Swamidass, 1998).

Over the last twenty years, through the use of various manufacturing technologies, U.S. manufacturers have:

1. Become more profitable
2. Become flexible
3. Reduced production batch sizes without a cost penalty
4. Become cost effective
5. The capability to bring new products to markets quicker
6. Reduced manufacturing lead time
7. Reduced inventories
8. Reduced the number of suppliers
9. Increased supply reliability
10. Increased the quality of the product, distribution system, and customer service
11. Improved the design process to make manufacturable designs
12. Become skilled in the use of teams for problem solving, design and development, and lead-time reduction.

GROWTH IN THE OUTPUT AND EXPORT OF MANUFACTURED GOODS

US manufactured exports have grown steadily since 1987; from $200 billion to $597 billion in 1998. While the gross domestic product (GDP) increased 300 percent between 1980 and 1987, manufactured exports increased 368 percent. U.S. exports of manufactured goods in 1991 were 12 percent of total world exports, which was 0.1 percent better than Japan. U.S. exports, as a percent of world exports, were better than Japan by the same margin in 1995. German exports as a percent of world exports dropped from 14.2 percent to 12.2 during the same period. Export performance of U.S. manufacturers is one indication of their competitiveness in the global market.

COMPETING IN THE FACE OF GLOBAL MANUFACTURING OVER-CAPACITY

The progress in manufacturing in the United States came at a time when developing nations, particularly in the Pacific Basin, added manufacturing capacity rapidly during the last two decades. China, Taiwan, Hong Kong, S. Korea, Malaysia, Thailand and Singapore, in addition to Japan expanded their manufacturing capacities substantially since the 1970s. The resulting over-capacity created a strong competition for markets among manufacturers. Only the best and most efficient have survived.

In the era of expanding global manufacturing capacity, U.S. manufacturers with relatively high labor wage rates have retained their competitiveness because of the improvements in manufacturing listed above.

STRATEGIC THINKING

In the last twenty years, strategic thinking has overtaken single-minded cost reduction and cost minimization in manufacturing. Consequently, the pursuit of cost, quality, flexibility, dependability and timeliness has replaced the single-minded cost reduction in manufacturing firms, which was the norm in manufacturing until the sixties and seventies. Now, manufacturers find competitive advantage through better design, improved customer satisfaction, quick response, faster new product introduction and other goals overshadowed in the past by the sole pursuit of cost reduction.

Strategic thinking in manufacturing filters down to the lower level in manufacturing when the corporate strategic planning process accepts manufacturing to be an integral part of strategic decisions that affect the competitiveness of the plant. This requires that the strategic business units (SBU) develop and execute an appropriate manufacturing strategy that is consistent with the overall strategy of the business. Once a manufacturing strategy is developed, it enables systematic and timely investments in manufacturing equipment and people to maintain the vitality of manufacturing and keep it competitive.

LEAN PRODUCTION

Lean production, in simple terms, refers to producing more with less. This popular term was introduced by Womack, Jones and Roos (1990) in their best-selling book *The Machine That Changed the World*. Lean production is the result of the use of several newer manufacturing developments that were adopted by US manufacturers in the last twenty years. The underlying principle is waste elimination in the use of all resources including people and time. While the foundational principles were imported from Japanese manufacturers, they have been Americanized, and smoothly adapted and integrated into US production.

DESIGN REDEFINED

Earlier, product design in the United States was dominated by design engineers' perspective of a product without sufficient input from the manufacturing function, suppliers and customers. Today, concurrent engineering, cross-functional teams, design for manufacturability, and customer inputs in product design are overshadowing the traditional approach to uncompetitive product design.

The increase in manufacturing capacity worldwide has spawned several new competitors for almost every product. One approach to attracting customers in this market is to increase the extent of customization in products, even in mass-produced products. An important development in this area called *mass customization* increases customization of mass-produced products. Mass customization begins with design and is executed by manufacturing.

SUPPLY CHAIN MANAGEMENT REPLACES MUNDANE PURCHASING

The role of suppliers has been greatly increased over the last twenty years. In the earlier part of this century, the tendency of manufacturers was to integrate vertically on the supply side, which ensured captive supply of numerous materials and components needed for assembly. General Motors and Ford Motor Corporation were good illustration of this phenomenon till the seventies. Since the eighties, the trend is to reduce the dependence on in-house supply of most components including technology intensive components. Only the components that represent the core competency of the company are not sourced from outside.

The increased use of outsourcing has escalated the dependence on suppliers. The result is to concentrate on a few good suppliers and integrate them more tightly with the manufacturing operation. The resulting supply chain provides competitive advantage.

HIGHER THRESHOLD FOR QUALITY

The quality of manufactured products across-the-board has improved over the last 20 years. In many industries, the minimum threshold for quality of the product is very high. Those manufacturers, who cannot meet or exceed this minimum threshold, cannot survive. The automobile and personal computer industries are very good examples of this phenomenon. Statistical quality control (SQC), JIT manufacturing, and total quality management (TQM) principles have been adopted by manufacturers from a variety of industries to produce consistently higher quality products.

Improved and consistent quality is made possible by: waste reduction, mistake-proofing, doing things right the first time, improved control of manufacturing processes instead of inspection of the output, reduction in rework and rejection, improved and frequent operator training, and immediate, collective quality problem solving in quality circles or similar teams.

IMPROVED COSTING AND PERFORMANCE MEASURES

Costing is the process of assigning costs to manufactured products. This process has remained unchanged for several decades, while manufacturing processes underwent dramatic changes. For example, when traditional unit-based costing systems were instituted at the turn of the century, labor input was a major component of the product cost.

A weakness of the approach is the use of direct labor cost to compute overhead costs. Today, with increased automation and labor productivity, in some automated plants, labor costs could be as low as five percent of the manufactured product cost. When direct labor costs are very low, the overhead cost reaches several hundred percent of direct labor cost. Consequently, a small change in direct labor cost could alter the total cost very substantially. The negative impacts of this are many. They include the reluctance of the management to invest in equipment and process technology because additional investment would increase the overhead burden rate.

For competitive manufacturing, the investment in equipment and process should be determined by the strategic and competitive priorities and realities. Therefore, traditional costing approaches have been criticized and other techniques such as activity-based costing (ABC) have received attention in recent years for internal control and management purposes.

Product costs based on ABC more accurately reflect the true cost of the product. Decisions based on more accurate costs provide better information for pricing the products, and do not hinder timely investment in automation and equipment.

Additionally, target costing has gained ground as the need for competitive pricing has increased. A product's price is determined using a study of the market and competitors before it is designed and marketed. Multiple iterations of the design are then undertaken until a viable product could be designed within the targeted price. A well-known example is that of Mercedez-Benz (MB). Until the early nineties, MB cars were built and then priced to cover all the incurred costs and desired profit. After one hundred years of this practice, due to increased competition in the luxury car markets and falling market share, target costing was instituted in this company in the early nineties. With in a few years, MB introduced several new products designed under the target costing principles and recovered.

The M-Class vehicles (sport utility vehicle) designed and produced by MB using target costing has been very successful and profitable.

NEW RESPECT FOR THE CUSTOMER

The customer was an after thought in traditional manufacturing. Japanese manufacturing principles embedded in JIT manufacturing and TQM have elevated the role of customer satisfaction and customer input into the design and manufacturing of products. Further, customer satisfaction earned through dependable, reliable products and services is now considered an important competitive advantage. The most notable aspect of customer satisfaction is that it cannot be easily and quickly replicated by competition. Manufacturers, who earn a reputation for sustained customer satisfaction, will be hard to dislodge from their markets.

COMPETITION IS NO LONGER LOCAL; GO GLOBAL

Increase in global manufacturing and an increased sensitivity to global markets defines the last two decades. The removal and reduction of restrictions to trade and investments have enhanced the global reach of manufacturers across national borders. This was particularly evident in Asia and the Pacific Rim countries.

Global reach of manufacturers makes domestic markets vulnerable to international competition. Manufacturers have responded with globalization of product design, international alliances, and international operations to cope with increased globalization of their markets. The net result is that product prices have remained depressed and manufacturers are forced to be more efficient to remain competitive. Inflation in the prices of manufactured goods has remained well under control in many markets over most of the world. This did not permit manufacturers to raise prices for years and kept the pressure on manufacturers to become more productive and efficient.

ENVIRONMENTAL ISSUES

Environmentally conscious manufacturing is changing the way design, search for raw materials and manufacturing are conducted. Further the rise of remanufacturing and reuse of durable goods pose new opportunities and threats to manufacturers.

Manufacturers such as GM and Ford are adopting environmental management systems (EMS) to improve their environmental record. Companies seek EMS to be seen as "green companies" by discriminating customers and to achieve cost efficiencies throughout the supply chain. Of the many EMS standards, ISO 14001 is the most widely accepted; it is similar to the more-widely known ISO 9000 standards for quality management.

ENTERPRISE RESOURCE PLANNING

Enterprise resource planning (ERP) is a relatively new development. It is essentially an integrated information and decision support systems accomplished by a data processing and information system, which enable the integration of manufacturing operations and other functions of the firm with suppliers and customers. ERP softwares permit a range of activities including advanced planning, paperless purchasing, MRP, lot and serial tracking, engineering change order management, Internet transactions, service management, finite capacity scheduling, sales management, the use of multiple currencies, subcontract processing, costing, accounting and financial management.

The promise of ERP is cost efficiency and improved effectiveness in complex organizations, which could be small or have multiple facilities spread across many nations. ERP is based on improved data handling, data warehousing, data analyses, and decision making for improved coordination of manufacturing as well as the entire business. Improved integration of activities, transactions, and information through ERP generates copious information for superior and faster planning and execution.

ERP softwares are identified with several global leaders such as SAP, BAAN, PeopleSoft, and so on. ERP implementation can be costly and time consuming. It is common to implement ERP in discrete functional modules to increase the success of ERP implementation. The successful use of ERP can increase efficiency, flexibility, and the planning capability of manufacturers.

Given the magnitude of the effort needed to deploy a full ERP, a new trend is evident in procuring ERP. Application service providers (ASPs) offer ERP outsourcing, where the ASP owns the software and keeps and maintains it at the site of the customer. The customers pay to use the applications and the services of the ASP. This lightens the load on the information technology department of the company.

AGILE-, QUICK-RESPONSE- AND VIRTUAL MANUFACTURING

Several new initiatives in manufacturing promise to make manufacturing more flexible and responsive. Agile manufacturing, quick response manufacturing (Suri, 1998), and virtual manufacturing initiatives in the US are aimed at supply chain efficiency and quick response to changing markets and new opportunities. Virtual manufacturing refers to new manufacturing entities created through very rapid integration of scattered resources in one or several firms. The rise of information technologies and the Internet fuels the growth of virtual manufacturing.

Business process reengineering (BPR), which gained momentum in the nineties is a major force in reducing wasted effort and resources in established manufacturing firms. BPR can revitalize established firms, remove rigidity in the manufacturing system and make it more and more flexible and responsive. Manufacturers increase the agility and the quickness of system response using BPR.

INTEGRATION: THE ROLE OF IT AND INTERNET

Integration will be a major theme in manufacturing in the years ahead. Integration reduces overhead costs, manufacturing lead times, increases flexibility to change and makes it possible to introduce new products more frequently.

The use of information technology (IT) is a proven way of increasing integration, which results in effectiveness and efficiency of manufacturing firms. Integration technologies are intranet, extranet, LAN, WAN, EDI, and bar codes. These technologies help integrate within and across the supply chain, design functions, distribution facilities, customers, subcontractors, factories, and manufacturing equipment. Manufacturers everywhere are using more and more integration technologies to become competitive.

The use of the Internet for the above purposes in a manufacturing firm reduces overhead costs. It also contributes to the integration of customers or potential customers, supply chain members, subcontractors, distributors, factories and many other entities. Marketing and administrative costs can be reduced substantially through integration. It makes manufacturers more competitive.

A STRONG NEW TREND: E-BUSINESS

The latter half of the nineties witnessed an Internet-based revolution take off. The Internet has spawned a host of new businesses and has enhanced the capabilities of established manufacturers.

With the help of Internet, e-business or e-commerce is permanently changing manufacturing firms. The role of e-business in manufacturing is positioned to grow at a very fast and unknown rate. So far, the effects of e-commerce on manufacturing have been very beneficial.

E-business enhances a manufacturer's competencies in (1) the accuracy of delivery estimates; (2) overnight order fulfillment; and (3) in providing customers with real-time, self-service information.

The meaning of e-business may vary from individual to individual. But, in general, it includes the following: (1) tight integration of various information systems of the firm, its customers and vendors; (2) order entry, (3) purchasing; and (4) extended or expanded customer service system.

How does one evaluate e-business? Effective evaluation of e-business is possible when it is included in strategic business goals based on how it can contribute to competitive advantage in the marketplace.

The Internet-based e-business and its use in manufacturing are relatively new phenomena. The use of Internet for increasing manufacturing efficiency and effectiveness is growing at an explosive rate. As the Internet matures, it can be used for progressively more and more complex purposes. A list of tasks for which Internet is used by manufacturing firms in ascending order of complexity is below:

• Dissemination of company information.
• Dissemination of product information.
• Dissemination of product specifications.
• Internal company or plant communication.
• To process customer request for samples and/or literature.
• Automated sales and order entries through the Internet.
• Bidding and receiving contracts over the Internet.
• Posting of order status on the Internet for coordination purposes.
• Product design and development.
• Training of employees and customers.

The success of e-business depends on order-fulfillment or e-fulfillment. E-fulfillment depends on "brick-and-mortar" functions such as inventory checking, order tracking, scheduling, etc.

A growing phenomenon is the rise of independent trading exchanges (ITEs), these are business-to-business (B2B) trading hubs. According to one estimate, there were 150 such hubs in early 1999 but there are more 1000 such sites in operation by early 2000. B2B businesses come in three types: horizontal, vertical and hub-based.

Horizontal websites deal mainly in low-value, non-specialized goods such as staplers; these sites host many buyers and sellers for the purchase of items that require very little industry expertise. Companies are exposed to vendors and customers they have not been aware of.

Vertical sites host many buyers and sellers but offer a high level of expertise in focused industries. Vertical B2B trade communities are industry-specific web sites which provide comprehensive sources of information, and enable interaction among members for buying and selling of goods over the Internet. The communities combine product information, directories, request for proposals, classifieds, industry news, electronic commerce, discussion forums, job opportunities, etc.

In hub-based sites, existing business relationship is taken online. The automotive and aerospace industries have exploited this approach where smaller suppliers can be integrated with larger trading partners.

The growth of e-business has created a legitimate debate comparing the use of electronic data interchange (EDI), which as been around for a longer time, and Internet-based e-business. The former uses value-added networks (VANs), which are expensive, complex and inflexible. E-business based on Internet technology reaches a wider community of users of all kinds than users of VANs. Further E-business over the Internet supports real-time communication rather than batch processing used with EDI. However, EDI transactions are suitable for large volume of transactions, and much more secure than the Internet.

Given the growing impact of e-business on manufacturers and their competitiveness, it is now imperative that manufacturers incorporate e-business in their strategic planning and top management must assess the impact of e-business, e-applications, and Internet commerce on the overall business.

THE FUTURE

For manufacturers, the future will be even more competitive than the last two decades. GM, Ford, and DaimlerChrysler AG "are creating jointly owned, Internet-based trading exchange that will be open to all car makers and their suppliers. . . . Fierce competition on the exchange will drive down prices of components" (McClenahen, 2000). Car companies no longer have to periodically browbeat supplier to reduce their price, the Internet-based exchange will do it.

Strategic flexibility of the type, where a manufacturer is able to change its configuration quickly, will be more common in manufacturing firms. For example, a new network hardware design and manufacturing company was started in San Diego, California in 1994. It designed proprietary network products and manufactured them in its own facilities. By the year 2000, the company decided to phase out all manufacturing in 12 months time and transfer all manufacturing to a contract electronic manufacturer/assembler (CEM/A). The company will continue to strengthen its design and engineering capabilities in the future. While it will take orders from customers, orders will be transmitted electronically to the CEM/A firm, who will fulfill the orders and ship them directly to customers. In this case, in about seven years, a manufacturing firm becomes a virtual manufacturing firm; it has the flexibility to change its structural configuration and strengthen its core competence while exploiting the core manufacturing competence of the largest CEM/A in its industry.

ABOUT THE BOOK

The following chapters of the book address major topics covered in this chapter.

REFERENCES

Aversa, J. (2000). "Productivity surges in '99." *ABC News.com: Business*, February 8.
McClenahen, J.S. (2000). "Connecting with the future." *Industry Week*, April 17.
Reuters (2000). "Greenspan: Economy stronger than expected." *Yahoo! News: Business*, March 22.

Reuters (2000). "Greenspan: Rising output curbs inflation." *Yahoo! News: Business*, April 7.
Reuters (2000). "Productivity soars in 4th quarter of 1999." *Yahoo! News: Business*, March 7.
Suri, R. (1998). *Quick Response Manufacturing*. Portland, OR: Productivity Press.
Swamidass, P.M. (1998). *Technology on the Factory Floor III: Technology Use and Training in U.S. Manufacturing Firms*. Washington, D.C.: Manufacturing Institute of the National Association of Manufacturers.
Womack, J.P. Jones, D.T., and Roos, D. (1990). *The Machine That Changed the World*. New York: Harper Perennial.

II. COMPETITIVE POSTURE

2. MANUFACTURING STRATEGY

PAUL M. SWAMIDASS

Auburn University, Auburn, AL, USA

NEIL R. DARLOW

Cranfield University, Cranfield, U.K.

ABSTRACT

Strategy defines the long-term direction of an organization in which its resources are matched to the economic climate and the expectations of clients and business owners. Manufacturing strategy is defined as "a sequence of decisions that, over time, enables a business unit to achieve a desired manufacturing structure, infrastructure, and set of specific capabilities" (Hayes and Wheelwright, 1984). Moreover, the pattern of decisions actually made constitutes realized manufacturing strategy, which may differ from the intended manufacturing strategy. The terms 'structure', 'infrastructure', and 'capabilities' will be subsequently explained. The primary aim of a manufacturing strategy is to ensure that all long-term manufacturing developments are congruent with the overall business strategy or competitive priorities of the firm. Read here about manufacturing strategy development in two different companies.

BACKGROUND

Wickham Skinner introduced the concept of manufacturing strategy and its implementation in the 1960s. The concept derives from corporate strategy literature of the early 1960s. From its ancient military origins, the word 'strategy' came to be defined and conceptualized in the 1960s in business literature, which often treated strategy in terms of long-term planning. The domain of strategy includes the following components:

Mission: Overriding premise in line with the values or expectations of
stakeholders.
Goal: General statement of aim or purpose.
Strategies: Broad categories or types of action to achieve objectives.
Tasks: Individual steps to implement strategies.

There are three levels of strategy of interest to a manufacturing company. At each level, the strategy contributes to the strategy immediately above it. The key questions that define strategies in descending order of scope are:

Corporate strategy: What set of businesses should we be in?
Business strategy: How should we compete in a given business?
Manufacturing strategy: How can the manufacturing function contribute to the
competitive advantage of the business?

Much of the literature on manufacturing strategy assumes the hierarchy shown above. Hayes and Wheelwright (1984) classify the role played by manufacturing strategy in firms into four categories or stages: internally neutral (Stage 1), externally neutral (Stage 2), internally supportive (Stage 3) and externally supportive (Stage 4). In Stage 1, manufacturing strategy seeks to minimize its negative contribution to the business, whereas in Stage 2, capital investments assist manufacturing to catch up with the industry norm. In Stage 3, manufacturing strategy and business strategy are made consistent by making investments in manufacturing in the light of business goals, and longer-term manufacturing developments are systematically addressed. Finally, in Stage 4, manufacturing strategy results in world-class manufacturing (WCM), where manufacturing is a competitive weapon.

THE ORIGIN OF MANUFACTURING STRATEGY THINKING

With the rise of business strategy thinking, Skinner reasoned that if manufacturing assumed a subordinate, reactive role in the corporate strategy, U.S. manufacturers will be ultimately uncompetitive (Skinner, 1969). He found that top management tended to distance themselves from the manufacturing function with the result that production management became an area for specialists with a limited view of the business. This led to unrealized competitive potential of the manufacturing function in such firms, and a lack of manufacturing people with a voice in top management decisions.

Skinner proposed a top-down approach to production management in which the manufacturing function and its managers are linked to business strategy and corporate strategy. He recognized the existence of 'tradeoffs' in manufacturing decisions that are unique to each business. Although tradeoffs remain an area of debate, Hayes and Pisano (1996) point out that competitive emphasis has changed over long periods of time, from low-cost production in the 1960s to the Japanese-led quality emphasis of the 1980s without sacrificing cost. In the nineties, cost, quality and flexibility have become collectively attainable strategic goals. To implement manufacturing strategy, Skinner recommended that manufacturing should be guided by a manufacturing task, which is a set of targets for productivity, service, quality and return on investment. These tasks reduce the plant's manufacturing strategy to a simple and practical set of goals and guidelines for implementation.

Factory focus

Skinner also developed the reasoning for the 'focused factory' (*FF*) concept (Skinner, 1974), which is widely used today. FF concentrates the resources of the plant to accomplish a limited manufacturing task. A focused factory may be large and complex, but the focus is in the mix of products made, processes used, and markets served. Skinner argued that attempting to serve too many purposes within the same manufacturing plant or system could lead to conflicts and incongruent objectives, resulting in inefficiency and ineffectiveness. There are several aspects of focus to consider (Hayes and Wheelwright, 1984). First, a factory can focus on manufacturing products in either low or high volumes. Second, focus may relate to the manufacturing processes used in the facility. Third, the facility may be customer focused, manufacturing products for a specific client or market. The focused factory concept is one of the most widely used principles in large scale factory reorganization for more than a decade. An extension of the principle can be found in cell manufacturing which is revolutionizing manufacturing everywhere. Manufacturing cells permit 'lean', 'agile' and 'flexible' manufacturing.

WHAT IS MANUFACTURING STRATEGY?

The *content* of manufacturing strategy sets priorities for a plant in five major areas so that the plant's products and processes are competitive, and these priorities guide key manufacturing decisions and tradeoffs. The areas are: (1) *Differentiation:* technological sophistication and product features; (2) *Flexibility* to modify products to suit customers, or to modify delivery volumes; (3) Manufacturing *cost*; (4) *Timeliness:* lead time, delivery reliability; and (5) *Quality:* conformance or perceived quality in regards to product performance, reliability, durability, etc. Hill (1985) recommends further clarifying these priorities as 'order-qualifying' or 'order-winning'. Order qualifiers refer to those attributes of a product that are required in order for entry to a market, whereas order winners differentiate a product by excelling in particular areas or offering features not available from the competition. Hill's approach allows for different industries having different order qualifiers and for the fact that an order winner at one point in time may become an order qualifier at a later date.

Manufacturing priorities of a plant defined by its manufacturing strategy have a major impact on decisions concerning all 'hard' and 'soft' manufacturing issues. The 'hard' or structural issues affected are: (1) *Capacity:* amount, timing, subcontracting; (2) *Facilities:* size, location, focus; (3) *Process technology:* equipment, automation, configuration; and (4) *Vertical integration:* make-or-buy, supplier policies. The infrastructural, or 'soft' issues affected are: (1) *Organization:* management structure; (2) *Quality policy:* monitoring, intervention, assurance; (3) *Production control:* decision rules, material control; (4) *Human resources:* skills, wage, management style; (5) *New products:* design for manufacture, flexibility; and (6) *Performance measurement and reward:* financial and non-financial systems.

The structural and infrastructural decisions play a dynamic role in shaping the development of specific long-term manufacturing capabilities. The decisions drive continual manufacturing improvement and reflect changes in manufacturing strategy. The 'hard' areas require capital investments in machinery and processes. The 'soft' areas require employee training and retraining.

HOW MANUFACTURING STRATEGY IS DEVELOPED

Manufacturing strategy process is concerned with how manufacturing strategy decisions are made and implemented. In the business strategy literature it is recognized that a company's strategy may not be completely planned, but may comprise various proportions of deliberate planning and emergent components (Mintzberg, 1978). However, planning plays a key role in manufacturing strategy making. A number of approaches to manufacturing strategy formulation have appeared in the literature, for example see Fine and Hax (1985), Miller (1988), Mills, Platts, Neely, Richards, Gregory and Bourne (1996), and Schroeder and Lahr (1990). A typical manufacturing strategy development and implementation process is illustrated in Figure 1.

The chief stages of the process in Figure 1 are:

1. *Define corporate objectives:* A top-down approach to manufacturing strategy is often taken, where manufacturing strategy is derived from corporate strategy. In this way, top management owns the manufacturing strategy. These corporate objectives may include a proliferation of new products, low costs, or a high level of product customization.
2. *Select product families:* Products made at the plant may be grouped in terms of their competitive requirements, followed by the formulation of a manufacturing strategy for each group. Manufacturing strategy for each group will be distinctive to suit the firm's business interests.
3. *Audit of the external conditions:* This audit addresses market requirements and competition in order to determine what is required from the firm's products to meet each of the strategic goals. This stage uses input from the marketing function of the company in order to view the competitive environment. Benchmarking may assist in understanding who the competitors are and what threat they represent.
4. *Audit of internal capabilities:* The internal audit considers the manufacturing capability of the firm in the context of the existing manufacturing strategy in the priorities outlined earlier, i.e. differentiation, flexibility, cost, timeliness, and quality. This stage assesses the state of the current manufacturing facilities, technology and infrastructure with respect to the intended manufacturing strategy.
5. *Analyze gap between actual and desired performance:* Gap analysis is a well-recognized concept from the business strategy literature. A comparison is carried out between the internal and external audits in order to assess how the manufacturing capabilities of the firm need to change to meet the competitive criteria, whilst remaining congruent with the corporate objectives and business strategy of the firm. This stage ensures a role for the manufacturing function in the competitive strategy.

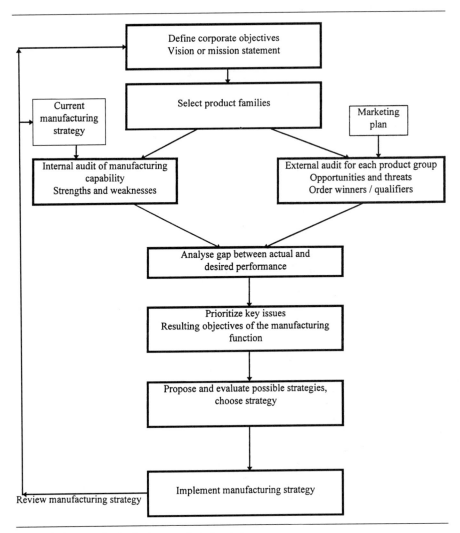

Figure 1. A process for developing manufacturing strategy.

6. *Prioritize key issues and define objectives of manufacturing:* After gap analysis, the manufacturing task is defined. This step uses the results of the gap analysis to translate the shortcomings of the manufacturing system (relative to the competitive strategy) into a set of tangible priorities and objectives.

7. *Choose manufacturing strategies:* The priorities and objectives from the previous stage are subject to specific action plans. This stage uses creativity techniques, such as brainstorming or mind mapping, to develop a set of action plans for manufacturing strategies that could be deployed. If multiple options are developed, one among them is chosen for implementation. The evaluation of alternatives may be done subjectively or by using computer models.

IMPLEMENTATION OF MANUFACTURING STRATEGY

Many consultants thrive in developing and assisting businesses in implementing a manufacturing strategy. Management processes and tools such as those used by Mills, Platts, Neely, Richards, Gregory and Bourne (1996) can be used as an aid to choosing and building appropriate manufacturing capabilities that will enhance and perhaps influence the competitive strategy of the company. Some plants use a steering group sponsored at a high level for developing and implementing manufacturing strategy. All key functions of the company including the manufacturing department should be involved in the process. Marketing, finance, human resources functions, and others should be part of the process. The following cases illustrate some aspects of the implementation of manufacturing strategy. Case 1 shows how cooperation between top management and cross-functional teams helped in identifying shortcomings in the manufacturing strategy, and then developed and implemented a revised one in several phases. In Case 2, a systematic process for developing manufacturing strategy leads to the development of implementation programs for the various areas of the plant.

Case 1: Specialty paper producer

This is a case of a company that perceived a need to change its manufacturing strategy due to a change in the competitive situation (Bennigson, 1996). For several years, it had occupied a niche in the market thanks to a unique production process developed internally, which enabled the company to manufacture a unique paper product. However, competitors were attracted to its market segment for its high profit margin potential, and the company began to lose its differentiation advantage. It then competed chiefly on price, but this strategy had limited success when other companies could offer better delivery timeliness with similar high quality. A steering group was initiated to investigate the company's manufacturing strategy. Members of manufacturing, sales, marketing, finance, management information systems, and research and development departments were involved in the steering group, and cross-functional teams were formed to carry out investigations.

The teams, together with top management, formulated the manufacturing strategy in the following steps. Firstly, the competitive challenge was clarified. Price was no longer an appropriate competitive criterion. Delivery reliability was the factor that was damaging the company's competitive edge, therefore customer service was identified as a problem area that required improvement. With this focus, machine scheduling was made the key issue. This was approached from the total company perspective, not simply as a manufacturing issue. Management found that many issues were impacted by machine scheduling. These were investigated from an external perspective, paying close attention to competitors.

Secondly, the current manufacturing strategy was identified and questioned. Management began to see clearly how the current manufacturing strategy had evolved and how it needed to be changed in the light of the new competitive challenge. Alternative approaches were examined for feasibility. Many 'sacred cows' were abandoned, such as scheduling to minimize set-ups rather than set-up time. Thirdly, analytical studies of the proposed changes to focused manufacturing facilities were done. These showed that by using 'focused factory' approaches and employing better machine scheduling, cost savings could be made whilst improving customer service. A goal was set to have the capability to improve delivery reliability performance by 300%. To work towards this goal, the cross-functional teams were reorganized to address ten tasks for implementation of the new customer-focused manufacturing strategy. These tasks included both the 'hard' issues, such as machine layout, and 'soft' issues such as information systems. The implementation of the manufacturing strategy took place in four phases, each of six months duration. Over a period of two years, improvement was achieved in the areas of delivery timeliness, inventory costs, and capital requirements, strengthening the competitive position of the company. The change in manufacturing strategy worked: They beat a chief competitor in the areas of low cost and volume flexibility, thus gaining market share.

The case highlights the following factors productive to manufacturing strategy development: (1) A market perspective on the business rather than an overtly technical view of the manufacturing system; (2) Cross-functional cooperation in the analysis; (3) Understanding by and involvement of all levels in the company; and (4) Good communication in the manufacturing strategy development process. These are contrasted with the pitfalls that can damage a manufacturing strategy review: (1) Internal, technical view of manufacturing strategy without an external, competitive view; (2) Involvement of only the manufacturing function in consultation and decisions; and (3) Lack of communication outside of the steering group.

Case 2: Packard electric

This case (Fine and Hax, 1985) describes the process of manufacturing strategy development in a division of a supplier of automotive parts. A strategic business unit manufacturing wire and cable, comprising three plants, was the focus of Fine and Hax's study. The division decided to develop a new manufacturing strategy when the chief customer, General Motors, required reduced costs, improved quality and better product development. Using the cost-delivery-quality-flexibility framework and the 'hard' and 'soft' issues described earlier, the first step in developing the manufacturing strategy was to look at the strategy hierarchy. Business strategy follows from corporate goals; the functional strategies are derived from business strategy. The manufacturing requirements to successfully implement the action programs of the business strategy were outlined. The requirements were: flexibility in capacity and processes for future demand, a changeover to new electronics technologies, and the development of a new packaging capability.

The current manufacturing strategy was then examined, broken down into the ten areas outlined earlier, and strengths and weaknesses of each policy were discussed. Product grouping was done using the *product-process life cycle* matrix of Hayes and Wheelwright (1984). The products were grouped according to their position on the product life cycle, from newly-introduced to mature, and additionally according to the market they served. This information was collected by the use of custom-developed forms about the products. A product-process matrix was again used to detect the degree of product or process focus at each of their three plants. A product was removed from a plant and reallocated if another plant was deemed more suitable to manufacture it. Next, given the capabilities of the current manufacturing system, long-term and short-term objectives were developed for the manufacturing function in each of the 'hard' and 'soft' decision areas. These objectives were developed into action programs for each decision category, such as quality management. The action programs were prioritized, with estimated cost and resources requirements and schedules. Responsible persons were identified for each program. Finally, the implementation programs included the development of new performance measurement systems, simulation modeling of capacity planning, and the introduction of a standardized decision process for make-or-buy decisions.

Key Concepts: Competitive Strategy; Content of Manufacturing Strategy; Differentiation; Flexibility; Infrastructure; Manufacturing Strategy Development Process.

Related Articles: Core Manufacturing Competencies; International Manufacturing; Supplier Partnership as Strategy.

REFERENCES AND OTHER SEMINAL WORKS

Bennigson, L.A. (1996). "Changing manufacturing strategy." *Production and Operations Management*, 5 (1), 91–102.

Fine, C.H. and A.C. Hax (1985). "Manufacturing strategy: a methodology and an illustration." *Interfaces*, 15 (6), 28–46.

Hayes, R.H. and S.C. Wheelwright (1984). *Restoring our competitive edge: competing through manufacturing*, Wiley, New York.

Hayes, R.H. and G.P. Pisano (1996). "Manufacturing strategy: at the intersection of two paradigm shifts." *Production and Operations Management*, (Spring), 25–41.

Hill, T.J. (1985). *Manufacturing strategy*, Macmillan, London.

Kinni, T.B. (1996). *America's Best: Industry Week's guide to world-class manufacturing plants*, Wiley, New York.

Leong, G.K. and P.T. Ward (1995). "The six Ps of manufacturing strategy." *International Journal of Operations and Production Management*, 15 (12), 32–45.

Miller, S.S. (1988). *Competitive manufacturing*, Van Nostrand Reinhold.

Mills, J., K.W. Platts, A.D. Neely, H. Richards, M. Gregory and M. Bourne (1996). *Creating a winning business formula*, Findlay, London.

Mintzberg, H. (1978). "Patterns in strategy formation." *Management Science*, 24 (9), 934–948.

Schroeder, R.G. and T.N. Lahr (1990). "Development of manufacturing strategy: a proven process." *Proceedings of the Joint Industry University Conference on Manufacturing Strategy*, 8–9 Jan., Ann Arbor, Michigan, 3–14.

Skinner, W. (1969). "Manufacturing—the missing link in corporate strategy." *Harvard Business Review*, 47 (3), 136–145.

Skinner, W. (1974). "The focused factory." *Harvard Business Review*, 52 (3), 113–119.

Spring, M. and R. Boaden (1997). "One more time: how do you win orders?: a critical reappraisal of the Hill manufacturing strategy framework." *International Journal of Operations and Production Management*, 17 (8), 757–779.

Womack, J., D. Jones and D. Roos (1990). *The machine that changed the world*, Rawson Associates, New York.

3. CORE MANUFACTURING COMPETENCIES

MORGAN SWINK

Michigan State University, Michigan, USA

ABSTRACT

A clear understanding of manufacturing competencies improves the formulation and implementation of manufacturing strategy for competitive advantage. Manufacturing competencies play three key roles in the formulation of strategy: (1) they clarify the differences between manufacturing means and manufacturing outcomes; (2) manufacturing competencies help maintain a strategic direction over time; and (3) they provide deeper insights for translating manufacturing policies into product attributes that produce competitive advantages. The following companies are discussed here: **Allegheny Ludlum Corp; GM; Hitachi Seiki; John Crane Limited; Nummi; Toyota.**

WHAT ARE COMPENCIES?

Recently, researchers have argued that "competencies" form the primary basis for competition between firms. It has been said that in the current business environment, the essence of strategy is to develop ". . . hard-to-imitate organizational capabilities that distinguish a company from its competitors in the eyes of its customers" (Stalk, Evans, and Schulman, 1992). Core competencies offered by a firm's manufacturing processes enable it to differentiate its products from competitors' products.

Core manufacturing competencies have not been well-defined. Hayes and Pisano (1996) suggest that competencies are *activities* that a firm can *do* better than its competitors. Further, competencies cannot be purchased. Competencies are organization specific; they are developed internally. The fact that they are difficult to imitate or transfer is what makes them valuable. Thus, competencies derive less from specific technologies or manufacturing facilities and more from manufacturing infrastructure composed of people, management and information systems, learning, and organizational focus. A manufacturing core competency is a very fundamental skill that is not easily copied.

HISTORY

Over the years, several strategic planning frameworks have been developed that involve the notions associated with manufacturing competencies. These frameworks have been very useful, yet they contain some inherent limitations. Strategic planning frameworks identify key manufacturing decision areas and stress the need for consistency among decisions affecting business strategy, competitive priorities, and manufacturing structure and infrastructure.

Making decisions using historical manufacturing strategy planning frameworks typically centered around "competitive priorities", such as cost, quality, dependability, flexibility, and service (Skinner, 1969; Wheelwright, 1978). These priorities are used to describe the content of a manufacturing strategy (Fine & Hax, 1985; Schroeder, Anderson, and Cleveland, 1986; Swamidass & Newell, 1987). Competitive priorities are too conceptually aggregated. Each priority is multi-facetted and complex, making its interpretation very much dependent on the researcher, strategy-maker, etc. Further, many different meanings and interpretations can be (and have been) attached to each term. For example, there are several terms and definitions for manufacturing flexibility (see the article on *Manufacturing Flexibility*, Gerwin, 1993; Upton, 1994).

The ubiquitous list of manufacturing priorities: cost, quality, dependability, and flexibility (see the article on *Manufacturing Strategy*), includes competencies and outcomes. Cost is a manufacturing outcome; flexibility is a manufacturing competency; the former refers to an *end*; the latter refers to a *means* to an end. Customers view manufacturing outcomes and manufacturing competencies differently. As Penrose (1959) and McGrath, Tsai, Venkataraman, and MacMillan (1996) point out, customers do not purchase a firm's competencies, per se (e.g., flexibility). Customers desire and purchase product and service attributes (e.g., delivery speed) that a firm creates by deploying its competencies.

OUTCOME VERSUS CORE COMPETENCIES
Manufacturing outcomes

Manufacturing outcomes are product attributes that reflect the cost, quality, and timing of production and service provided by the operation (Corbett and Van Wassenhove, 1993; Chase, et al., 1992). These dimensions of manufacturing performance are corollaries to the familiar marketing dimensions of price, product, place, and promotion. A number of typical manufacturing outcomes are defined in Table 1.

Table 1 Examples of manufacturing outcomes

Development cost:	The cost to design, test and develop production processes for a new product.
Production/transfer cost:	The cost to make and deliver the product, including the cost to return or replace the item if necessary.
Product conformance:	Degree to which a product meets pre-established specifications, and the accuracy of the delivered order to the specified mix and quantities of products.
Superior mfg. technology:	Manufacturing processes, which are unique or superior to competitors' manufacturing processes.
Order status information:	The availability and accuracy of real-time information about the location and status of an order shipment.
Mfg. process information:	The availability and accuracy of data regarding manufacturing performance or process parameters.
Order processing time:	The time required for the customer and the supplier to communicate and agree upon the order specifications and to place the order (i.e., ease of ordering).
Development time:	The time required to create, design, and introduce a new product into manufacturing, including the time to develop and ramp-up needed manufacturing processes.
Production/transfer time:	The time required to produce and transfer the complete contents of the customer's order, including the time to return or replace defective products.
Lead time variance:	Reduced variance between the scheduled delivery date and the actual, agreed upon date.

Let us elaborate on of the items in Table 1. One of the outcomes included in Table 1 is superior manufacturing technology. This refers to the outcomes resulting from the *use* of various manufacturing technology. An example of a superior use of manufacturing technology is provided by R&R Engineering, Inc., a small manufacturer of bent bolts and custom made wire forms. This firm has perfected the use of various planetary thread rolling machines to make products at faster rates and with higher conformance quality than any of its competitors, who continue to use circular die rollers and flat die rollers. The complexity of the planetary threaders makes it very difficult to calibrate and adjust the machines for peak performance. Realizing this, the firm's managers invested a great deal of time and effort into experimenting with the equipment and understanding its capabilities. Competitors have also purchased planetary threaders, but have been unable to use them as effectively. R&R managers, through a unique understanding of the equipment's competencies and have successfully integrated them into their operations, they have distinct quality advantages and can deliver products in half the time required by competitors.

Manufacturing competencies

While manufacturing competencies span a wide range of attributes, seven core competencies address steady state and growth aspects of manufacturing performance. Steady state competencies are known at any given point in time and contribute to desired manufacturing outcomes. Growth competencies enable improvement in manufacturing outcomes over time, or they contribute to the development of new steady state competencies. The various manufacturing competencies are described in Table 2.

Table 2 Core manufacturing competencies and examples

	Competencies Important for Growth
Improvement:	The ability to incrementally increase manufacturing performance using existing resources.
Example:	The ability to identify and remove non-value-adding activities.
Innovation:	The ability to create and implement unique manufacturing processes that radically improve manufacturing performance
Example:	The ability to apply new technologies or methods to solve problems.
Integration:	The ability to incorporate new products or processes into the operation.
Examples:	The ability to introduce and manufacture new products quickly.
	The ability to easily adjust processes to incorporate product design changes
	The ability to adjust smoothly to changes in product mix over the long term.
	Competencies Important in a Steady State
Acuity:	The ability to understand, acquire, develop, and convey valuable information and insights regarding products or processes.
Example:	The ability to assist both internal groups and customers in problem solving (e.g., in new product development, design for manufacturability, quality improvement, etc.).
Control:	The ability to direct and regulate operating processes.
Example:	The ability to determine the causes of adverse effects and remedy undesired variations in manufacturing outcomes.
Agility:	The ability to easily move from one manufacturing state to another.
Example:	The ability to manufacture a variety of products, over a short time span, without modifying facilities.
Responsiveness:	The ability to react to changes in input or output requirements in a timely manner.
Example:	The ability to accommodate raw material substitutions or variations.

In the framework in Table 2, growth in manufacturing effectiveness result from *three growth related competencies:*

- improvement
- innovation
- integration.

Steady state competencies result from:

- acuity
- control
- agility
- responsiveness.

The seven components of manufacturing competencies are elaborated below.

Improvement: Improvement relates to the ability to steadily increase the efficiency and productivity of existing manufacturing resources over time. The NUMMI plant, established as a joint venture between General Motors and Toyota, provides an example of superlative improvement competencies. Adler (1993) documents the steady performance improvement that resulted at the plant as a result of increased worker motivation, learning and problem solving, waste reduction, and work standardization.

Innovation: Innovation refers to the ability to radically improve manufacturing performance through the creation and implementation of new resources, methods, or technologies (Schroeder, Scudder, and Elmm, 1989). Innovation stems from the awareness of technological developments, plus the abilities to adapt and apply technology in ways that meet needs or create opportunities. Innovation competencies are vividly illustrated in the early development of flexible manufacturing systems by Hitachi Seiki (Hayes, 1990). The company cultivated the intellectual assets and organizational skills it needed to successfully apply a burgeoning micro-processor technology to machine tool systems. Technical and R&D expertise were important to the success of these developments. However, equally crucial were the contributions of manufacturing experts who understood the myriad production issues that needed to be addressed (e.g., scheduling, materials handling, etc.). Numerous other stories in the popular press recount the advantages manufacturing firms enjoy due to innovative developments of superior, often proprietary, processing competencies.

Integration: Integration is the ability to easily expand an operation to incorporate a wider range of products or process technologies. Upton (1994) discusses the ability to quickly manufacture new designs at John Crane Limited. The company's proficiency at introducing custom mechanical seal designs into an existing mix of manufactured components greatly enhanced its ability to meet unique customer needs. The related abilities to quickly introduce and utilize new processes or equipment are important for firms that compete in dynamic environments involving rapidly changing process technologies.

Acuity: Acuity refers to the insights of operations managers regarding process competencies and performance. These insights derive from high quality operations data and from abilities to translate internal or external customer needs into manufacturing specifications. Allegheny Ludlum Corporation, a specialty steel manufacturer, provides an example of superior acuity in manufacturing (March, 1985). The company developed acuity through extensive process modeling and experimentation. Information systems rapidly and frequently provided in-depth data regarding productivity, utilization, yields, rejects, and operating variances. In addition, manufacturing personnel had close ties to key customers. Allegheny Ludlum used its extensive process information coupled with a keen understanding of customer needs for evaluating strategic alternatives. In doing so, the company achieved a high level of financial success even during periods of industry-wide recession.

Control: Control is the ability to direct and regulate operating processes. A necessary requirement for control is feedback, a property that permits comparisons of actual output values to desired output values. An illustration of control is found in statistical process control techniques. Statistical process control tools are used to analyze and understand process variables, to determine a process's capability to perform with respect to those variables (Gitlow, Gitlow, Oppenheim, and Oppenheim, 1989). In a larger sense, control refers to managers' abilities to understand and reduce sources of unwanted variation in a process.

Agility: Agility is the ability to move from one manufacturing state to another with very little cost or penalty. It refers to manufacturing's ability to produce a wide range of products using a fixed set of resources. Agile processes are able to switch process set-ups quickly and efficiently, so that non-value-added time is minimized and so that smaller production runs are economical. Referring to John Crane Limited, Upton (1994) relates agility (which he names "mobility") to the ability of the company to maintain a wide offering of products without having to maintain a large finished goods inventory. Agility also involves the ability to produce a wide range of products in a range of quantities.

Responsiveness: Responsiveness refers to the ability to quickly adjust manufacturing processes to deal with changes in inputs, changes in resources, or changes in output requirements. For example, a responsive process can accommodate variations in the quality of raw materials or the uptime of equipment. Similarly, responsive processes can shift work schedules, job sequences, or physical routings to deal with unexpected changes in customer needs. In these ways, responsive processes are robust under conditions of input or demand variations.

In conclusion, a manufacturer may use a combination of core competencies in manufacturing, which cannot be easily replicated by others, for strategic advantage.

Key Concepts: Competitive Priorities; Generic Strategies; Manufacturing Objectives; Manufacturing Outcomes.

Related Articles: Manufacturing Flexibility; Manufacturing Strategy.

REFERENCES AND BIBLIOGRAPHY

Adler, P.S. (1993). "Time-and-Motion Regained." *Harvard Business Review,* January–February, 97–108.

Chase, R.B.K., R. Kumar and W.E. Youngdahl (1992). "Service-Based Manufacturing: The Service Factory." *Production and Operations Management,* 1 (2), 175–184.

Corbett, C. and L. Van Wassehnove (1993). "Trade-Offs? What Trade-Offs? Competence and Competitiveness in Manufacturing Strategy." *California Management Review,* Summer, 107–122.

Fine, C.H. and A.C. Hax (1985). "Manufacturing Strategy: A Methodology and an Illustration." *Interfaces,* 15 (6), 28–46.

Garvin, D.A. (1993). "Manufacturing Strategic Planning." *California Management Review,* Summer, 85–106.

Gerwin, D. (1993). "Manufacturing Flexibility: A Strategic Perspective." *Management Science,* 39 (4), 395–410.

Gitlow, H., S. Gitlow, A. Oppenheim and R. Oppenheim (1989). *Tools and Methods for the Improvement of Quality,* Irwin: Homewood, IL.

Hayes, R. (1990). "Hitachi Seiki (Abridged)", Case No. 9-690-067, *Harvard Business School,* Boston: MA.

Hayes, R. and G.P. Pisano (1994). "Beyond World Class: The New Manufacturing Strategy." *Harvard Business Review,* January–February, 77–86.

Hill, T. (1983) "Manufacturing's Strategic Role." *Journal of the Operational Research Society,* 34 (9), 853–860.

Kim, Y. and J. Lee (1993). "Manufacturing Strategy and Production Systems: An Integrated Framework." *Journal of Operations Management,* 11 (1), 3–15.

Kotha, S. and D. Orne (1989). "Generic Manufacturing Strategies: A Conceptual Synthesis." *Strategic Management Journal,* 10 (3), 211–231.

March, A. (1985). "Allegheny Ludlum Steel Corporation." Case No. 9-686-087, *Harvard Business School,* Boston: MA.

McGrath, R.G., M. Tsai, S. Venkataraman and I.C. MacMillan (1996). "Innovation, Competitive Advantage and Rent: A Model and Test." *Management Science*, 42 (3), 389–403.

Mintzberg, H. (1988). "Generic Strategies: Toward a Comprehensive Framework." *Advances in Strategic Management*, 5, 1–67.

Penrose, E. (1959). *The Theory of the Growth of the Firm*, Wiley, New York.

Porter, M.E. (1980). *Competitive Strategy: Techniques for Analyzing Industries and Competitors*, Free Press, New York.

Schroeder, R.G., J.C. Anderson and G. Cleveland (1986). "The Content of Manufacturing Strategy: An Empirical Study." *Journal of Operations Management*, 6 (3–4), 405–415.

Schroeder, R.G., G.D Scudder and D.R. Elmm (1989). "Innovation in Manufacturing." *Journal of Operations Management*, 8 (1), 1–15.

Skinner, W. (1969). "Manufacturing—Missing Link in Corporate Strategy." *Harvard Business Review*, 47 (May–June), 136–145.

Skinner, W. (1992). "Missing the Links in Manufacturing Strategy." in: C. Voss (Ed)., *Manufacturing Strategy: Process and Content*, Chapman-Hall, London, 13–25.

Stalk, G., P. Evans and L.E. Schulman (1992). "Competing on Capabilities: The New Rules of Corporate Strategy." *Harvard Business Review*, March–April, 58–69.

Swamidass, P.M. and W.T. Newell (1987). "Manufacturing Strategy, Environmental Uncertainty and Performance: A Path Analytic Model." *Management Science*, 33 (4), 509–524.

Swink, M. and M. Way (1995). "Manufacturing Strategy: Propositions, Current Research, Renewed Directions." *International Journal of Operations and Production Management*, 15 (7), 4–27.

Upton, D.M. (1994). "The Management of Manufacturing Flexibility." *California Management Review*, Winter, 72–89.

Wheelwright, S.C. (1978). "Reflecting Corporate Strategy in Manufacturing Decisions." *Business Horizons*, 21 (1), 57–66.

III. COMPETITIVE CUSTOMER SERVICE

4. CUSTOMER SERVICE, SATISFACTION, AND SUCCESS

STANLEY E. FAWCETT

Brigham Young University, Utah, USA

M. BIXBY COOPER

Michigan State University, Michigan, USA

ABSTRACT

The strategic benefit of customer-success initiatives is that they move the firm closer to the position of indispensability. The critical component of these efforts is a knowledge of the entire supply chain, which can be used to help key customers improve their competitive position. For example, one manager pointed out that a particular supplier was practically indispensable because of the tremendous value provided by the supplier. The manager noted that the supplier knew more about the industry than his firm, and more importantly, actively shared that knowledge to help his firm be more successful. By promoting state-of-the-art strategic initiatives, sharing operational tools, and facilitating new merchandising techniques, the supplier had helped its customer compete more effectively against larger rivals. Loyalty to the supplier was a natural result. Customer-success strategies recognize the importance of knowledge, and they use knowledge to increase switching costs through unique product/service offerings.

Of course, using knowledge to turn customers into winners can be a difficult process both in gaining the requisite knowledge and expertise as well as in translating that expertise into competitive advantage for the customer. The firm must obtain vital information about downstream requirements that can then be used to tailor the firm's product/service packages to deliver exceptional value. At the same time, the firm takes on the role of consultant to its customers, educating them in areas where they lack needed skills or knowledge. When a good relationship exists between the firm and its customers, this educational role is not only natural and straightforward but also genuinely appreciated. When close relationships do not exist, informing customers that they "are not always right" is difficult and can be highly uncomfortable. It is, however, better to turn down a customer request for a product that will not meet the customer's real needs than to deliver a product that will ultimately lead to dissatisfaction. The short-term struggle to explain how a product might be inadequate as well as what product/service package might better fulfill the customer's requirements is preferable to selling a product that dissatisfies the customer later. It is very important, however, not to use customer success as an excuse to needlessly oversell customers on higher margin products.

Finally, customer-success strategies must recognize that the buyer/supplier relationship should yield competitive advantage, or profit to both firms. Mutual benefit is particularly important given the experience of many firms that adopted "customer-delight-at-any-cost" approaches to keeping customers satisfied. These firms established selective relationships with key accounts that proved to be highly unprofitable because of the expense of meeting "excessive" service requests. The fact that many such relationships are cost ineffective can be discovered only if activity-based costing systems are adopted. The bottom line is that being a of preferred supplier is only beneficial when it can be done at a reasonable profit.

Customer service practices of the following companies are discussed here: **Electro; Hershey; IBM; Nuskin; Xerox.**

HISTORY

For most of this century, firms have focused on enhancing the performance of internal service-related activities or processes with the expectation that internal excellence would somehow translate into satisfied customers. This approach frequently left customers disenchanted; however, a lack of adequate alternatives often meant that customers returned to sellers despite their dissatisfaction. Only as new alternatives arrived in the marketplace—in response to global competition—did companies begin to perceive the gaps that existed between the product/service packages that they offered and the desires of their customers. When the gaps became large enough, customers opted for competitors' products. Haas (1987) referred to this point of customer departure as a *strategic breakpoint*.

As competitive pressures increased during the 1980s, customers throughout the supply chain became more vocal in expressing their displeasure with existing service levels. As customers demanded higher levels of service, more and more companies began to pursue customer satisfaction. By 1990, the term customer satisfaction had all but replaced customer service in the trade press as well as in the managerial lexicon. Expectation theory emerged as the guiding principle in companies' efforts to provide higher levels of satisfaction (Zeithaml, et al., 1996). Customers were satisfied when their actual experience with a supplier's product/service package met apriori expectations. Any gap between expectations and experience led to dissatisfaction. The key to achieving high levels of satisfaction was to understand customers' needs so that the firm could develop and deliver distinctive product/service packages that would meet those needs.

Unfortunately, many initiatives designed to provide real customer satisfaction degenerated into efforts to convince customers that the company was truly interested in them and their satisfaction. One important aspect of communicating this concern to customers was to get them involved in sharing expectations as well as feelings about their experience with the company and its products. By eliciting customer feedback, companies generated expectations that the feedback would be used to improve both products and the buying experience. From the customers perspective, companies appeared to systematically ignore the feedback they had received. This resulted in cynical, dissatisfied, and increasingly vocal customers.

Further, even for those companies that assiduously sought to incorporate customer feedback into the way they conducted business, the challenge of meeting expectations was magnified by the fact that competitors were constantly evolving and improving. New products and services are constantly introduced in today's global marketplace, which has the effect of obsoleting "old" technologies and processes with little or no notice. To combat this challenge, many companies embarked on the quest to provide high levels of customer delight. The new mission was to exceed customer expectations. Yet, going beyond customer expectations can be both difficult and expensive. Many companies have found that exceeding customer expectations have been unprofitable (Bowersox, et al., 1995). Today, the focus of leading companies is to mesh internal excellence with the ability to deliver valued satisfactions. For these firms, customer success, the ability to improve the customer's competitive advantage, is the ultimate strategic objective. The distinctive characteristics of customer service, satisfaction, and success are defined below:

- Customer Service: Customer service focuses on what the firm can do; that is, customer service measures internal service levels. The firm hopes that by performing well along these internal measures customers will be pleased.
- Customer Satisfaction: The measures of customer satisfaction are externally oriented. The firm seeks feedback from its key customers and uses this feedback to design its measurement system.
- Customer Success: Customer success focuses on helping customers succeed. Customer success requires an intimate understanding of the value-added potential of the entire supply chain. The firm uses this knowledge to help its customers meet the needs of customers further down the supply chain.

IMPLEMENTATION

A key to the design and implementation of an appropriate customer service strategy is to understand the benefits and challenges inherent in different service strategies. Table 1 highlights some of the pertinent issues related to the implementation of service,

Table 1 Customer service, satisfaction, and success fundamentals

Approach	Focus	Issues
Customer Service	Meet internally defined standards.	• Fail to understand what customers value. • Expend resources in wrong areas. • Measure performance inappropriately. • Fail to deliver more than mediocre service. • Operational emphasis leads to service gaps.
Customer Satisfaction	Meet customer-driven expectations.	• Ignore operating realities while overlooking operating innovations. • Constant competitor benchmarking leads to product/service proliferation and inefficiency. • Maintain unprofitable relationships. • Vulnerable to new products and processes. • Focus on historical needs of customer does not help customer meet new market exigencies.
Customer Success	Help customers meet their competitive requirements.	• Limited resources require that "customers of choice" be selected; that is, customer success is inherently a resource intensive strategy

satisfaction, and success strategies. The primary benefit of traditional customer service strategies is that they are relatively easy to implement since they are internally oriented. Objectives, measures, and procedures all originate and reside within the firm. This internal emphasis often meant that managers did not really understand customers' needs and desires. For example, Does a firm's internal measure of quality match the customer's definition of quality? It is not uncommon for a manufacturer and a customer to measure quality differently or to have different quality standards. One manufacturer was disappointed at the number of shipments that were returned as unacceptable by a sister division of the same firm. The two divisions used different quality standards in their internal operations. When firms do not understand customer requirements, they frequently "dissipate resources on things viewed as unimportant by customers" (Stock and Lambert, 1992). Traditional service strategies can easily lead to service gaps, which represent an invitation for competitors to enter the market and "steal" valuable customers (Parasuraman, et al., 1985).

The main benefit of customer-satisfaction strategies is that they incorporate customer input into both product and process designs in order to remove service gaps. Typical approaches to gathering customer feedback include surveys, panels, focus groups, in-depth personal interviews, and shadowing. In recent years, senior executives have begun spending more time with key customers to gain a better understanding of customer requirements. Because these approaches are costly in terms of both time and money, relatively few companies systematically collect reliable and valid data from customers. Even fewer really incorporate the data that they do collect into product and process design decisions. Other pitfalls of satisfaction strategies is that they often lead to unnecessary product proliferation. A leading consumer products manufacturer recently asked shoppers to try and find a given product on the store shelves. Even when given a sample of the product to facilitate the search, many shoppers were unable to find the product and indicated that they were confused by the multitude of similar products on the shelf. The manufacturer decided that it needed to simply its product lines. Customer satisfaction strategies can also lead to myopic, "me too" product/service strategies, preventing important innovation. Customer-satisfaction strategies attempt to use information to bridge the gap between the firm and its customers. (Stewart, 1997).

The principal benefit of customer-success strategies is that they are forward looking. That is, they consider the needs and requirements of the entire supply chain. This perspective enables a firm to actually help its customers become more competitive. Unfortunately, customer-success strategies are inherently resource intensive and can only be effectively implemented with a few "key accounts" or "customers of choice." The challenge of implementation is twofold: first, to identify attractive opportunities where the firm's capabilities match the market needs of key customers and second, to coordinate the firm's value-added efforts so that the firm can efficiently deliver increased value to the selected customers. Even as management begins to evaluate market opportunities, it must consider marketing and operations implications to determine whether the needs of specific customers can really be met efficiently and effectively. This evaluation process requires that managers realize that not all customers are equal and that individual customers have different needs that require different resource commitments.

Figure 1 illustrates a matrix alignment approach that is useful in helping the firm identify opportunities where a customer-success strategy would be appropriate. The vertical axis represents the important success factors of potential customers of choice. These success factors are determined by evaluating the entire supply chain to determine what imperatives drive the success of the customer. The key is to find what does the customer's customer define as important? For firm's at the end of the supply chain, the question translates into, how can we improve the operations or lives of key customers so that they feel not just satisfied but feel more successful. The horizontal axis focuses

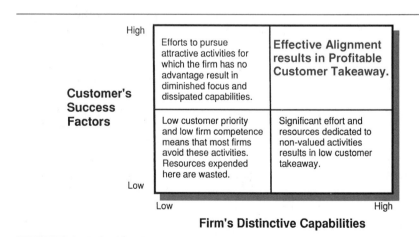

Figure 1. The alignment matrix.

on the firm's distinctive capabilities; that is, what does the firm do particularly well? In the matrix, where the customer's success factors and the firm's distinctive capabilities come together, an opportunity for the successful implementation of a customer-success strategy exists.

Having identified key customers and what constitutes real value to them, the task is to use the diverse value-added activities of the firm to profitably create and deliver the sought-after value. To do this, "value-added" must be clearly defined and communicated, and each part of the organization must understand: (1) its role in the value-added process; (2) the value that is added by other areas of the firm; and (3) how decisions in one area of the firm impact the other areas as well as the overall value-added process. The following questions can help the firm explicitly assess where it is in its quest to achieve customer success. These questions also help direct implementation efforts.

- Does the firm still operate from a service orientation that is driven by a set of internal measures of activity performance?
- If the firm has adopted a true customer-satisfaction orientation, are the systems and measures in place to really know what customers want and need?
- Does the firm really understand the supply chain and the competitive needs of the different members of the supply chain? What imperatives are changing the competitive dynamics of the supply chain? Is the firm providing value to other supply chain members that will enhance the overall competitiveness of the supply chain? Can the firm's product/service package be provided by another member of the supply chain in a way that threatens the firm's active participation through role shifting?
- Does the firm know how profitable different customer relationships are? Where is the value created that leads to profitable customer success? How could communication and coordination be improved to deliver more profitable customer value?

THE USE OF TECHNOLOGY

The impact of customer success initiatives depends heavily on the attitude of management and the training and measurement systems that are put into place. Technology is not as critical or central to these programs. However, the more the firm moves toward a success-driven strategy, the greater the opportunity to utilize technology in the development and delivery of tailored product/service packages. For example, a primary impediment to more effective customer-satisfaction initiatives is the inability to accurately and inexpensively capture customer perceptions in a timely manner. Information technologies such as electronic data interchange (EDI) provide a direct, low-cost, and real-time vehicle for sharing customer satisfaction information both in terms of expectations as well as actual performance levels. Internet "chat rooms" and e-mail can also facilitate the low-cost, real-time sharing of information between a firm and its key customers. The key to getting real value from these information technologies is to motivate people to use them on a frequent and consistent basis. These technologies only work when the proper incentives and training are provided early in the implementation of the customer-success strategy.

Information technologies can also play an important role in the design and management of supply chain alliances, providing the firm with a better opportunity to analyze and understand supply chain dynamics. The knowledge gained from this analysis is the key to accurately identifying customer success factors. Moreover, better enterprise resource planning (ERP) and inventory tracking systems help the firm reduce cycle times and make more accurate delivery promises. That is, when the managers know exactly what products are available in inventory as well as what current production schedules look like, they can make realistic delivery promises based on information and not the guesswork of a sales person. Companies such as IBM, Hershey, and Nuskin have all been able to improve customer responsiveness through the use of ERP systems. Hershey has also been able to standardize its customer order and invoicing procedures via its ERP system to provide more customer friendly interactions. Bar coding and satellite positioning systems also provide more accurate information regarding expected and actual delivery times. These technologies allow firms to track, expedite, and reroute shipments in response to customer inquiries and requests. Each of these technological advances allows the firm to be more creative in the products and services that it offers. They also help the firm get closer to customers than ever before. This connectivity helps to eliminate surprises and allows better responsiveness when they do occur. In effect, technology is altering both the product/service packages that can now be offered to important customers and the relationship with them.

ARE ALL CUSTOMERS TREATED ALIKE?

The foremost timing issue related to customer satisfaction strategies is the determination of which type of strategy a firm will employ with each customer in its client portfolio. Quite simply, service excellence is essential in today's intensely competitive global marketplace. Remember, any service gap is an open door for the competition, existing and potential, to enter the market and capture market share. Efforts to go beyond service excellence require greater resource intensity and should therefore be evaluated carefully with particular attention being paid to understanding customers' needs. The choice of a customer service, satisfaction, or success strategy depends on the following issues: customer expectations, customer leverage, the value of the relationship, the number and diversity of the customers, the firm's resource base, and the breadth of firm's product line. An "ABC" classification of customers might be useful. "A" customers would be likely candidates for implementation of a customer success strategy—the intensity of these relationships almost assures that a firm will lack the resources to offer dedicated, tailored product/services to more than a few non-"A" customers. Likewise, most "B" customers would be potential targets for focused satisfaction strategies where customer input is actively sought and utilized to meet expressed customer requirements. Finally, the sheer number of "C" customers generally means providing high levels of service excellence is the best a firm can do with constrained resources. As data capturing technologies improve and firms become more adept at using database mining software, "C" level customers may become candidates for more tailored satisfaction strategies.

SOME EXAMPLES

The rationale for devoting resources to achieve high levels of customer satisfaction has always been the belief that satisfaction and loyalty move in tandem—satisfied customers are loyal customers. Existing evidence, however, refutes this assumption for all but the most satisfied customers (Jones and Sasser, 1995). In fact, Xerox found that largely satisfied customers—those who marked a four on a five-point scale where five is equal to completely satisfied—are six times more likely to defect to the competition than those who are completely satisfied. Generally speaking, the reality is that customers in the 1990s are fickle and really do not trust the companies that promise to meet their needs. Only when a company truly excels at something that customers value can it expect to achieve some semblance of loyalty, which leads to profitable repeat business.

Figure 2 shows the relationship between satisfaction and loyalty. When a company's service offering is markedly lower than the competition, customers simply do not come back. Repeat business and loyalty begin to appear as the service offering enters the zone where it is comparable to competitors' offerings. However, customers will defect if it is convenient to do so. Only when the firm achieves an intrinsically higher level of service or a truly distinctive product do customers exhibit strong loyalty. The challenge is that most products and services can be replicated relatively easily by the competition within a short period of time—frequently less than one year. Thus, most distinctive advantages are transitory if they are not constantly upgraded. Firms must learn to make their own product/service offerings obsolete with improved product/service offerings.

This predicament underlies the emergence of customer success as a viable strategy. As one CEO notes, "We turn our customers into winners. Their success is cash in our bank. Our customer is our most important partner in cooperation—his customer benefits from this as well" (Ginsburg and Miller, 1992). Unfortunately, relatively few companies are willing to invest the needed time, effort, brainpower, and capital necessary to develop truly distinctive capabilities and match them to customers' important success factors. In reality, among all firms, three out of four still operate using traditional customer service definitions—regardless of the management rhetoric concerning the

Figure 2. The satisfaction/loyalty conundrum.

importance of customers. Another 20 percent have truly implemented customer satisfaction programs that utilize a systematic approach to gathering and incorporating customer feedback. Only a very small percentage of today's firms understand that customer success is an investment in long-term profitability. The end result is that customer success companies achieve levels of loyalty and profitability that are generally envied by the competition.

The following case illustrates the perspective and power of a well-implemented customer success strategy. Electro, a manufacturer of electrical components used in the assembly of consumer durable products such as household appliances and automobiles, adopted a customer-success philosophy a little over three years ago. The firm's customer orientation was brought to life and communicated through the slogan, "Pride in helping customers compete!" The phrase was quickly disseminated throughout the company and can now be seen hanging on walls in reception areas, corporate offices, and throughout the firm's manufacturing facilities. More importantly, Electro provided the needed resources and training to make the slogan a reality. As a result, Electro is now a preferred supplier to the largest appliance manufacturer in the U.S. as well as to two of the largest automobile manufacturers in the world. One corporate official summarized Electro's philosophy as follows:

"We feel it is no longer adequate to simply provide our customer good service since that has become the standard in our industry. It isn't even adequate to satisfy them if that means meeting their expectations since their past experience may lead them to expect certain problems to arise with their suppliers. Instead, we feel that the most appropriate thing for us to do is to do our best in making our customers better competitors in their industries. If they are more successful due to our ability to provide them with better products, more timely delivery, lower total costs, or whatever, then they will gain market share and grow. Of course, when they grow, we grow!"

LESSONS

All too often, customer-service initiatives are viewed with tremendous skepticism by the firm's employees and customers alike. Employees consider such programs to be among top management's favorite "flavor-of-the-month" programs—if they can avoid making any big changes for a while, the initiative will go away, having been replaced by the next management fad. Customers, likewise, are inundated with statements that "customers are our reason for being" or "no one ever won an argument with a customer" while at the same time experiencing poor customer service time and time again. The evidence suggests that a relatively small set of impediments prevent the majority of firms from providing the high levels of customer service, satisfaction, and success that they desire. These impediments begin with a lack of true commitment from top management. Once real managerial commitment is in place, the other impediments, which follow below, can be addressed.

• Adversarial worker/manager relations that get translated to the company's products and customers.

- Inadequate training, especially with respect to how each job impacts customer success.
- Incongruent performance measures that do not reinforce appropriate customer-success behavior.
- Lack of information flow to employees to help them learn; i.e., what worked and why!
- Compensation and reward systems that fail to recognize the customer relationship as critical.

As commitment to customers increases and as managers recognize and overcome the principal impediments to the implementation of an effective customer fulfillment strategy, their firms will be better positioned for long-term success. As noted by one executive, customer success is the new competitive imperative, "The only sure way to grow is to share in the growth of our customers. If we are the preferred supplier and they are their customer's choice, we can jointly identify breakthrough opportunities. Jointly, we can outperform the competition."

Key Concepts: ABC classification; Customer satisfaction; Customer success; Customers of choice; Expectation theory; Preferred supplier; Product proliferation; Service orientation.

Related Articles: Performance Excellence: The Malcolm Baldrige National Quality Award Criteria; The Implications of Deming's Approach; Total Quality Management.

REFERENCES

Bowersox, D.J., R.J. Calantone, S.R. Clinton, D.J. Closs, M.B. Cooper, C.L. Droge, S.E. Fawcett, R. Frankel, D.J. Frayer, E.A. Morash, L.M. Rinehart, and J.M Schmitz (1995). *World Class Logistics: The Challenge of Managing Continuous Change.* Council of Logistics Management, Oak Brook, IL.

Fawcett, S.E. and S.A. Fawcett (1995). "The Firm as a Value-Added System: Integrating Logistics, Operations, and Purchasing." *International Journal of Physical Distribution and Logistics Management*, 25 (3), 24–42.

Fierman, J. (1995). "Americans Can't Get No Satisfaction." *Fortune* (December 11), 186–194.

Ginsburg, I. and N. Miller (1992). "Value-Driven Management." *Business Horizons* (May–June), 23–27.

Haas, E.A. (1987). "Breakthrough Manufacturing." *Harvard Business Review* (March–April), 75–81.

Jones, T.O. and W.E. Sasser, Jr. (1995). "Why Satisfied Customers Defect." *Harvard Business Review* (November–December), 88–99.

Ohmae, K. (1988). "Getting Back to Strategy." *Harvard Business Review*, 66 (6), 149–156.

Parasuraman, A., V.A. Zeithaml and L.L. Berry (1985). "A Conceptual Model of Service Quality and Its Implications for Future Research." *Journal of Marketing*, 49 (4), 41–50.

Stewart, T. (1995). "After all you've done for you customers, why are they still NOT HAPPY?" *Fortune* (December 11), 178–18.

Stewart, T.A. (1997). "A Satisfied Customer Isn't Enough." *Fortune*, 136 (July 21), 112–113.

Stock, J. and D. Lambert (1992). "Becoming a "World Class" Company With Logistics Service Quality." *International Journal of Logistics Management*, 3 (1), 73–80.

Zeithaml, V.A., L.L. Berry, and A. Parasuraman (1996). "The Behavioral Consequences of Service Quality." *Journal of Marketing*, 60, 31–46.

IV. DEVELOPING COMPETITIVE PROCESSES

5. BUSINESS PROCESS REENGINEERING AND MANUFACTURING

TIMOTHY L. SMUNT

Wake Forest University, Winston-Salem, NC, USA

ABSTRACT

The ability to link tasks electronically can both reduce costs and increase customer service levels, mainly due to the fact that fewer handoffs are required and the work–in-process moves smoothly through the system. The work–in-process can be a manufactured item or the information that is required to support the physical transformation activities. These information processes are sometimes referred to as administrative processes and may reflect the majority of expenditures for the firm.

As we find the direct costs shrinking to an average of less than 25% of the cost of goods sold in today's environment, the reduction of costs in administrative processes can often have the largest impact on competitiveness. A number of firms, including **Texas Instruments**, have also used a process focus to address manufacturing capabilities and technology development over time. Rather than delegate the responsibilities for these important issues to more narrowly focused departments such as "research and development" (R&D) or "manufacturing engineering," companies are reorganizing and combining personnel from a number of departments, including manufacturing, engineering, R&D and marketing to work in a process mode on these dynamic, long-term issues. Reengineering practices of the following companies are included here: **Ford; Taco Bell; Texas Instruments.**

Figure 1. Process focus.

DESCRIPTION

This article considers operations from a process perspective, rather than from the traditional functional point of view. Such a consideration is important as an increasing number of firms are implementing reengineering projects in order to improve their systems' effectiveness. This article first provides a description of the process perspective, then provides implementation guidelines using reengineering principles as a basis for analysis. Finally, examples are given of how firms have made the transition from an operations functional structure to one where a process model is used.

A *process* can be defined as the set of tasks required to transform a set of inputs into a set of outputs for a customer. Customers can be either *internal* or *external*. External customers are ones to whom a company sells their final products, sometimes called *end items*. Internal customers are the people within the organization who receive reports or other types of information that are produced by a process but are not given to the external customers.

Most U.S. and Western European companies have been organized around functional expertise, rather than processes. Processes typically require the completion of tasks from a number of functional areas; however, they can become inefficient or ineffective when handoffs must be made across functions. (See Figure 1 for an illustration of the process focus.) As an example, consider the order fulfillment process that takes a customer order as the initial input and ends with the delivery of a product to the customer. In a functionally-oriented organization, the customer order is entered by employees reporting to a sales or marketing manager. The order then proceeds to a production planning department (as depicted in Figure 2) which then schedules the manufacture of the various components and subassemblies in order to complete the end item. Purchasing utilizes the information from production planning to place orders with suppliers for material and parts that are acquired from the firm's vendors. After the end item becomes available in finished goods inventory, the distribution department determines whether the product will be shipped to satellite distribution centers or directly to external customers.

Note that, in this functionally oriented system, responsibilities for the total process are separated among many different departments. Incentives and coordinating mechanisms must be carefully designed and implemented in order to provide smooth completion of tasks and high service levels to the external customers. In a manufacturing company, complex information systems are typically implemented to provide a focal point for the various interdepartmental activities. Material Requirements Planning (MRP) systems and their extensions to MRP II or Enterprise Requirements Planning (ERP)

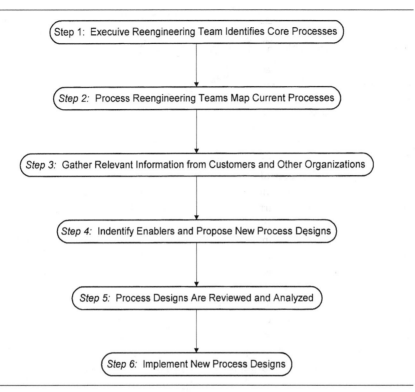

Figure 2. Major steps in implementing a process focus.

systems, which provide additional financial and capacity planning information, are the most common types in use today.

These information systems are often very expensive to either design or purchase and to implement, but are necessary for control in any organization of reasonable size. Even though the information systems provide a link across functional areas, communication between any two departments can remain difficult and may result in delays and errors. A process perspective, on the other hand, attempts to resolve the inherent communication and incentive problems of separate functional departments. These are often referred to as *silos*, since the managers and employees within a department tend to behave narrowly, although they have a deep set of functional skills. The process focus requires that the appropriate personnel from each of the functions required to produce a product be reassigned to work as a team, whose sole responsibility is to provide the necessary steps for a particular product or set of products.

HISTORY

While the process perspective can be traced to the origins of industrial engineering, it has recently come to light as an important contemporary management issue by

Michael Hammer and James Champy through their best selling management book, *Reengineering the Corporation: A Manisfesto for Business Revolution* (Hammer and Champy, 1993). Two other notable books, by James Harrington and Thomas Davenport, have outlined the process approach in fair detail and provide valuable reading in this area (Harrington, 1991 and Davenport, 1993).

The major reasons cited for the recent ability to change from a functional organization structure to a process one is the fact that information systems, including both hardware and software, have finally matured to the extent that they can enable such a change in focus. Previously, communication across functional areas could only be accomplished through a complex reporting structure using various levels of managers reporting up and across a pyramid, with the CEO or president of an organization at the top and the lower level managers and employees at the bottom. The use of email, intranets, electronic white boards, etc. have provided the impetus to consider the dissolution of functional departments and reallocation of appropriate personnel to process teams that handle all, or subsets, of tasks required to produce an end item.

REENGINEERING

Redesigning the company or parts of it into a set of processes typically requires that a number of steps are followed over a period of one to two years. Figure 2 shows the major implementation steps.

The first step is to identify core processes of the firm, especially the ones that have the potential for the greatest improvement in competitiveness. A top management team, sometimes called an executive reengineering team, is responsible for the initial identification of the core processes of the firm and for assigning managers and other personnel to process reengineering teams. The executive reengineering team may be specific about the goals or visions for each of the processes or may choose to let the process teams themselves identify the goals after an initial analysis of the current process design or lack thereof.

The second step requires that the process reengineering teams become familiar with the current problems and current practices of their assigned process. One non-trivial analysis here is the determination of process boundaries. It is often unclear where a process starts or stops, and to what degree there is overlap with other processes. However, unless the process teams can reach agreement on the process boundaries, any future analyses may be redundant. Interaction with the executive reengineering team is often necessary at some point in this step in order to obtain further definition.

The third step requires that the process teams gather information from both their customers and other organizations in order to determine the types of change necessary for an effective process design. Since any process should be designed specifically to meet customers' needs, the customers may be interviewed, surveys may be performed, discussions may be had with field sales representatives, or records of warranty work may be reviewed to better understand the direction of the process design. Additionally, benchmarking of similar processes should be made as possible. While the benchmarking of a direct competitor's process may not be feasible, many firms from other industries would likely have similar processes and might be willing to cooperate in this regard. Consultants are often used in this step since they have access to a variety of firms and industries and can provide information on best practices.

The fourth step requires that "enablers of change" are identified and tentative process designs are proposed. Enablers of change include both information technology (IT) enablers and other cultural enablers, including organizational and human resource enablers (Teng et al., 1996). IT enablers could include email, electronic white boards or other file sharing mechanisms, digital imaging technology, laptop computers, local area networks and expert systems. Organizational enablers include the use of case teams, case managers, new budgeting and resource allocation methods, and new management reporting relationships. The intent of the organizational enablers is to enhance employee empowerment so that delays caused by layers of vertical management approvals are drastically reduced. Human resource enablers include compensation systems that are based on team performance, training programs to enhance cross-functional skills, work assignment rotations, and advancement criteria that are related to ability rather than performance. Using "ability" as a criterion provide incentives to employees for increasing the breadth of their skill-set rather than just performing well on a narrow set of tasks.

Process designs are then tentatively proposed taking advantage of the most effective enablers available to the design team and company. It should be recognized that constraints often exist in companies with regard to enablers, either due to budget constraints or the current culture. For example, while the use of electronic white boards and shared data bases might be effective enablers for a process design, the current state of IT at a particular company may not allow quick adoption of these enablers. Or, perhaps, the current corporate culture is highly control-oriented and certain changes to the approval process might be met with stiff resistance from key managers. While the optimal process design would necessitate advancements in these areas, process designers may need to accept such constraints in the short-term in order to make any progress toward moving to a process orientation.

The fifth step calls for the tentative process designs to be reviewed by both the executive reengineering team, customers, and the personnel, who will be involved in the new design. Analysis of the proposed process design can be made at this step, and revisions to the process may be incorporated as needed.

The last step, implementing the new process design, requires top management support and financial resources for success. The changes that result from moving towards a process focus can be radical, affecting the fundamental precepts on how an organization works. Such change must be fully supported by the top management with both their time and the company's resources. There are a number of case studies that indicate that the lack of such support can be fatal to the successful implementation of a process focus and the specific designs determined by the process teams (Champy, 1995 and Hall et al., 1993).

TECHNOLOGY NEEDS

The availability of new information technology and other process technologies can greatly enhance the ability for a company to take a process focus. Essentially, the use of such enabling technology allows an organization to rewrite the "rules" of doing work and what is possible in meeting the customers' needs. The use of fast computers, networks, the internet, intranets, group decision-making software, expert systems, shared data bases, video conferencing, bar code scanning, and other technology provides advantages on a number of dimensions. The speed of the process flow increases, timeliness of information improves, work can be accomplished in a number of different locations, authority can be more easily decentralized, and communications between the company with both customers and suppliers can be vastly enhanced. For example, the use of shared data bases allows information to appear simultaneously in many places, reducing delays in communication and improving the accuracy of information reporting. The use of expert systems provide the capability of lower-skilled generalists to do the work of a number of specialists. While it may be possible to move from a functional focus to a process focus without the use of some type of enabling technology, the benefits of doing so will likely be much smaller.

ANALYTICAL TOOLS

Process mapping and related tools are the most prevalent methods used in process design. Additionally, computer simulation is rapidly becoming an analysis tool used to determine both the efficiency and efficacy of new process designs for both manufacturing and administrative processes.

A typical process map (sometimes called a process flow chart) is shown in Figure 3. While any symbols can be used to represent tasks, flows of material and flows of information, the ones used here are widely accepted across the world and provide a common vocabulary for discussion of the current and proposed designs. Also shown in Figure 3 are the three typical perspectives of processes, i.e. "what we think it is," "what we want it to be," and "what it really is." In my experience of mapping current processes for organizations, managers are always shocked to see the visual representation of how a process is currently working. Much of the reason for the complexity of the current systems is due to the fact that the processes are run using a functional organizational structure, causing the creation of a number of "work arounds" to deal with the inherent inefficiencies of communicating across functional silos.

An important analysis of the current process design requires the determination of *value-added* activities. A value-added activity or task is one, when viewed by the customer, is necessary to provide the output that the customer expects. Note that the definition of value-added relates to the customer's expectation, not the company's. While some activities may not be value-added to the customer, a company may still choose to keep them for control or communication purposes. The objective, however, is to eliminate as many non-value added activities as possible in order to reduce costs and time required to complete the process. See Figure 4 for an example of a value-added analysis.

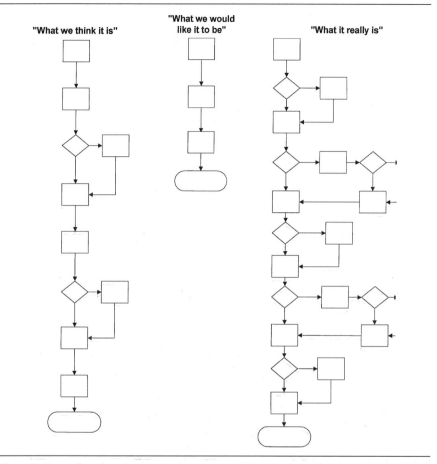

Figure 3. Process flow charts (process maps).

Another common tool for analyzing process designs is an enhancement of the simple process map to one that includes both time and function attributes. These process maps are often called time-function maps or deployment flowcharts, and an example is depicted in Figure 5. The major objective is to illustrate the cumulative time required for completing activities in a sequence and to indicate the functional expertise required at each step.

The concept of mapping a process has been extended to user-friendly computer simulation programs. Many new software packages provide for easy mapping of the process on the screen, with simple menu-driven input windows for customization of task times, assembly tasks, worker assignments and the like. The main advantage of the use of one of these software packages is that more realistic conditions can be rapidly tested without the need for expensive "laboratory" experiments of proposed process designs.

ACTIVITY (TASK)	TIME	VALUE-ADDED ?
Order taken by salesperson and entered onto order form.	1 hour	Yes
Order form taken back to car and placed into car organizer recently purchased from "office super store."	10 minutes	No
Waited in car as salesperson made visits to other customers in her area.	4 days	No
Overnight delivery to central order entry department at headquarters.	1 days	No
Order entered into computer system.	10 minutes	No
Computer system notes missing data.	1 second	No
Order awaits review by supervisor	1 day	No
Supervisor reviews order form and calls saleperson for missing data.	30 minutes	No
Order again entered into computer system with full information; confirmation number assigned.	10 minutes	No

Figure 4. Value added analysis.

RESULTS

The major results of a process focus can be categorized in two ways, i.e. in terms of *process efficiency* and in terms of *process effectiveness*. Process efficiency can be measured in terms of *processing time*, including specific measures like cycle time per unit or volume of transactions per employee. Other process efficiency measures include *resources expended, value-added*, and *wait time*. Effectiveness measures are those that measure the extent to which outputs of the process meet the needs of the customers. They include measures of *accuracy* (e.g. number of customer complaints or number of errors found in inspection before delivery), measures of *timeliness* (e.g. time to solve problem and closeness to product rollover promise), measures of dependability (e.g. percentage of missed deliveries by promised due date), measures of *responsiveness* (e.g. number of rush orders completed), and measures of *market* and *financial* success, including market share, number of lost customers, return on investment and earnings per share.

CASES AND EXAMPLES

Companies typically consider the move to a process focus when one or more of the above measures are indicating clearly poor performance, although a few companies have done so as a preemptive action. A number of case studies have been reported over

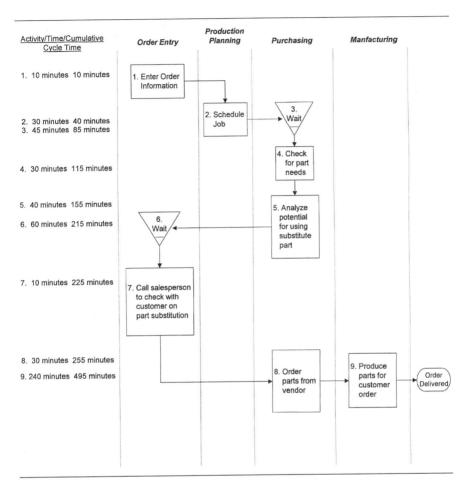

Figure 5. Time function maps.

the past few years on successes and results achieved from using a process focus. The Taco Bell and Ford Motor Company Accounts Payable examples are most notable.

The *Taco Bell* case (Hammer and Champy, 1993) is an excellent example of a business that was in danger of losing a market share war, but was turned around by radical changes in its management and operations. While IT was used to provide better information flow within the organization, the major enabler from an operations perspective was the centralization of food preparation. Using what they called a "K–minus" approach for the elimination of the kitchen from each restaurant location, both food preparation costs and timeliness were improved. Today, the food is mostly prepared at central "kitchens" (factories) and shipped to the individual restaurant locations for warming and final assembly to customer order. Less space is then needed at the restaurants for food preparation, allowing more seating area and, thus, larger sales volumes at the same facility cost. There is also less material waste, lower production costs, and higher consistency. Through these and other changes, Taco Bell's growth went from a negative 16% per year to a growth in sales from $500 million to $3 billion in just 10 years.

The *Ford Motor Company* Accounts Payable example (Hammer, 1990) provides an excellent example of how IT can change the rules and provide both reductions in cost and increased effectiveness. In this early implementation of reengineering concepts, Ford found that their accounts payable department used five times more employees than did an equivalent Mazda department. Ford instituted what they called "invoiceless processing" in order to eliminate paperwork that was frequently in error and to place the responsibility of shipment conformation on the receiving dock. The goods were checked on the receiving dock when they arrived, and if the order matched the shipment, the receiving clerk noted such on the computer system. Checks are promptly sent to the vendors. At the same time, inaccurate shipments were sent back to the vendor. In this way, many of the old checks and balances were eliminated along with the need for numerous accounts payable clerks. The use of IT in this situation enabled the process to maintain control at a much lower cost and higher effectiveness.

CAVEATS

There are many other examples of successful reengineering projects that have improved the operations in numerous companies by changing the firm's structure from a functional to process perspective. At the same time, even more failures are reported. The failures can be blamed on a number of factors, including lack of top management commitment, lack of resources for implementation, lack of will power of top and middle managers in making radical changes, and a poor understanding of analyzing and designing processes. This article provides an overview of the issues and tools used in focusing operations towards a process orientation.

Key Concepts: Benchmarking; Core Processes of the Firm; Enablers of Change; Process Effectiveness; Process Efficiency; Process Map; Value-added Activities.

Related Articles: Enterprise Resource Planning; Total Quality Management.

REFERENCES AND BIBLIOGRAPHY

Champy, J.A. (1995). *Reengineering Management: The Mandate for New Leadership*, Harper Collins, London.
Davenport, T.H. (1993). *Process Innovation: Reengineering Work Through Information Technology*, Harvard Business School Press, Massachusetts.
Hall, G., J. Rosenthal, and J. Wade (1993). "How to Make Reengineering Really Work." *Harvard Business Review* (November–December), 71 (6), 119–131.
Hammer, M. (1990). "Reengineering Work: Don't Automate, Obliterate." *Harvard Business Review* (July–August), 68 (4), 104–102.
Hammer, M. (1996). *Beyond Reengineering*, Harper Business, New York.
Hammer, M. and J. Champy (1993). *Reengineering the Corporation: A Manifestor for Business Revolution*, Harper Business, New York.
Hammer, M. and S. Stanton (1995). *The Reengineering Revolution*, Harper Business, New York.
Harrington, D.H.J. (1991). *Business Process Improvement: The Breakthrough Strategy for Total Quality, Productivity, and Competitiveness*, McGraw-Hill, New York.
Johansson, H.J., P. McHugh, A.J. Pendlebury, and W.A. Wheeler, III (1993). *Business Process Reengineering: BreakPoint Strategies for Market Dominance*, Wiley & Sons, New York.

Majchrzak, A. and Q. Wang (1996). "Breaking the Functional Mind Set in Process Organizations." *Harvard Business Review* (September–October), 74 (5), 93–99.

Reeves, E. and E. Torrey, Editors (1994). *Beyond the Basics of Reengineering: Survival Tactics for the '90s*, Industrial Engineering and Management Press, Georgia.

Rohm, C.E. (1992/1993). "New England Telephone Opens Customer Service Lines to Change." *National Productivity Review*, (Winter) 12 (1), 73–82.

Shapiro, B., V.K. Ranga, and J.J. Sviokla (1992). "Staple Yourself to an Order." *Harvard Business Review* (July–August), 70 (4), 113–122.

Short, J.E. and N. Venkatraman (1992). "Beyond Business Process Redesign: Redefining Baxter's Business Network." *Sloan Management Review*, (Fall), 34 (1), 7–20.

Tent, J.T.C., V. Grover, and K.D. Fiedler (1996). "Developing Strategic Perspectives on Business Process Reengineering: From Process Reconfiguration to Organizational Change." Omega, *International Journal of Management Science*, 24 (3), 271–294.

Terez, T. (1990). "A Manager's Guidelines for Implementing Successful Operational Changes." *Industrial Management*, 32 (4), 18–20.

6. THE EVOLUTION OF ENTERPRISE RESOURCE PLANNING

V. SRIDHARAN,
R. LAWRENCE LaFORGE
Clemson University, Clemson, SC, USA

ABSTRACT

Today, many Enterprise Resource Planning (ERP) systems are run on a network of personal computers. Their features include a client/server architecture supported by a distributed relational database system with query and reporting capabilities, electronic data interchange capability to communicate with both suppliers and customers, decision support systems for managers, a graphical user interface, and standard application programming interfaces.

In its 1998 software survey, APICS identified over one hundred products/ vendors, indicating the prevalence of ERP systems. A small but representative sample of software companies (and their products) that market ERP systems include SAP America (R/3), Bann Company (Bann ERP), Peoplesoft, J.D. Edwards, MAPICS Inc. (MAPICS XA), Marcam Solutions Inc. (PRISM) and others. Information about these companies are accessible through the Internet.

The science and practice of manufacturing planning and control has evolved over the years from MRP to MRP II to ERP systems. This evolution has been necessary to meet the changing needs of businesses, perhaps due to the changing economic and market conditions, and to bring the state-of-the art technology to the work place. The evolution has also brought about a resource-view of the corporation and a process-based view of its activities. Consequently, it has produced a tight integration of the activities of the various business functions such as engineering, operations, marketing, finance, and human resource management. The net effect of this evolution is to make enterprises more lean and agile.

HISTORY

Material Requirements Planning (MRP) systems have been in vogue since the early 1970s. The MRP approach provided much-needed forward visibility and the ability to co-ordinate plans for dependent-demand inventory items. MRP-based systems, however, focused mainly on planning materials and organizing shop floor activities. Thus, their scope was limited to the manufacturing function. In addition to its narrow focus, the MRP approach also had several weaknesses including the absence of a formal feedback mechanism, especially to handle situations when the material plan was found to be infeasible due to a shortage of capacity.

Recognition of the weaknesses of the MRP approach led to a cycle of improvements. With these developments came recognition of the need to take a more holistic approach encompassing the entire organization. To some extent, concurrent advances in the area of computing technology (both hardware and software) greatly facilitated the rapid growth of a new integrated approach for manufacturing planning. Over time, the viability of the master production schedule (MPS) was ensured by incorporating the so-called "feedback loop" between shop floor execution and production planning activities. Furthermore, sophisticated approaches for testing the capacity-feasibility of both the master production schedule and MRP-generated material plans were incorporated. This led to the development of rough-cut capacity planning approaches and capacity requirements planning. In turn, these developments resulted in improved execution of the plan and facilitated basing financial planning on the detailed manufacturing plan, leading to a new cadre of systems known as manufacturing resource planning systems.

A *Manufacturing Resource Planning* (also known as MRP II) system is essentially a business planning system. In addition to enhancing the planning of manufacturing activities it also integrates information systems across departments. In an enterprise implementing MRP II, manufacturing and marketing managers, financial officers and engineers are linked to a company-wide information system. Each manager has access to information relating to his functional area of management as well as to the information pertaining to all other aspects of the business. In reality, to produce higher quality products and provide excellent customer service this integration is clearly mandatory. For example, the sales department needs the production schedule to promise realistic delivery dates to customers, and finance needs the shipment schedule to project cash flow.

TRANSITION FROM MRP TO MRP II

Manufacturing planning and control activities are closely related to the activities of other functional areas such as accounting and finance, marketing and sales, product/process engineering and design, purchasing, and materials management.

Traditionally, each function within an organization had its own way of doing things and its own databases. Consequently communication between the various functional areas has not always been perfect. However, such separation of the activities across functions is artificial. In businesses all activities are interrelated, and constitute the whole rather than a collection of different functions.

Therefore, the next logical step to MRP was to combine the manufacturing activities with those of finance, marketing, purchasing and engineering through a common database. This was the leap from MRP to MRP II, known as Manufacturing Resource Planning.

The quality of the major inputs to manufacturing planning, the MPS, bills of material (BOM) and inventory record information is not determined solely by manufacturing. These inputs are prepared, shared and updated by other functions within the organization as well. To start with, the MPS is a statement of planned production and thus is the basis for making delivery promises. It forms the basis for coordinating the activities of sales and production departments. Any changes or updates by sales need to be approved by manufacturing and vice versa. In addition, whereas marketing is charged with creating the demand, manufacturing is responsible to fulfill it. Therefore, any marketing activity that may influence future demand needs to be communicated to and confirmed by manufacturing.

Any changes in the BOM will have to be agreed upon by both engineering and manufacturing to assure the feasibility of tolerances, impact of product revisions and new product introductions (where marketing also is involved) on the current shop floor system. The impact of changes in the BOM can be seen on material routings and lead times as well, which are used in material and capacity planning. Changes in agreements with suppliers affect purchase quantities and delivery lead times for purchased items.

Finally, the accounting/finance functions should use the same data as manufacturing while converting the materials, units produced, and activities into dollars. The MPS converted to dollars represents the revenue projection; purchase orders converted to dollars represent the cost of materials; shop floor activities (represented in work orders) converted to dollars reflect the labor and overhead costs. Discrepancies in the information used by manufacturing and finance/accounting is therefore unacceptable.

Once we use the MRP II logic, it is easy to realize that there is one physical system in operation in a company, and there is no justification for having more than one information system representing different dimensions of the physical system. Multiple systems lead to discrepancies and errors over time. The information system should also be unique and reflect the actual physical system. Thus, MRP systems evolved into MRP II when a common database was shared by all functions, and any changes and updates by one functional area immediately became visible to the rest of the organization.

With the advent of faster and cheaper computers it became possible to provide MRP II systems with the ability to simulate and thus provide a "what if" capability. Thus, modern MRP II systems can be used to simulate what would happen if various decisions were implemented, without changing the actual database. This makes it possible to see the impact of changes in the schedule on capacity requirements, the impact of schedule changes on material requirements, and/or the impact of design changes on customer responsiveness.

MRP II FEATURES

Interfunction coordination

As already mentioned, MRP II systems grew incrementally from material requirements planning (MRP) systems. One of the goals was to take a resource-based view of the firm. That is to say, engineering, marketing, manufacturing, distribution, and finance are all resources that must be managed in a coherent fashion to ensure consistency of purpose over time. This warranted improved coordination among the various functions.

Consider for example the activities of marketing and manufacturing. Recall that the master production schedule is a key input for planning material requirements for manufacturing. Given a master schedule, customer order promising is regulated via the so-called Available-To-Promise (ATP) logic to achieve coordination between production and sales. MRP II systems extend this similar integration and coordination between operations and other functions of the business such as human resource management, finance, and engineering.

Closed loop planning

Another enhanced functionality of MRP II is the "closing the loop" feature to ensure capacity-feasibility of material plans. APICS, The Educational Society for Resource Management, defines a closed-loop system as follows.

"A system built around material requirements that includes the additional planning functions of sales and operations (production planning, master production scheduling, and capacity requirements planning). . . . The term 'closed-loop' implies that each of these elements is included in the overall system, but also that feedback is provided by the execution functions so that the planning can be kept valid at all times." (APICS Dictionary, 8th ed. 1995)

This is accomplished at two levels—first at the master production scheduling (MPS) level, and again after the detailed material plans are developed. Figure 1 presents a schematic of a closed-loop system. As indicated by the figure, first a trial MPS is prepared. Then, the feasibility of the trial MPS is verified using the so-called rough-cut capacity planning techniques. The available techniques range from the simple (which use historical load factors to project time-aggregated total capacity requirements at each work center) to the complex (which produce time-phased workloads at each work center using product routing, bill of material, and lead time data). If adequate resources are not available to meet the MPS, the MPS may be revised to achieve feasibility. Thus, feasibility of the final MPS is ensured.

Notice, however, that a feasible MPS cannot guarantee feasibility of the detailed material plans developed by MRP. Hence, MRP II systems also provide users the capability to check the feasibility of the detailed material plans. This is accomplished by means of what is called Capacity Requirements Planning (CRP). The CRP module

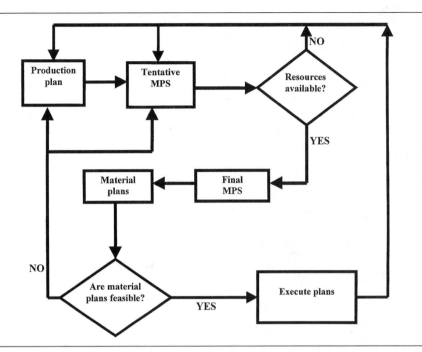

Figure 1. Closed loop planning.

verifies capacity feasibility taking into account; (1) the time-phased material plan; (2) inventory on-hand; (3) current status of work-in-progress in the shop; and (4) demand for service parts. Obviously the CRP module is highly data- and computation-intensive. The benefits, however, are realistic plans that can be achieved and better customer service via increased on-time delivery rate. The loop is closed by providing the necessary feed back of the status of plan execution to relevant planning modules.

What-if analysis capability

MRP II systems also provide the capability to conduct simulations to study and evaluate alternative decision options. Thus the effect of operations decisions on cash flow or working capital requirements can be easily examined. This capability to examine the consequences of each course of action without actually disturbing the system greatly enhances the quality of the decisions made. In turn, costs can be reduced without sacrificing customer service.

MRP II TO ERP SYSTEMS

Both MRP and MRP II systems focus at the plant production level. The reader may be aware of recent economic trends such as global competition; multiple plant-locations; geographically diverse, international, network of operations; and global net-work of suppliers. In addition, the markets are also becoming global. These changes in economic and market environments are complicated by the diversity of the political, cultural, and financial conditions in different parts of the globe. One consequence of these developments has been the need for instant access to information pertaining to customer needs, operations capabilities, and sourcing and distribution options. Almost simultaneously the computer hardware and software technologies have also been making rapid advances in terms of faster and cheaper processors, barcode scanners, computer graphics, networking, electronic mail, Internet, and the world-wide-web. Many software companies have realized the need and the potential for improvement in the design and practice of manufacturing planning and control systems.

The net effect of the new era has been to change the focus from plant level to an enterprise-wide view. This means an even tighter integration of the various business functions such as marketing and sales, accounting and finance, human resources, purchasing, distribution, and design and engineering with the traditional operations activities of the entire enterprise. The ability to handle multiple languages and currencies is also becoming important in view of the global nature of operations. This led to what is now called *Enterprise Resource Planning* (ERP) systems.

The main goal of ERP systems is to enable business processes that support customer needs. ERP accomplishes this goal by developing and presenting an integrated view of the supply chain. Such a view exposes system bottlenecks and constraints. The core view of ERP systems, however, remains resource-based. That is to say, ERP systems view the business as a collection of interconnected resources, whose optimization produces a superior performance. In turn, this approach enables accurate forecasting and realistic planning.

ERP systems tend to address the present-day needs for distributed planning and control. ERP systems also cater to the unique needs of personnel at different layers of the organization, such as individual divisions, locations, and machine centers. A typical ERP system provides the following functionality:

Engineering:
    Item part number control
    Bill of material control
Engineering Change & Documentation Control
    Routing
    Estimating
    Design engineering
Purchasing
    Vendor performance
    Purchase order management
    Subcontracting
Material
    Inventory control
    Master production scheduling

Material requirements planning
Lot tracking
Rough-cut capacity planning
Manufacturing
Capacity requirements planning
Shop floor control
Finite scheduling
Project control
Human Resources
Human resource information system
Payroll
Costing
Job costing
Cash flow analysis
Actual costs
Standard costs
Work breakdown structure
Finance
Accounts receivable
Accounts payable
General ledger
Multi-company consolidation
Foreign currency conversion
Marketing/Sales
Sales order management
Order configurator
Billing/Invoicing
Full sales analysis
Commission calculation/reporting
Sales forecasting/rollups
Quoting

IMPLEMENTATION OF ERP

ERP implementation can be a very long process requiring much financial resources and involuntary disruption of work. The results from the field are mixed. It is wise to evaluate the options for ERP software and implementation services provided by software vendors or independent consultants. Poor understanding of the nature of ERP software systems, the company's needs, its existing software, and the capability of vendors and service providers could prove very costly in the end.

It has been a common practice to install ERP systems in modules. Almost all vendors make ERP systems in easily manageable modules. The benefits of ERP, its integrative capability, and its use by competitors are forcing non-ERP users to take a close look at ERP systems.

Key Concepts: Bill-Of-Materials (BOM); Capacity Planning; Closed-Loop Planing; Manufacturing Resource Planning; MRP; MRPII; Process View.

Related Articles: Agile Manufacturing; Business Process Reengineering and Manufacturing; Supply Chain Management: Competing Through Integration; Virtual Manufacturing.

REFERENCES

APICS Dictionary (1995). 8th Edition, APICS, Falls Church, VA.
APICS The Performance Advantage (1998), APICS, Falls Church, VA, June.
APICS The Performance Advantage (1998), APICS, Falls Church, VA, September.
Vollmann T.E., W.L. Berry, and D.C. Whybark (1997). Manufacturing Planning and Control Systems, 4th Edition, Irwin/McGraw Hill.

V. COMPETING ON QUALITY

7. TOTAL QUALITY MANAGEMENT

R. NAT NATARAJAN

Tennessee Technological University, Cookesville, TN, USA

ABSTRACT

Collis P. Huntington, owner of Newport News Shipbuilding, engraved in 1917 the company's motto on the side of the building: "We shall build good ships here; at a profit, if we can, at a loss if we must, but always build good ships." (Dobyns and Crawford-Mason, 1991: 11). Five thousand years before Huntington, Egyptian inspectors checked the work of masons who dressed the stones of the pyramids, and the ancient Chinese had a department of the government to establish and maintain quality standards (Juran, 1993). These examples show that, meeting the standards, whether they were set by kings or merchants, was important. Ever since businesses have been engaged in the production of goods and services, quality has been a concern to the producer and the user. In that sense, quality has never gone out of fashion and never will. But what has changed over the years is the meaning of the term "quality," its significance to organizations, and how it is managed.

In the present day context, the word "quality" has come to mean more than it did in the past. It means meeting or exceeding the requirements, expectations, and needs of the customers—even those needs which are latent and not articulated by the customers. In recent years, a pragmatic and comprehensive system for managing quality called Total Quality Management (TQM) has evolved. Many organizations have successfully developed and implemented such systems, with dramatic improvements in performance. Many different experts and the practices of diverse organizations have contributed to the principles of TQM. Generally, TOTAL quality management implies performance excellence throughout the total system—including design, production, distribution, service, and the involvement of all categories of employees, customers, and suppliers in the quality initiative. The practices of the following companies are included here: **Caterpillar; Dana Corporation; General Electric Company; General Motors-Nummi Plant; Hewlett-Packard; Highland Park Plant, Ford; IBM; Komatsu; Motorola; Toyota; Xerox.**

WHAT IS TQM?

The following are essential characteristics and core values are common to most effective TQM systems.

Customer Focus: This puts the customer first for real. The essence of customer focus is identifying the external and internal customers, their needs and expectations, and doing whatever it takes to satisfy them. Under TQM, the scope of the customer extends to those who are not directly involved in using the product but may have a legitimate concern such as the regulators, consumer organizations, the community, and the general public, whose concern over safety, health, the environment, and consumer protection are taken seriously by TQM.

Active Involvement and Support of the Top Management: The top management of the organization must actively demonstrate by their deeds, and not just words, their dedication to total quality. Their leadership must be informed and visible.

Active Involvement of All Employees: Top management must lead everyone, from line workers to clerical employees to professionals and managers, to participate and become actively involved in the TQM process. Such involvement must be supported by policies for training, empowerment, performance measurement, recognition, and reward.

Prevention Emphasis: The emphasis of TQM is on prevention of defects and errors, rather than after-the-fact detection and reaction, such as inspection. Defects are proactively eliminated by designing quality into the product and the process.

Continuous Improvement and Learning: This basic philosophy includes constant efforts to identify and eliminate non-value adding activities, and to continuously improve product/service, processes, and all the inputs. It also includes education, training, and upgrading of the skills of human resources.

Management by Fact: This means using facts and data to solve problems, understand sources of variation, and uncover root causes. Analytical tools are used, both intensively and extensively throughout the organization, to collect, communicate, analyze, and share data.

Business Planning and Performance Measurement: Quality plans and objectives are integrated with overall business strategies and other objectives. Performance measurement, which includes comparative and competitive benchmarking, is aligned with quality objectives and organizational goals.

Collaborative Relationships: This refers to partnerships and alliances with suppliers, customers, educational institutions, and other organizations.

It is emphasized that any one of the above characteristics by itself does not imply TQM. They all interact and constitute a total system. Organizations will not realize the full benefits of TQM if they only pick and choose some of the above characteristics and core values for implementation. In order to fully understand what these characteristics imply, we need to consider the evolution of the terms "total," "quality," and "management" in TQM.

HISTORY

Before industrial revolution, skilled craftsmen produced products in small quantities customized to meet the needs of individual customers. Generally, a small group of craftsmen were involved in product design, acquisition of the inputs, production process, and interaction with the customer. They were also responsible for the quality of what they produced. There were no quality inspectors as such. All that changed with the advent of mass production in factories to serve the needs of the mass markets. To make mass production more efficient, division of labor and task specialization were introduced. This led to organizations structured along functional specialties like design, production, marketing, and accounting. Consequently, production activities were separated from the task of inspecting the quality of what was produced. Quality meant conformance to certain standards, which were set mostly by the producers themselves. End-of-line inspection, which weeded out the products that did not meet the standards was the primary means for assuring quality.

It is important to note how quality was managed in such a set-up. Of course, it was not economical for cost and time reasons, to inspect each and every item that was produced. But, in many instances, very rudimentary inspection procedures or no procedures were followed. This was particularly true of mass production systems, where consumer products were produced. In Ford's Highland Park plant, which had one of the most advanced forms of mass production for its time, finished automobiles were rarely inspected, and no Model T was ever road-tested (Womack, Jones, and Roos, 1991). Any defective product revealed by inspection was sent to a separate rework line, because rectifying the defects on the assembly line sacrificed production. How did customers' needs figure in all of this? According to Henry Ford, "They can have any color they want, as long as it is black." Because mass produced products were competing on price, mass production systems focused on raising productivity and reducing costs by automation and reduction of direct labor. Quality did not receive the same emphasis as cost in mass production.

In contrast, in batch-oriented job shop type manufacturing systems, product quality was considered important. Even in those organizations, if customers received defective products, the inspectors, rather than the operators, were questioned. The more important question—Why was the product made defective?—was never raised. In other words, the source of chronic waste in the system was not attacked. Such practices also fostered the belief that quality can be inspected in and that inspectors were responsible for quality! In countries where the mass production systems were not widespread because of the smaller size of the markets, quality became a competitive factor in certain niche industries, e.g., optical equipment and cameras in Germany and watches in Switzerland.

However, the situation was quite different when it came to production of military goods, in the U.S. the customer was the U.S. Department of Defense. Here, exacting standards of performance for the products were applied and were strictly enforced. Often, it was quality attained at any cost. This was true in many industrialized countries of the world including U.S., Japan, and, until recently, the former Soviet Union. Juran observes: "Japanese quality was all military until 1945. Their toys were shoddy, but their torpedoes were superior." (Juran, 1993, 43). Because of the attention given to quality and reliability in the case of hardware and materiel, important techniques were developed for sampling and inspection in production of military goods. World War II provided further stimulus to these efforts, resulting in the development and application of more advanced techniques. In fact, the origins of the now popular ISO 9000 standards could be traced to the standards developed by the U.S. Department of Defense.

From the 1930s onwards, along with advances in methods for process control and inspection, there were significant advances in measurement, gauging, testing, and other technical aspects of quality. Quality control had emerged as a separate function but with a narrow focus and managed by technically oriented specialists. In this era, the word "total" would mean application of quality control techniques across all product lines of a firm. This was the era of quality with a small "q."

From small "q" to big "Q"

Things began to change in the 1950s. As the economies of the industrialized countries recovered from the devastation of the war and began to grow, competition among suppliers increased, which increased the choices for the consumers. There were other major forces and trends worldwide: greater complexity and precision of products; product safety and liability litigation; government regulation of quality; and the rise of consumerist movements. In this emerging environment, the firms were forced to address the requirements of the customers and the regulators. Now, satisfying the customers required the coordination of all the activities that had a bearing on customer satisfaction but may take place in separate functional silos. For instance, the impact of design on conformance quality, i.e., meeting the design specifications during production, was recognized. It became imperative that customers' viewpoint was reflected in all the activities and quality had to be managed across the functions from design to marketing.

Dr. A.V. Feingenbaum was one such quality professional who had first-hand experience in dealing with such across-the-company cooperation issues at General Electric

Company in the U.S. He originated the concept of Total Quality Control (TQC) to describe the broader scope of quality assurance function. In Western companies, with their functionally and professionally oriented specialization, it was the QC departments that designed and ran the quality programs. Top management was only peripherally involved, and the rank and file in the organization did not play any role in such programs.

By 1960s, many Japanese companies also had TQC or Company Wide Quality Control (CWQC) systems as they were called, but the term "total" in TQC had a very different meaning in Japan. It meant involvement of every one in the hierarchy, from top management to the production worker on the shop floor, and the clerical worker in the office. Dr. Juran's lectures on managing quality in 1954 had convinced many Japanese top managers of the importance of top management's responsibility. As a result, they no longer viewed quality as a technical function that could be left to the specialists. Another important development was due to the efforts of Dr. Ishikawa who insisted on involving all employees in studying and promoting QC (Ishikawa, 1985). In 1962, he developed the quality control circles, small groups of Japanese workers who met voluntarily to discuss ways to improve their own work and the system. His approach was to provide easy-to-use analytical tools—including his own innovation, the cause-and-effect or fishbone diagram—that all workers could use to analyze and solve problems. He also persuaded Japanese management to not only support the activities of quality control circles but also incorporate their suggestions in the total quality effort. He also made them think about co-workers and colleagues as internal customers.

In Summary, the momentum generated by the following factors had a profound impact, not only on product quality but on cost as well: (1) top management's involvement and leadership for quality; (2) lowering the barriers between departments; (3) relying on workers' brainpower for quality improvement; (4) adopting the philosophy of *kaizen* or continuous improvement of the product, the process, and the inputs; and (5) focusing on the needs of external and internal customers. Many Japanese companies, by the 1970s, were able to achieve rapid improvements in the quality of their products, and, more significantly, could deliver them consistently at lower costs to the market (Shiba, Graham, and Walden, 1993).

TQM, STRATEGY, AND COMPETITIVENESS

From the 1970s, global competition has intensified due to the revolutions in communication, transportation, and reduction in barriers to trade. Threats to human safety, health and environment have become major concerns for citizens, governments, and corporations.

Quality may be necessary for the survival of a firm without guaranteeing its success. But the TQM perspective can provide the firm with new strategic options for business (Belohlav, 1995). To illustrate, a few years ago, companies like Xerox, Motorola, IBM, and the Big Three automakers in the U.S. found themselves in serious trouble against competitors from Japan, whose product quality and customer acceptance levels were much higher. For these firms, quality improvement became a necessity, not a choice. In these recovering companies, quality defined their business strategy. At Xerox, quality improvement thrust drove the entire strategy, and major organizational changes that followed. For Motorola, the new strategy meant that high quality would not only differentiate its products in the market place, but it would also make the company a low cost leader in the industry. Top priority given to quality in the overall strategy gave Motorola a significant competitive edge. It had, in fact, set the industry standards for quality, forcing its competitors to play catch up. Sometimes, a perceived quality edge can be sustained even after the competitors have caught up. For instance, General Motor's Geo Prizm and Toyota's Corollas were produced in the same NUMMI plant in California, a joint venture between GM and Toyota. Because of similar designs, same production processes, and workers, the design and conformance quality of these model vehicles were nearly identical. Yet, for many years, the Corollas outsold the Geo Prizms, and commanded a premium price because of the differences in customers' perceived quality.

Other implications of TQM

Increased expectations of customers regarding not only quality but regarding cost, responsiveness, and flexibility as well, are challenging companies to design, produce, and deliver products better, cheaper, and faster. According to Dr. Curt Reimann, the former Director of the Malcolm Baldrige National Quality Award in the U.S., "Consumers now have choices from around the world. Choices may be made on the basis of price; on the basis of features, variety, and service; on the basis of responsiveness; on the basis of quality. All of the factors in purchase decisions—price, features, variety, services, responsiveness, and quality, not just quality alone—are addressed in an integrated way in total quality management." (Dobyns and Crawford-Mason, 1991: 93).

   In some cases, quality improvement becomes a prerequisite for the implementation of strategies that are not necessarily built around quality. Manufacturers trying to implement Just-In-Time (JIT) production have to first achieve stable and capable manufacturing processes. The Pull system of production in JIT treats the next process as the internal customer. In JIT manufacturing, suppliers deliver defect-free parts directly to the line when incoming inspection is eliminated.

IMPLEMENTATION

Quality awards as a framework

TQM implementation can be based on many different frameworks. Many companies have used the approaches advocated by one of the quality gurus, i.e., Deming, Juran, Crosby, Feingenbaum, Ishikawa or other experts to launch their TQM efforts. Some companies have been more eclectic, integrating what they thought were the best ideas of the different experts and developing their own approaches to TQM.

   In 1951, Japan instituted the Deming Prize to recognize companies that were successful in implementing Company Wide Quality Control, the Japanese term for TQM. In recent years, several countries ranging from Australia to Mauritius, have created national awards to promote quality. In the U.S., apart from the Baldrige award at the national level, many states have their own award programs that are patterned after the

Baldrige award. The European Quality Award was developed for recognizing excellence in European companies. The criteria for these awards provide a non-prescriptive framework for TQM. Many companies have used these criteria for self-assessment. In many countries, a lot of prestige is attached to these awards. The winners are recognized in public ceremonies, and the results achieved by these companies and the methods they used to achieve them are widely disseminated. In some cases like the Baldrige, the winners are obligated to share non-proprietary information regarding their quality strategies, systems, and practices with others. Such award-winning companies can be viewed as role models. Indeed, they have often stimulated other companies, both large and small, to initiate TQM.

ISO quality standards

The requirements of the ISO 9000 series of quality standards can also serve as a framework for developing a quality system. Though compliance with ISO 9000 is voluntary, many customers are requiring that their suppliers be certified. However, the registration process usually deals with only a part of overall business operations and therefore does not represent a system in the sense that most TQM experts like to emphasize. The criteria are largely process focused and do not include many of the essentials of TQM, such as personal leadership by the top management, integration of quality goals with business goals, achieving rapid rates of improvement in quality, emphasis on overall performance and competitive positioning, participation and empowerment of the workforce, and benchmarking (Reimann and Hertz, 1993). As they stand now, the criteria provide only minimum requirements for a quality system and therefore, certification should be viewed only as the beginning and not the end of the evolution towards TQM.

CRITICAL SUCCESS FACTORS AND TOOLS

There is no single correct way to implement TQM. Each organization has to develop a customized system that is tailored to its culture, history, and the industry it is in. However, the companies that have succeeded with TQM share some common traits discussed below:

Leadership

Top managers in these firms promote emphasis on quality, establish quality goals, enlarge business plans to include those goals, and provide resources for achieving the goals. They are personally involved in education, training, and recognition. They are accessible and have routine contact with employees, suppliers, and customers. Juran considers these roles of top management in TQM nondelegable (Juran, 1995). A strong leadership system that is not dependent on any one individual is often evident. Written policy, mission, and other documented statements of quality-based values provide clear and consistent communications. Through their personal roles, the senior leaders serve as role models reinforcing the values and expectations.

Systematic processes such as Management-by-Policy (Hoshi-Kanri) or Policy Deployment (PD) are used to deploy company strategies by considering the relevant ends, the means and measures to achieve those ends at each level of management (Shiba,

Graham, and Walden, 1993). These processes reassure people that the organization has a strategy for the future, and make clear how they fit into the strategy. Based on leadership team's vision and active participation of all employees, strategy and action plans are generated. The plans cascade down hierarchically with progressively more detailed information on the means to achieve the goals. At each level, priorities (*Hoshin*) are developed and targets are set to focus on areas identified for improvement. Thus company strategy is made meaningful for all employees in light of their own responsibilities. These methods are oriented towards both results and improvement. At Hewlett-Packard, PD is called a planning and implementation methodology that is driven by data and supported by documentation.

Senior leaders also provide strategic directions in other ways. For instance, dimensions of quality critical to customers are identified. These dimensions are used to set clearly defined customer satisfaction and internal quality objectives and priorities. Aggressive targets are set. These targets go beyond incremental improvements, and look at the possibility of making large gains, and get the workforce to think about different processes. Strong drivers such as cycle time reduction or other targets are used to focus the efforts. Best practices within and outside their industry are benchmarked, and the results are used for improvements. Flatter organizational structures are created that allowed more authority at lower levels. Senior managers act as coaches rather than bosses. Cross-functional management processes and interdepartmental improvement teams are used. Organizational culture with respect to quality practices—e.g., identifying defects over prevention of defects is changed.

Customer focus

Customer is the focal point of any TQM effort. Identifying their requirements and expectations is the very important first step in customer satisfaction, retention, and building a relationship with the customer. Successful companies use a variety of strategies and technologies such as market surveys, focus groups, feedback from employees in contact with the customers, trade shows, toll-free lines, electronic mail and bulletin boards for listening to "the voice of the customer." The choice depending on the type and size of customer segments. To be effective, such learning and listening strategies have to be applied continuously and tied to the overall business strategy. Customer-driven quality is more than just meeting specifications. It implies adapting and responding quickly to the changing and emerging customer requirements. It demands an awareness of developments in technology and competitive product offerings.

Quality Function Deployment (QFD) is a methodology designed to ensure that all major requirements of customers are identified and subsequently met or exceeded through the resulting design of the product and its manufacturing. QFD can be viewed as a set of communication and translation tools for making quality customer-driven—it translates the voice of the customer into the relevant characteristics of the product, parts, and process. QFD tries to eliminate the gap between what the customer wants in a new product, and what the product is capable of delivering. "Customer" could be internal, external, present or future, or it can be any set of requirements, such as ISO 9000 standards. QFD acts as a vehicle that facilitates inter-functional communication and simultaneous engineering involving marketing, design, and manufacturing. Through a series of cascading matrices (called House of Quality), the voice of the customer (the "what's") are translated into technical specifications (the "how's") to meet, successively, the requirements of product design, parts, process, and production. Often competitive assessment of "what's"—"where do we stand relative to competitors"—is also taken into account. The "how's" of one stage become the "what's" for the next stage. This can later lead to optimization of the values of the technical parameters ("how much"). QFD can also help speed new products to the market—it enabled Komatsu to introduce eleven new products in two and a half years, shocking its competitor Caterpillar, which managed only one to two per year!

Customer satisfaction is measured and tracked using a mixture of hard and soft measures tailored to different market segments. The results are then compared to key competitors and industry averages. A system for keeping customers satisfied is to provide easy means for their complaints and prompt resolution of complaints. TQM companies accumulate information on customers in a central database so that this intelligence can be used to drive improvements.

Continuous improvement

Continuous improvement or *kaizen* is one of the pillars of TQM. Kaizen is the Japanese term for continual improvement involving everyone—both managers and workers. In manufacturing, kaizen means finding and eliminating waste in machinery, labor, or production methods. Such an improvement can be either incremental or breakthrough ("reengineering") in nature.

Applied to manufacturing processes, improvements start with controlling the variation in the quality characteristic of the output. In this context, a landmark development took place in the 1920s, when Dr. Walter A. Shewhart of Bell Telephone Laboratories developed the theory of statistical process control (SPC). It was significant because it focused not on the output of the process, but on the monitoring, and, more importantly, on the improvement of the production process. His study of different processes led to the conclusion that all manufacturing processes exhibit variation. He identified two components: a steady component, which appeared to be inherent in the process, and an intermittent component. Shewhart attributed inherent variation, currently called common cause or systemic variation, to chance and undiscoverable causes, and intermittent variation to assignable or special causes. He also developed, based on statistical methodology, a graphical device, which became known as control chart—a misnomer because in reality it does not "control" any aspect of the process—to monitor the variation in the process and signal the presence of special causes.

His major finding was that assignable causes could be economically discovered and removed with a tenacious diagnostic program, but common causes could not be economically discovered or removed without making basic changes in the process (Shainin and Shainin, 1988). To date, this insight remains fundamental to monitoring and improving the quality of manufacturing processes. Deming's lectures to the Japanese industrialists in 1950, which served as a catalyst for the quality revolution in that country, and his theory of management were based on Shewhart's theory (Deming, 1986).

The manufacturing process has to be made stable and predictable, and brought under statistical control by identifying and removing the special causes of variation. Only then does it make sense to find ways to reduce the systemic variation, i.e., to improve the process. Once the process is improved, it is standardized and documented to enable transfer of knowledge and consistent application. Another cycle of improvement, commonly referred to as the Plan, Do, Check, Act (PDCA) Cycle, starts all over again.

Mistake-proofing

In many modern manufacturing systems using high speed production, the use of SPC is not enough to prevent defects. The signals that control charts provide are often too late to take any preventive action. In such cases, Shigeo Shingo's innovation of mistake-proofing (Poka-Yoke) can be used. Mistake-proofing is a proactive and generally inexpensive technique for building quality into the manufacturing process and preventing problems (Shingo and Robinson, 1990). Examples are manufacturing or setup activities designed to prevent errors that could lead to product defects. For example, in an assembly operation, if every correct part is not used, a sensing device detects that a part was unused and shuts down the operation, thereby preventing advancement of the incomplete assembly to the next station.

Taguchi method

As the evolution from small "q" to big "Q" took place, some important concepts and tools were developed. Of particular relevance to manufacturing are two methodologies that are used at the design (off-line) stage. One is the set of tools to evaluate and improve the manufacturability of design. For instance, reducing the number of parts and increasing the use of common and standard parts across designs, improve manufacturability and ease of assembly. The other methodology is quality engineering, attributed to Dr. Genichi Taguchi. It consists of off-line quality control methods applied at the product and process design stages in the product development cycle. This concept, developed by Dr. Taguchi, encompasses three phases of product design: system design, parameter design, and tolerance design. The goal is to reduce quality loss by reducing the variability of the product's characteristic during the parameter design phase of product development. The application of such techniques upstream at the design stage prevents poor quality occurring downstream at the production stage.

Six sigma

Another powerful concept, pioneered by Motorola, and currently used to drive improvements in many companies, is *six sigma*. Here process variations are reduced to half of the design tolerance, and process mean could shift as much as 1.5 times the process standard deviation from the target to the process mean. This limits the defect rate to 3.4 defects per million. In addition, if the process mean is centered on target, the defect rate is reduced to 2 defects per billion. This lofty goal is attained by reducing process variations due to insufficient product design margin, inadequate process control, and less than optimum parts and materials. Six Sigma could be applied to everything a company does, including administrative activities such as filing, typing, and document preparation.

Motorola launched the *six sigma* program because it found that a four-sigma (defect rate of 6210 per million) manufacturer, who spends an excess of 10% of sales revenues on internal and external repair, cannot compete against a six sigma manufacturer, who spends less than 1%. Motorola also stressed reductions in cycle times in all elements of its business, and set a goal of ten-fold improvements in cycle time to be achieved in five years. They found the goals of six sigma and cycle time reduction to be mutually supportive—cycle times were reduced when fewer mistakes were made.

Benchmarking

Often an important question to be addressed in TQM is *what* should be improved. This can be answered by competitive and comparative benchmarking. Benchmarking is the continuous systematic search for, and implementation of, best practices which lead to superior performance. First, key processes of strategic importance are identified. Then, for each key process, the best competitors or the best-in-class companies are bench-marked, gaps in performance are assessed, priorities are developed and improvement initiatives launched to close the gap.

Employee involvement

Involvement of all employees is absolutely critical to the success of TQM efforts. Continuous improvement and learning cannot take place if employees are not trained and motivated to improve processes. Since the knowledge of production workers plays a crucial role in identifying and eliminating special causes of process variation, training and empowering the workers involved in the process and creating an organizational climate in which they can apply their knowledge is essential. Systemic variations are beyond the control of workers who work within the system created by the management.

For a long time, managerial implications were not recognized with respect to quality because process control was considered to be the domain of the quality control (QC) specialist. Now, top management in TQM-oriented companies view their employees as internal customers, and consider it their responsibility to provide them with the tools and the authority to take actions to improve customer satisfaction and work processes. A systematic approach is used to solicit suggestions from all employees. This approach also recognizes and rewards them for their ideas. Employees are empowered to operate in autonomous, self-directed teams, engage in problem solving, participate in projects like benchmarking and mistake-proofing, and manage their own work. Companies like Motorola, and Dana Corporation have set up corporate universities and made significant investments in employee training and education.

DOES TQM WORK?

Many studies now provide the evidence that generally TQM works and has a positive impact on: (1) internal measures ranging from productivity to inventory turns;

(2) market measures such as stock prices, market share, revenue growth, customer satisfaction; and (3) bottom line measures such as profitability, reduction in costs, and in legal liabilities (Hiam, 1993). Deming Prize winners have demonstrated profit levels twice those of other Japanese companies. Studies conducted by the National Institute of Standards and Technology (NIST) in the U.S. in 1995 and 1996 showed that, over time, the stocks of Baldrige award-winning companies outperformed the market index by 4 to 1. The studies mentioned above also indicate that TQM has produced better results at some companies than others. This is due to the differences in the approaches used and implementation.

There are a number of challenges faced by companies that implement TQM. As companies learn to make use of self-directed work teams in TQM, they are likely to meet with resistance because it involves transfer of work and responsibility from specialists and supervisors to the work force. Empowerment of workers has to be managed and accompanied by proper training. In the short term, this could be seen as adding to the costs. In many companies, there are a myriad of initiatives such as team building, SPC, and ISO 9000, without any linkage to the overall strategy. Without coordination by top management and strong linkages, they will be ineffective.

Globalilzation creates a problem because of the differences in quality systems in business units dispersed over the globe. Geographic distribution of people also precludes maximum interaction. Such companies have to balance central coordination and standardization of quality management with the local requirements and autonomy. TQM process is vulnerable to sharp downturns in the market and the economy, but TQM puts firms in a far better position to recover because of the superior management systems it creates. During downturns in the market, the commitment of top management to TQM can become weak; layoffs, not quality, may be seen as the way to cut costs, but they often lead to loss of vital knowledge about processes and customers. TQM can complement downsizing if the latter is based on customer satisfaction and continuous improvement needs, and is not a thoughtless reaction to reduce head count. Mergers, acquisitions, outsourcing, and takeovers are often accompanied by leadership and organizational changes and can create uncertainty about the future and direction of TQM within organizations.

Key Concepts: Benchmarking; Control Chart; Cycle Time; Deming Prize; Design for Manufacturability; Employee Empowerment; *Hoshin Kanri*; Integrated Quality Control; ISO 9000; ISO 9000/QS 9000 Quality Standards; Just-in-Time Manufacturing (JIT); *Kaizen*; Malcolm Baldrige National Quality Award; Management-by Policy; Mistake-proofing; Plan–Do–Check–Act (PDCA); *Poka-Yoke*; Policy Deployment; Pull System; Quality Circles; Quality Function Deployment (QFD); Six Sigma; Statistical Process Control (SPC); Taguchi Method.

Related Articles: The Implications of Deming's Approach; Just-in-Time Manufacturing; Lean Manufacturing Implementation; Total Productive Maintenance.

REFERENCES

Belohlav, J.A. (1995). "Quality, Strategy, and Competitiveness." *The Death and Life of the American Quality Movement*. Edited by R.E. Cole. Oxford University Press, New York, 43–58.

Deming, W.E. (1986). *Out of the Crisis*. MIT Center for Advanced Engineering Study. Massachusetts.

Dobyns, L. and C. Crawford-Mason (1991). *Quality or Else: The Revolution in World Business*. Houghton Mifflin, New York.

Godfrey, A.B. (1993). "Ten Areas for Future Research in Total Quality Management." *Quality Management Journal*, October, 47–70.

Hiam, A. (1993). "Does Quality Work? A Review of Relevant Studies." *Conference Board Report Number 1043*. The Conference Board, Inc., New York, 7–38.

Ishikawa, K. (1985). *What is Total Quality Control? The Japanese Way*. Translated by David J. Lu. Prentice-Hall, New Jersey.

Juran, J.M. (1993). "Made in U.S.A. A Renaissance in Quality." *Harvard Business Review*, July–August, 42–50.

Juran, J.M. (1995). "Summary, Trends, and Prognosis." *A History of Managing for Quality*. Edited by J.M. Juran, ASQ Quality Press, Wisconsin, 603–655.

Reimann, C.W. and H. Hertz (1993). "The Malcolm Baldrige National Quality Award and ISO 9000 Registration: Understanding Their Many Important Differences." *ASTM Standardization News*, November, 42–51.

Shainin, D. and P.D. Shainin (1988). "Statistical Process Control." *Juran's Quality Control Handbook*. Edited by J.M. Juran and F.M. Gryna (4th ed.). McGraw-Hill, New York. Section 24, 24.1–24.40.

Shiba, S.A. Graham and D. Walden (1993). *A New American TQM: Four Practical Revolutions in Management*. Productivity Press, Oregon, 3-30, 411–460.

Shingo, S. and A.G. Robinson (1990). *Modern Approaches to Manufacturing Improvements: The Shingo System*. Productivity Press, Massachusetts.

Wever, G.H. (1997). *Strategic Environmental Management: Using TQEM and ISO 14000 for Competitive Advantage*. John Wiley, New York.

Womack, J.P., D.T. Jones, and D. Roos (1991). *The Machine that Changed the World: The Story of Lean Production*. Harper Collins, New York, 37–38, 91–93.

8. THE IMPLICATIONS OF DEMING'S APPROACH

ELISABETH J. UMBLE

Texas A&M University, College Station, Texas, USA

ABSTRACT

Many current management concepts can be traced back to Deming's teachings. Quality was his central theme, and most "Total Quality Management" (TQM) programs are based on his beliefs about quality. He emphasized the need to break down barriers between management and labor and to effectively utilize employees to make processes work better. This eventually led to powerful concepts like worker empowerment and cross-functional teams. Highly effective manufacturing systems, such as "Just-in-Time Manufacturing," would not have been possible if not for the quality-based improvements that resulted in reliable processes as well as the supply of high quality products from vendors. And, Deming's philosophy about the need to overhaul and improve business processes laid the foundations for the current trend toward "reengineering." Deming's theories on quality can be found in *Quality, Productivity, and Competitive Position* (1982), *Out of the Crisis* (1982), and *The New Economics for Industry, Government, and Education* (1993).

In this article read how the following organizations have gainfully used selected ideas of Demings: **Army Ordnance; Ford; Graphic Enterprises; Nashua Corporation; Ryan Transport Management Systems; War Department; War Production Board; Xerox Corporation.**

WHO WAS DEMING?

W. Edwards Deming (1900–1993) was a statistician, physicist, consultant, and teacher. He is best known for developing a system of Statistical Quality Control, although, his contributions far exceed the contribution of those techniques. He contended that quality is the responsibility of top management and that high quality costs less than poor quality. He also championed the idea that quality must be built into the product at all stages of design, development and production in order to achieve excellence.

Deming was one of the earliest well-known proponents of quality, and his philosophy concerning the managerial changes necessary to achieve quality changed management practices in both manufacturing and service industries worldwide. His work illuminated and extended that of earlier proponents of quality control and productivity. Deming's many contributions to quality improvement and the corresponding managerial changes, that are a prerequisite for quality, include his famous 14 Points, his ideas based on Statistical Process Control, and his System of Profound Knowledge.

HISTORY

Quality has always been important in manufacturing. Before the industrial revolution, skilled craftsmen served both as manufacturers and inspectors, and their pride of workmanship inspired them to build quality into their products. By the turn of the twentieth century, mass production techniques had become firmly entrenched in manufacturing. In 1911, Frederick Taylor published *Principles of Scientific Management*, substantially changing the nature of quality assurance. By focusing on production efficiency and decomposing jobs into small work tasks, the assembly line changed the holistic nature of manufacturing. The ensuing specialization necessitated the development of independent "quality control" departments that used inspection to ensure that products were manufactured correctly. Thus, "inspecting out" defective items became the chief means of ensuring quality.

Statistical Quality Control (SQC), the forerunner of today's Total Quality Management, had its beginnings in the mid-1920s at the Western Electric plant of the Bell System. Walter Shewhart, a Bell Laboratories physicist, designed the original version of SQC to help eliminate defects in the mass production of telephone exchanges and sets. In 1924, Shewhart made the first sketch of a modern "control chart." The new technique was subsequently developed in various memoranda and articles; and in 1931, Shewhart published a book on statistical quality control which bore the title *Economic Control of Quality of Manufactured Product*. This book provided a precise, measurable definition of quality and promoted statistical techniques, such as control charts, for evaluating production processes and improving quality. In the early thirties, Bell Systems, in collaboration with the American Society for Testing and Materials (ATSM), the American Standards Association (ASA), and the American Society for Mechanical Engineers (ASME), undertook to popularize the new statistical methods in the United States. But, despite their support, the rate of adoption of the new techniques proved to be extremely slow.

The initial reluctance of American industry to adopt statistical quality control was rapidly overcome during World War II. The armed services became large consumers of U.S. industrial output, and consequently, influenced the adoption of statistical quality control in two ways. First, the armed services themselves adopted scientifically designed sampling inspection procedures. The initial step in the development of military sampling inspection procedures occurred shortly after America's entry into the war, when a group of Bell Lab engineers were brought to Washington to develop a sampling inspection program for Army Ordnance. Second, the military established educational programs for military and industrial personnel. At the request of the War Department, the American Standards Association developed concise statements of American control-chart practices and published materials for training courses. Between 1943 and 1945, representatives from 810 organizations attended one or more of the 33 intensive courses on SQC offered by the Office of Production Research and Development of the War Production Board. Among those attending were faculty from 43 different educational institutions who wished to prepare themselves to teach quality control.

After World War II, many of the SQC techniques, which had been developed and applied in the production of military goods, could have been readily applied in the consumer goods sector. However, many U.S. executives stubbornly resisted the implementation of the new techniques. Deming personally experienced this resistance when he delivered a series of seminars on SQC for auto industry executives at Stanford University in 1945. Because of such experiences, Deming became somewhat bitter toward U.S. management.

In 1946, under the auspices of the Economic and Scientific Section of the U.S. Department of War, Deming spent two months in Japan assisting U.S. occupation forces with studies of nutrition, agricultural production, housing, fisheries, and so forth. Quality control concepts were first introduced to Japanese industry in the form of an occupation forces order to communications equipment manufacturers. During this and many subsequent visits to Japan, Deming discussed his ideas on quality control with members of the Union of Japanese Scientists and Engineers (JUSE). In late 1949, Deming was invited to Japan to teach statistical methods for industry. He preached a very simple message: *controlling the process, rather than inspecting items that come from the process*, is the key to quality; process control depends on a knowledge of statistical process control concepts; and total quality management requires the participation and training of all members of the organization. The Japanese took Deming's message to heart. By the 1980s, many Japanese companies that adopted Deming's approach were outperforming their U.S. competitors.

Eventually, organizations around the world saw the benefits of implementing Deming's philosophy. While Deming is not solely responsible for quality improvement in Japan or the U.S., he played a key role in increasing awareness of the process of quality control and the need to improve. Few people have changed worldwide business practices as positively and completely as did Deming. Today, he is regarded as a national hero in Japan and has been honored by the Japanese with the world-famous Deming Prize for Quality which recognizes successful efforts in instituting company-wide statistical quality control principles.

STATISTICAL PROCESS CONTROL

Shewhart (1931) established his criteria for determining when numerical data are in statistical control. He demonstrated how graphs can be used to identify fluctuations in the process and to determine when a process is exhibiting more than simple random variation. This kind of analysis can be applied to distinguish common causes of trouble that occur due to the natural variability of the system from local sources of trouble, which Shewhart called assignable causes of variation. Local sources of trouble must be eliminated before managerial innovations leading to improved productivity can be achieved through removal of common causes.

Some consider Deming's most significant contribution was recognizing the tremendous potential of Shewhart's work and the use of statistical tools for the continuous improvement of production processes and the delivery of a quality product. He emphasized that variability in manufacturing and service processes can be traced to either common causes or special causes (Shewhart's assignable causes). Common cause variability is that variability naturally inherent in the system. Common cause variability is not necessarily acceptable, but it cannot be removed by employees attempting to do better. Problems generated by common cause variability can only be addressed by the management by improving the system. On the other hand, special cause variability is that variability beyond the natural variability of the process. Such variability can be addressed by employees.

Deming believed that 85% of the variation in a process is due to common causes and only 15% is due to special causes. Deming stressed the importance of clearly identifying the source of the variability. For example, if an employee tries to compensate for variability without any statistical guidance, the employee will probably either overadjust or underadjust to the variability. This will not cure the problem and may actually increase variability. And when employees' best efforts fail to produce improvement, they may become confused and demoralized, eroding their creative spirit and pride of workmanship, and making future process improvements unlikely.

Deming relied on control charts to describe both the natural variability of the system and to detect the existence of special cause of variations such as broken tool, worker fatigue, etc. A control chart is a time-ordered plot of a sample statistic—usually a sample mean, range, proportion, or number of defects—and upper and lower control limits that reflect the magnitude of common cause variation present in the system.

Suppose we want to use a control chart to monitor a process that produces one-inch diameter bolts. Upper and lower control chart limits for the mean diameter of bolts can be developed from previous production runs. These limits describe the normal amount of variability in the bolt diameters. Bolts are then periodically sampled from future production runs and the average bolt diameters from each sample are plotted on the control chart. As long as average bolt diameters fall within the control limits and exhibit no particular pattern, the process is considered to be in statistical control. However, this only means that the process is acting as it has in the past. A process that is in statistical control may still produce defective items; it is stable only in the sense that the amount of variability is predictable. A sample mean diameter beyond the specified limits can be attributed to special causes that increase the amount of variability in the system.

DEMING'S 14 POINTS FOR MANAGEMENT

Deming's goal was to make companies more competitive and keep them in business. Because he believed 85% of the problems in a system to be the responsibility of management, he offered 14 Points for management as the basis for transforming industry. These can be applied to large or small organizations, service or manufacturing industries, or divisions within a company. The 14 Points are:

1. Create constancy of purpose toward improvement of product and service with a plan to become competitive and stay in business.
2. Adopt the new philosophy of quality.
3. Cease dependence on inspection to achieve quality. Require instead statistical evidence that quality is built in.
4. End the practice of awarding business on the basis of price tag. Consider quality as well as price. Move toward a single supplier for an item, based on a long-term relationship of trust.
5. Improve constantly and forever the system of planning, production, and service, to improve quality and productivity and constantly decrease costs.
6. Institute modern methods of training on the job.
7. Institute modern methods of supervision that emphasize quality rather than sheer numbers.
8. Drive out fear.
9. Break down barriers between departments.
10. Eliminate slogans, exhortations, and targets for the work force.
11. Eliminate numerical quotas for the work force and numerical goals for management.
12. Remove barriers that rob workers (and management) of their right to pride of workmanship.
13. Institute a vigorous program of education and retraining.
14. Put everyone in the company to work to accomplish the transformation.

Deming believed that the consumer is the most important part of the production line. He defined quality as the translation of future needs of the consumer into measurable characteristics so that a product or service can be designed and produced to give satisfaction at a price the user will pay and that will make the user better off in the future. Low quality costs money because it means rework or scrap; wastes material, labor, and machine capacity; increases the cost of the product; and causes a loss of sellable product. As quality improves, costs will decrease and productivity will increase, resulting in greater market share, more jobs, and long-term survival. Deming argued that high quality is achieved by measuring, controlling, and improving the processes involved in the design, production, and delivery of goods or services.

In his 14 Points, Deming highlighted some key business practices that cost companies dearly. These practices include the awarding of contracts to the lowest bidder while ignoring quality, creating obstacles that rob workers of pride of quality workmanship, and trying to inspect quality into products. For example, Deming points out that, in most practical cases, "all or none" sampling will minimize inspection costs. For a process in control, cost is minimized with no sampling, when proportion defective is less than the break-even proportion, which is calculated as cost to inspect divided by cost to repair. Further cost is minimized with 100 percent sampling, when proportion defective exceeds the break-even proportion. Because properly motivated workers are necessary for the production of high quality products, Deming stressed worker pride and satisfaction rather than slogans and numerical goals. Furthermore, since the system, rather than the worker, is the root cause of most problems, Deming's overall approach focused on improving the system, which is management's responsibility.

The Deming cycle (Plan, Do, Check, Act; also known as PDCA) is a methodology for continuous improvement. This methodology was originally developed in *Statistical Method from the Viewpoint of Quality Control* (1939) by Shewhart. Deming suggested that this procedure should be followed for the improvement of any stage of production, and as a procedure for finding a special cause of variations detected by statistical signals. The Plan stage involves studying the current situation, gathering data, and planning for improvement. The Do stage consists of implementing the plan on a trial basis. The Check stage is designed to determine if the trial plan is working and to see if any further problems or opportunities have been discovered. The Act stage consists of implementing the final plan. This leads back to the Plan stage for further diagnosis and improvement. The cycle is never ending. Deming used this Plan, Do, Check, Act cycle as the basis for accomplishing his 14th Point by putting everyone in the company to work to accomplish the transformation.

THE SYSTEM OF PROFOUND KNOWLEDGE

Deming described a System of Profound Knowledge (SOPK) which provides insights into the process of effectively transforming organizations. The overriding theme of SOPK is that the prevailing style of management must first undergo a transformation. There are four interdependent parts of SOPK. The first part of SOPK is an appreciation of the system and systematic thinking. A system is a set of functions or activities within an organization that work together to achieve organizational goals. To run any system, managers must understand the interrelationships among all subsystems and the people that work in them.

The second part of SOPK is knowledge of theory of variation. A production process contains many sources of variation. The complex interaction of the variations in materials, tools, machines, operators, and the environment cannot be understood in isolation. However, the combined effect of all sources of variation can be examined statistically.

The third part of SOPK is theory of knowledge—a branch of philosophy concerned with the presuppositions, nature, and scope of knowledge. Deming emphasized that there is no knowledge without theory and that experience alone cannot establish a theory. Managers thus have a responsibility to learn and apply theory. The final part of SOPK is psychology. Psychology helps managers understand people. Managers must recognize differences between people and use psychology to nurture each individual's positive innate attributes. The 14 Points for management follow naturally as an application of Deming's SOPK.

SUCCESSFUL IMPLEMENTATIONS

For over 40 years, Dr. Deming served as a world-renowned consultant in statistics and quality. Companies that have successfully implemented Deming's approach to quality management have experienced reduced inventories, decreased costs, increased profits and improved worker morale and labor relations. Two examples of companies that have benefited from Deming's philosophies follow.

In the book, *Quality, Productivity, and Competitive Position*, Deming offers the following example. A superintendent at a plant knew there were problems with a certain production line. His only explanation was that the work force of 24 people made a lot of mistakes. The first step of a consulting statistician was to obtain data from inspection and plot the fraction defective day by day over a six-week period. This plot (a run chart) showed statistical control with stable random variation above and below the average. This meant that any substantial improvement had to come from actions on the system, which is the responsibility of management. What could management do? The statistician suggested that perhaps the people on the job and the inspector did not fully understand what constituted acceptable work. Operational definitions were developed and posted for everyone to see. The result was a substantial reduction in the proportion defective. The gains were immediate and the costs were essentially zero, all accomplished with the same work force and no investment in new machinery. The next step was to reduce the proportion defective through better incoming materials and better maintenance of equipment, both the responsibility of management.

The case of Graphic Enterprises

A more recent illustration of a successful implementation of Deming's approach is found in Graphic Enterprises, Incorporated of Detroit, Michigan. The company was on the brink of bankruptcy. Following Deming's principles, the company decided to train and empower action teams to pursue special tasks. These teams implemented many improvements. The company began to monitor and control humidity levels in its storage room to keep paper from becoming sticky and from acquiring a static charge. A customer survey log was instituted to follow up every job, and, internal statistics were used to keep track of the quality of maintenance. As a result, the company became profitable, and sales growth jumped to between 6% and 10% per year.

The case of Ford Motor Company

Deming visited Ford Motor Company in 1981 to meet with its president, Donald Petersen and other company officials. Subsequently, Deming gave seminars for top executives and met with various employee groups, suggesting changes in accordance with his 14 Points. Ford managers visited Nashua Corporation, the first American company to incorporate Deming's philosophy, to learn how statistical methods were used there, and chief executives from many of Ford's major suppliers visited Japan. The 14 Points became the basis for a transformation in Ford's management.

Ford's quality commitment was embodied in its "Guiding Principles": quality comes first, customers are the focus of everything we do, continuous improvement is essential to our success, employee involvement is our way of life, dealers and suppliers are our partners, and integrity is never compromised. Management at Ford endeavored to create an environment where everyone could contribute to continuous improvement. Subsequent to the adoption of these ideas, over the period 1982 to 1984, direct labor productivity at Ford increased by 13%, setup time improved by 80%, inventory turns increased from 1.9 to 4.0, and customer service was substantially improved.

The case of Ruan Transportation Management System

Ruan Transportation Management Systems followed Deming's prescription to trim expenses and improve profits. After the federal government deregulated the trucking industry in 1980, Ruan faced new low-cost competitors. To compete, Ruan decided to develop the "mega truck," the first truck designed to last one million miles. Ruan introduced quality circles to help with the development of the truck. At first, the quality circles didn't work because management had no idea what they were supposed to be doing. Then, company executives decided to learn and apply Deming's methods and SPC. Top executives made quality integral to the company's mission.

Employees at all levels were trained to effectively investigate and analyze problems, and they began collecting detailed data about how the company worked. For example, Ruan trucks were experiencing an affliction known as engine "pitting." The pitting was caused by insufficient radiator coolant; the manufacturers claimed that Ruan's maintenance workers weren't filling the radiators properly, and therefore would not honor warranties. Then a team of maintenance workers created a detailed chart of when engine problems were occurring. It turned out that things got worse when the weather changed significantly. The employee team discovered that the company was using a radiator hose that expanded and contracted with changes in temperature, causing coolant leaks. The company purchased a new kind of flexible clamp that kept the hoses properly sealed, which solved the pitting.

Deming's ideas also enabled Ruan employees to determine a major cause of excessive maintenance costs. The employees began keeping statistics on maintenance expenditures. One problem was: headlights were burning out too quickly. The company equipped all trucks with long-life halogen headlights, resulting in substantial savings.

The Deming philosophy eventually led to Ruan's "added value" emphasis. By charting every move, employees gained understanding of what they did, and what their purpose in the company was. The data showed that many workers were underutilized and could use their time providing additional services to clients. By more effectively utilizing employees, Ruan was able to provide services such as driver training, fuel purchasing, fleet dispatching, and vehicle routing. In 1993, these activities accounted for about half of the $300 million in revenue the company was expected to generate.

LESSONS

A change in managerial mindset is required in order to successfully implement Deming's ideas. Companies that undertake programs to improve quality without changing the way the company is managed are doomed to failure. Thus, Deming would work with a company only on the invitation of top management, only if top management was willing to actively participate, and only on a long-term basis. Deming noted that the biggest obstacle to improving quality was the lack of constancy of purpose. Improvement takes time, determination, and commitment to change. Workers must perceive that the push for quality is more than just another gimmick or slogan that will have little or no impact on the way the company is run.

Layoffs and labor difficulties are possible symptoms of a failure in the implementation of Deming's concept of quality. Xerox Corporation, considered to be a national leader in total quality, embraced "Leadership Through Quality," a process aimed at providing innovative products and services that fully satisfy customers. The movement depended on employee involvement and empowerment. Employee support was eroded when Xerox decided in 1993 to cut ten percent of the company's document processing work force worldwide. Workers said they feared for their jobs; such fear is disruptive and stifles future improvements. Deming contended that fear is a symptom of failure in hiring, training, supervision, and motivation, and can adversely affect the ability to produce a quality product that the market is willing to pay for.

Finally, in order to realize Deming's benefits of quality improvements (the cycle in which improved quality leads to lower costs, resulting in lower prices that increase the market share, ultimately, leading to more jobs), quality must be related to other business objectives, plans, and functions. Quality must be integrated with other dimensions of business that deliver value to the customer, such as on-time delivery, low cost, and increased flexibility.

Key Concepts: All or None Sampling; Assignable Causes of Variation; Common Causes of Variations; Deming Cycle; Deming Prize; Deming's 14 Points for Management; Integrated Quality Control; ISO-9000/QS-9000; Malcom Baldridge Award Criterion; Natural Variability; Performance Excellence; Quality Assurance; Quality Management Systems; Quality Standards; Sampling in Quality Control; Special Causes of Variations; Statistical Process Control; Statistical Quality Control; Variability in Manufacturing.

Related Articles: Just-in-Time Manufacturing; Lean Manufacturing Implementation; Performance Excellence: Malcolm Baldridge National Quality Award Criteria; Total Productive Maintenance; Total Quality Management.

REFERENCES

Deming, W. Edwards (1982). *Out of the Crisis*, Center for Advanced Engineering Study, Massachusetts Institute of Technology, Cambridge, Massachusetts.
Deming, W. Edwards (1982). *Quality, Productivity, and Competitive Position*, Center for Advanced Engineering Study, Massachusetts Institute of Technology, Cambridge, Massachusetts.

Deming, W. Edwards (1993). *The New Economics for Industry, Government, and Education*, Center for Advanced Engineering Study, Massachusetts Institute of Technology, Cambridge, Massachusetts.

Shewhart, W.A. (1931). *Economic Control of Quality of Manufactured Product*, E. Van Nostrand Company, New York.

Shewhart, W.A. (1939). *Statistical Method from the Viewpoint of Quality Control*, Graduate School, Department of Agriculture, Washington.

Taylor, Frederick (1911). *Principles of Scientific Management*, Harper and Row, New York.

VI. THE RISE OF WORK TEAMS

9. TEAMS: DESIGN AND IMPLEMENTATION

JOHN K. MCCREERY

North Carolina State University, Raleigh, North Carolina, USA

MATTHEW C. BLOOM

University of Notre Dame, Notre Dame, Indiana, USA

ABSTRACT

The decision to use work teams should be placed in the broader context of the organization's competitive and manufacturing strategies. The overall premise is that, for work teams to be truly effective within the organization, they must be geared to provide capabilities or outcomes that are valued by the firm as a whole. In other words, teams should add value to the firm's products and services in ways that are meaningful to the marketplace. This implies an outward-looking view of team design: teams should be empowered to take actions that, either directly or indirectly, support the organization's sources of competitive advantage.

Depending on its charter and design, manufacturing work teams may provide improvements in operational efficiency, quality, delivery, and flexibility. In turn, the organization may exploit these improvements through either low price or product differentiation strategies.

The use of teams as a way to organize people and conduct work is not a new phenomenon in business. What is new is the dramatic increase in the use of teams and the variety of purposes for which they are used. Teams are becoming pervasive in manufacturing organizations because they provide a variety of benefits (Katzenbach and Smith, 1993). While teams certainly have their place in the modern manufacturing organization, they should not be formed indiscriminately. If not used properly, teams may fail to provide the benefits hoped for and may even be harmful to manufacturing performance (Campion, Medsker and Higgs; 1993).

HISTORY

For many years, managers thought of teams as primarily a productivity tool. The old admonition that "many hands make the work light" served as the guiding principle: teams could simply do more work. This view was later expanded somewhat and viewed teams as a vehicle to induce desireable behaviors. The idea was that people who work as part of a team would be more likely to help each other, offer advice and counsel, or provide informal, on-the-job training to less experienced workers. The central focus of teams was to concentrate more people on a particular task.

In the decades after World War II, conventional wisdom shifted toward seeing teams as a way to increase quality. Quality circles (QCs) first came into use in Japan in the early 1960's (Ishikawa, 1989) and were tremendously successful in fostering quality improvements. Over the next decade, as word of Japan's success with QCs spread, the concept was introduced in the U.S. by managers who realized that a great deal of untapped knowledge resided in the minds of workers. Teams were seen as a mechanism to tap this store of knowledge because the attention of a group of workers could be focused on day-to-day work activities. Since workers know these situations best, it was reasoned, why not get them to think about these situations more broadly?

The success of these early efforts suggested that teams might be a way to increase innovation, better meet customer needs, and facilitate change (Osburn, Moran, Musselwhite and Zenger; 1990). "Teams can solve all your problems" was an often—heard rallying cry. Today, views are moderating as stories of team failures come in (Hackman, 1998). Attention is turning toward seeking to understand when teams can be helpful, and how they should be structured for optimum performance.

TEAM DESIGN

A number of key implementation issues must be considered when designing teams. The important one being the type of team needed to suit the task.

Types of teams

Casual observation tells us that teams come in many shapes and sizes. Although there are many ways to classify teams (Goodman, Devadas and Hughson, 1988; Guzzo and Dickson, 1996), we suggest Table 1 as a useful way to think about teams in a manufacturing context. Table 1 presents what we might call "pure types". In practice, it is likely that teams will exhibit the characteristics of more than one type, but the typology in Table 1 helps suggest what the principal features of teams are, and how they can be used to facilitate manufacturing performance.

Ad-hoc teams

At one end of the typology are teams that are primarily concerned with solving temporary problems, addressing specific issues, or recommending remedial actions

Table 1 Team design continuum

	TEAM TYPE	TEAM CHARACTERISTICS
Increasing ↑ SCOPE OF RESPONSIBILITY — DEGREE OF AUTONOMY ↓ *Decreasing*	**Autonomous Work Teams**	♦ Team makes all decisions or have veto power regarding the pace and flow of their work. ♦ Team determines the best way to structure and complete all phases of their work.
	Self-directed Work Teams	♦ Employees supervise their own work. ♦ Make decisions about the pace and flow of work. ♦ May make decisions about the best way to get work done
	Quality of Work Life Initiatives; Problem Solving Teams; Quality Circles	♦ Team identifies work problems or important work issues. ♦ Team makes recommendations or suggestions to management for resolving work problems. ♦ Team does not make final decisions.
	Consultative Teams	♦ As required, management informally seeks worker input and opinions about work issues.
	Ad-hoc Teams	♦ As needed, teams of employees form temporary groups to analyze work issues and recommend remedial action or changes.

to changing conditions. These teams are characterized by irregular team activities and a focus on temporary versus on-going issues and activities. Ad-hoc teams meet sporadically or for short, intense periods of time. Sometimes, the team is created to deal with a specific problem and may disband after a solution has been tendered and implemented. Other times, ad-hoc teams use people from on-going teams, but it may meet occasionally. The normal work responsibilities of the individual team members are usually not highly interdependent; members can perform their routine duties without relying on the team. Team activity is centered around specific, one-time or short-term issues. Team members may come from different functional areas to ensure that the team is complete and balanced.

Consultative teams

These are somewhat akin to ad-hoc teams. The primary responsibility of consultative teams is to provide information for other decision makers. For example, a consultative team may be used as a sounding board to test proposed changes, an example is an internal focus group. Or, managers can use the team to acquire important information, or to get the ideas or perspectives of workers about an issue or problem. Here, the team is a resource for information, data, ideas, opinions, and the like. Again, these teams can be temporary or more permanent. In either case, team activities are irregular and do not comprise the daily or normal work responsibilities of the team members.

Quality of work life initiatives, problem solving teams, or quality circles

Quality circles (QC) became popular in the late 1970's and are still around in various forms. They represent a higher level of team functioning in several respects. First, members are usually permanent; temporary QC teams are rare. Second, QC teams meet regularly and team activities are a routine part of team members' work responsibilities. Third, these teams are usually organized around a specific part of the work process or a clearly defined segment of work. For example, an automobile manufacturer might organize a QC around its body painting group, engine assembly group, etc. Members usually come from the same or very similar functional areas. Fourth, QC teams are expressly formed to *identify* work problems or important work issues. This means, for example, they can be used to uncover hidden problems in manufacturing processes or diagnose the root cause of symptoms such as consistent defects. Fifth, QC teams usually focus on a wide range of issues and problems that are common to a particular area. However, QC teams make recommendations to management for resolving work problems; employees do not make final decisions. Finally, QC teams usually have a long-term mandate, and have stable membership (team members do not change often).

Self-directed work teams

Self-directed teams represent a key point of demarcation in the typology because they consist of employees who supervise their own work. Here, the team makes decisions about such things as the pace and flow of work. An important distinction between these and the other teams discussed so far is the scope of decision making. Self-directed teams are usually given specific performance targets or final output goals, and then it is left to the team to determine what specific tasks or duties need to be done, when they should be conducted, and how workers, materials, and equipment should be utilized. They may be comprised of members from different functional areas (so-called cross-functional teams), but the work of these members is highly interdependent. Functioning within the team is integral to performing the normal work responsibilities of all members. That is, members rely on the team to perform most of their duties. These teams are permanent and have stable membership.

Autonomous work teams

Autonomous work teams are the most complex form and represent the high-end of the typology. These are truly self-contained teams. Organizations set broad performance targets (e.g., meet customer needs), and the team decides not only what must be done to meet the goal, but how to measure the goal itself. Autonomous work teams make all decisions about work, or have veto power regarding the best way to structure and complete all phases of their work. They set interim goals, establish performance standards, develop regulatory policies, and manage resources such as budgets, equipment, and materials. Often, autonomous teams select new members, discipline poorly performing members, and may even set human resource policies. They most often consist of members from a broad spectrum of functional areas, and usually include members from management, professional, or technical levels. That is, team members are usually highly skilled and deal with complex tasks. Autonomous work teams can be accurately thought of as mini-organizations in that most of their activities are self-contained, independent, and not subject to significant external controls.

IMPLEMENTATION

Motivational issues

A fact often ignored by managers is that successful implementation of work teams requires paying attention to the human element: teams are made up of people. Issues such as member support, communication, and celebration are important because they are integral to properly motivating people to perform well on teams.

One factor, which seems to be more important, as one moves up the typology from ad hoc to autonomous work teams is the notion of trust. Sometimes called altruistic attachments, teams seem to function better when the members trust each other. The idea of a combat unit in war times is perhaps one of the best examples of trust and its importance for teams. Team members must trust that others are competent to do the job and also trust others will be responsible, work hard, and seek the good of the entire team and not act in their own self-interest. Training (often called team building) is an important step during the initial formation of teams because it can help create the basis for ongoing trust among members. Likewise, setting up compensation policies to reward cooperation and trust (e.g., using team versus individual incentive pay plans) can foster trust among members. Developing trust often means managers taking a more laissez faire approach to supervising the team, opting to let teams develop their own internal controls consistent with the team's design and objectives. However it is achieved, trust is of central importance.

Another important factor is extensive communication. Centralized information flows may inhibit teams because they cannot get the information they need. Restricted communication flows may also disrupt the natural cooperative process because members cannot create the synergy or rhythm that facilitates good teamwork. Higher performing teams often learn how to do things best "on-the-fly," learning as they go. If communication is blocked, this learning cannot take place. Extensive communication also ensures all team members are apprised of important events, opportunities, threats, and the like. Communication is important for knowledge sharing, mentoring, and other interpersonal activities that strengthen team bonds, create skilled workers, and facilitate progress toward team goals. Training may be important in this area too. High performing teams are characterized by fine-grained information transfers where non-proceduralized or proprietary information is exchanged among members as a course of normal activity. Unless members are good at communicating, this information transfer will be less effective. Finally, communication and trust are connected: if team members do not trust each other, communication is unlikely to occur at all.

The importance of trust and communication point to a third motivational factor: keep the rules imposed from outside the team to a minimum. Especially for self-directed or autonomous teams, rules imposed from the outside cramp team functioning or force the team to operate in ineffective or inefficient ways. As much as is practical, and consistent with the team's design and objectives, the team needs freedom to find its own way, room to improvise, and the opportunity to develop its own performance norms. Consider that, in most societies, the worst form of punishment is not death, but expulsion from the group. This suggests that group or team norms tend to have a much stronger influence on the behavior of members than do rules imposed from outside the group.

The fourth motivational factor critical to successful team implementation is a shared, goal-oriented mindset among team members. Team members need to share a common understanding about what goals the team is pursuing so that their efforts can be directed appropriately. Managers should set goals that are specific, challenging yet attainable, meaningful, and relevant (O'Leary-Kelly, Martocchio, and Frink; 1994). With a clear idea of what the team's purpose is, members tend to perform better. This is probably the best way to exert external control on higher level teams. Set a goal, and let the self-directed or autonomous team take over.

The final motivational factor is that a successful team is intrinsically rewarding to many, if not most, people. People like being part of a successful team. The team needs to celebrate its success. This builds strong member-to-member support, fosters trust, enhances a common mindset, and encourages hard work in the future. Organizational rewards should be tied to behavior that helps the group. Recognition programs are one example, but cash compensation itself should be tied to factors that reward team performance. Team failures can often be attributed, at least in part, to a mismatched reward system that did not support team goals and teamwork (Hackman, 1998). Failing to recognize the satisfying aspects of teamwork per se may leave important motivational potential on the table. Encouraging esprit de corps can be an important precursor to high performance teams.

Individual characteristics

Creating a team out of a group of individuals is no simple task. Each potential member comes into the team with his or her own set of particular skills, beliefs, and personality traits. While an overly high level of homogeneity within a team is likely to stifle creativity and diversity of ideas, nonetheless there are certain skills and traits deemed necessary for all team members (Dunphy and Bryant, 1996). At the most fundamental level, all work team members must come into the team with a set of basic skills. These include reading comprehension, rudimentary writing and speaking skills, basic computational ability, and logical thinking. These basic skills are necessary in virtually any modern manufacturing setting, and are needed for members within any type of team design.

In addition, certain technical skills are required. These skills depend greatly on the particulars of the production tasks that will be performed by the team. A wide variety of fabrication, process monitoring and control, and product assembly tasks are included in this skill group. These skills build upon the set of basic skills, and are likely to be at least partially product- and process-specific.

A more general set of team management skills or traits are also required by all team members. These include the ability to communicate well with others both within and outside the team, the willingness to trust management and other team members to perform their jobs fairly and well, and the desire to be a cooperative team member. Further, each team member needs a positive attitude toward his or her duties and responsibilities as a member of the team. In essence, regardless of each team member's background, experiences, and beliefs, he or she needs to be willing to be a "team player".

Leadership

Leadership may be exercised within the team by certain members, depending on how the team is designed. This leadership may be formally delineated as part of the team design, or may come about informally as the team matures. Either way, the team management skills and traits mentioned above are also important here. Good communication skills, trustworthiness and a trusting attitude, cooperation, and a positive attitude are all critical for a team member to exhibit effective leadership (Mohrman, Cohen and Mohrman, 1995). In addition, technical skill proficiency, the ability to train team members, and the capability to effectively interface with other organizational units may be necessary, depending on the organizational context within which the team operates.

Leaders outside the boundary of the team have a more difficult role to play. These are often former supervisory and first-line managerial personnel. Their management styles under a more conventional hierarchical organizational structure may be at odds with what is needed in a team-based environment. They need the ability to delegate responsibilities to the team, and then allow the team to perform with minimal interference. The team needs the leeway to find its own solutions to issues and problems as much as is practical. The former supervisor/manager is now more of a coach, giving guidance when requested and providing support and encouragement to the team. The external leader must also be able to effectively manage relationships at the team's boundary, thereby allowing the team to operate without an undue level of influence or interference from the organization or other external constituencies.

Rewards

Performance expectations change when workers are configured into teams. Under more traditional hierarchical reporting structures, individual worker accomplishments are often the primary or sole basis for determining compensation. Typical measures used at the level of the individual worker include the worker's productivity or work output as compared to job standards, the worker's quality performance as measured by scrap and rework levels, and the worker's time-based performance as measured against planned schedules for the worker's station. However, the basis for compensation of individuals organized in teams must be adjusted to reflect their new set of responsibilities and accomplishments (Lawler and Cohen, 1992).

Along with rewards for individual performance, systems for measuring and rewarding team-based performance should be implemented. Using the individual worker performance measure examples from above, analogous team-based performance measures may be appropriate, such as the team's overall level of work output, levels of scrap and rework generated by the team, and the time-based performance of the team versus its production schedule. In addition, depending on the design of the team, other team-based measures may be appropriate. Examples of these are: (1) the number of quality problems solved by the team; (2) the number and impact of manufacturing process and product design changes identified and implemented by the team; and (3) improvements in stated customer satisfaction if the team has a direct effect on customer perceptions.

Whatever performance measures are chosen, for the team-based compensation scheme to be successful over the long term, it must be tied to the stated strategic objectives that first motivated the management to institute the team structures. This, however, is complicated by the fact that strategic objectives are often cross-functional in nature. As such, team performance may not be the sole determinant of whether the strategic objectives are met. Therefore, team performance must be measured against team-specific targets that support these strategic objectives, independent of whether these objectives are in fact attained by the organization as a whole.

Regardless of the particulars of the team-based performance criteria, they should possess a few vital characteristics. First, the performance criteria should be attainable by the team. The performance targets should challenge the team to perform at a high level of competence and effectiveness, but not be so challenging that the team cannot achieve them. Second, the performance criteria should be unambiguous. All team members should have a clear understanding of what is required of them. Third, the criteria should be within the domain and control of the team. Setting performance criteria that forces the team to rely significantly on the actions of other organizational units undermines the motivational aspect of the reward system. And fourth, the team criteria should apply equally to all members of the team. While individual performance criteria may also be utilized with team members, there should be a clear delineation between individual and team performance criteria.

BENEFITS OF TEAMS

Work teams, when successful, can enhance manufacturing's ability to support the competitive needs of the organization (see, for example, Katzenbach and Smith, 1993). At the most fundamental level, work teams may increase the quantity of products being built. Teams that share production tasks are often able to coordinate their actions, which increases the total amount of output per unit time. The result can be a dramatic increase in plant-wide efficiency. This in turn may be exploited by the organization that competes on the basis of low price.

Beyond efficiency improvements teams improve efficiency through the discovery of better ways to perform production tasks. Process streamlining, setup reduction, and waste reduction initiatives may all be enhanced in a cooperative team environment.

An addition to achieving higher output levels, teams also improve the quality of the products being built (Banker, Field, Schroeder and Sinha, 1996). Improvements in conformance quality are possible as teams share experience and knowledge of production methods. This may result in lower scrap and rework rates, as well as reductions in manufacturing-related field failures. Teams also have the potential to add to product quality through suggestions for product improvements. At the manufacturing level, the nature of the improvements are likely to be related to changes in fabrication or assembly methods, equipment, tooling, inspection methods, or testing procedures. At the product design level, teams may collectively suggest design changes that make the product easier to produce and test, and improve product durability and reliability.

Third, teams make a strategic impact on time-based competition. With improvements in productivity come the ability for manufacturing processes to reduce internal delays between the receipt of an order and its subsequent shipment. In make-to-order and assemble-to-order environments, reductions in cycle time can yield a significant competitive advantage. In make-to-stock environments the benefit of rapid delivery allows manufacturing to satisfy the changing demands of the distribution channel pipeline while minimizing inventories.

Finally, teams may support the organization that competes on the basis of flexibility. Specifically, teams enhance the ability of manufacturing to provide a greater mix of products, both existing and newly introduced. Teams that schedule their own work and develop highly cross-trained team members are better able to adjust to market-driven changes in demand. Likewise, empowered teams have the potential to participate in the design of new products, resulting in smoother product releases and shorter product ramp-up times.

Not all types of teams have the breadth of skills and responsibility necessary to deliver all of these strategic benefits to the organization. The next section discusses when different types of work teams should be employed, depending on other characteristics of the manufacturing environment.

WHERE TO USE TEAMS?

Teams can be used in a wide variety of manufacturing settings. While work teams have wide applicability in manufacturing, there is a need for research to identify the proper fit between team design and key characteristics of the manufacturing environment. This section will offer some thoughts and suggestions along these lines.

Referring to the typology of Table 1, it can be seen that team types at the low end of the spectrum—Ad Hoc, Consultative, Quality of Work Life, Problem Solving, QC—can be successfully implemented in virtually any manufacturing setting. The Ad-Hoc and Consultative types of teams focus on providing information, identifying problems, and suggesting solutions for a wide variety of manufacturing-related situations. These teams are not usually permanent in nature, and they do not require full-time participation from team members. Because of the transitory nature of these teams, there is no pressing long-term need for team members to work together in a highly coordinated manner. Hence, the particular characteristics of team members' normal, non-team based jobs are not of great importance here. Therefore, these types of teams can be successfully deployed in any production environment, be it job shop, batch manufacturing, line assembly, or continuous flow.

Similarly, the Quality of Work Life, Problem Solving, and QC types of teams have wide applicability in manufacturing. While these types of teams tend to be formed on a permanent basis, they nonetheless focus their efforts on distinct, identifiable issues such as quality improvement and process modification. These teams are not primarily engaged in ongoing, highly interactive sharing of work duties and responsibilities. As with the ad-hoc and consultative teams, these team types can be successfully employed in virtually any manufacturing setting.

In contrast to the universal applicability of low-end team types of Table 1, caution needs to be used when implementing team types at the high end of the spectrum, namely Self-Directed and Autonomous Teams. As discussed earlier in this article, these team types are significantly different than the others in that they focus on coordinating and managing ongoing activities directly related to building products. Because they are directly engaged in production tasks on an ongoing basis, successful implementation depends on the nature of the work they perform.

The key issue to examine here is the degree of interdependence among team members. Because these types of teams are designed to use a high degree of coordination among their members, activities must be such that members can perform better as a team than as individuals. To increase the probability of this occurring, there needs to be significant associations and dependencies among tasks to be performed by the team, so that team members can share their knowledge, expertise, or workloads. If the tasks are not dependent or somewhat related in the types of skills and abilities called for, there is little to gain by calling the group of workers a team.

Assuming an acceptable degree of interdependence among the team members, the focus of activities undertaken by self-directed and autonomous teams may vary depending on the particulars of the manufacturing environment. One team activity that should be done with caution is the detailed scheduling of its own work.

In line assembly and continuous flow environments with high levels of capital intensity, well balanced workloads and high process utilization rates are often critical to good performance. In this environment, teams may hinder manufacturing performance if their decentralized, localized work scheduling efforts degrade the workload balance or utilization rates. In contrast, manufacturing environments with lower levels of capital intensity found in batch manufacturing environments are more likely to see performance improvements through teams making work scheduling decisions. Similarly, job shop environments with moderately high levels of capital intensity but without the necessity for well-balanced workloads and high equipment utilization rates are also likely to improve their performance through teams performing their own scheduling of work.

High-end teams participate in activities related to equipment upkeep, maintenance, and improvement. Again, the ability of the team to successfully undertake these types of activities is somewhat dependent on the nature of the equipment used in the plant. Shops with highly complex, expensive equipment are less likely to have teams heavily involved in upkeep and maintenance. Further, in these environments, process improvements will often include modifications to equipment design or operation, and teams may lack the specialized technical skills necessary to determine and execute these modifications. This is often the case in continuous flow and automated, line assembly processes. This situation may also occur in job shops, where equipment tend to be more general purpose than in continuous flow and line assembly plants, but can nonetheless be complex to maintain and improve upon.

In general, manufacturing environments that require centralized scheduling/coordination and high levels of capital intensity may need to use prudence when designing and implementing self-directed and autonomous teams.

CONCLUSIONS

As discussed in the Benefits section of this article, there are a variety of benefits to be gained through the use of work teams. However, there are also some potential pitfalls to consider. A major concern when making the decision to implement manufacturing work teams is for the management to be realistic in their expectations. Despite what is often written in the popular press, teams will not solve all production-related problems. A mismatch between team design, team membership characteristics, and team objectives will likely result in mediocre or poor team performance.

In addition, a new management style is necessary in the team-based organization. The "command-and-control" approach common in manufacturing firms of the past must give way to a more supportive, flexible, and adaptive management style. To gain the full benefit of teams, they must be empowered to achieve their goals and performance targets without undue constraint from management. A reasonable approach for management is to direct teams by setting performance measures and specific targets, yet allow teams to decide specifically how they will achieve their performance targets.

Even assuming work teams are performing at a high level, there is a team life cycle issue to consider. Teams have a tendency to plateau in their effectiveness over time, as the luster and excitement of working as a coordinated unit starts to wear off. A current problem facing many manufacturing organizations is the need to renew the energy and enthusiasm of the mature team (Harrington-Mackin, 1996). Methods such as changing the team's objectives, increasing or otherwise modifying the team's performance targets, and bringing in new team members are all being attempted by team-based organizations. It remains to be seen if these, or other methods, will consistently revitalize mature teams.

Finally, it must be recognized that there is a cost/benefit tradeoff associated with teams. There are costs in terms of time, money, and effort expended to create a cohesive team out of a group of unique individuals. Further, these costs continue in some fashion throughout the life of the team. For the team to add net value to the organization, the benefits generated from the team must be greater than the team's startup and ongoing costs.

Key Concepts: Ad-Hoc teams; Autonomous work teams; Consultative Teams; Quality of work life Initiatives; Self-directed work teams.

Related Articles: Concurrent Engineering; Product Development and Concurrent Engineering.

REFERENCES

Banker, R.D., J.M. Field, R.G. Schroeder, and K.K. Sinha (1996). "Impact of Work Teams on Manufacturing Performance." *Academy of Management Journal*, 39 (4), 867–890.

Champion, M.A., G.J. Medsker, and A.C. Higgs (1993). "Relations Between Work Group Characteristics and Effectiveness: Implications for Designing Effective Work Groups." *Personnel Psychology*, 46, 823–850.

Cohen, S.G., G.E. Ledford, Jr., and G.M. Spreitzer (1996). "A Predictive Model of Managing Work Team Effectiveness." *Human Relations*, 49 (5), 643–676.

Dunphy, D. and B. Bryant (1996). "Teams: Panaceas or Prescriptions for Improved Performance." *Human Relations*, 49 (5), 677–699.

Goodman, P.S., R. Devadas, and T.L.G. Hughson (1988). "Groups and Productivity: Analyzing the Effectiveness of Self-Managing Teams." In *Productivity in Organizations*, J.P. Campbell et al. (ed.), Jossey-Bass, San Francisco, CA.

Guzzo, R.A. and M.W. Dickson (1996). "Teams in Organizations: Recent Research on Performance and Effectiveness." *Annual Review of Psychology*, 47, 307–338.

Hackman, J.R. (1998). "Why Teams Don't Work." In *Theory and Research on Small Groups*, R.S. Tindale et al. (ed.), Plenum Press, New York, NY.

Harrington-Mackin, D. (1996). *Keeping the Team Going: A Tool Kit to Renew & Refuel Your Workplace Teams*, AMACOM/American Management Association, New York, NY.

Ishikawa, I. (1989). *Introduction to Quality Control, 3rd edition*, 3A Corporation, Tokyo, Japan.

Katzenbach, J.R. and D.K. Smith (1993). *The Wisdom of Teams: Creating the High-Performance Organization*, Harvard Business School Press, Boston, MA.

Lawler, E.E. and S.G. Cohen (1992). "Designing pay systems for teams." *American Compensation Association Journal*, 1, 6–18.

Mohrman, S.A., S.G. Cohen, and A.M. Mohrman, Jr. (1995). *Designing Team-Based Organizations: New Forms for Knowledge Work*, Jossey-Bass, San Francisco, CA.

O'Leary-Kelly, A.M., J.J. Martocchio, and D.D. Frink (1994). "A Review of the Influence of Group Goals on Group Performance." *Academy of Management Journal*, 37 (5), 1285–1301.

Osburn, J.D., L. Moran, E. Musselwhite, and J.H. Zenger (1990). *Self-directed Work Teams: The New American Challenge*, Irwin Publishing, Chicago, IL.

VII. COMPETING ON FLEXIBILITY AND AUTOMATION

10. FLEXIBLE AUTOMATION

KATHRYN E. STECKE
The University of Michigan, Ann Arbor, Michigan, USA

RODNEY P. PARKER
The University of Michigan, Ann Arbor, Michigan, USA

ABSTRACT

Flexible automation was hailed as a remedy for the competitive challenges that modern manufacturing was encountering through rising quality standards, shortened product life cycles, and greater demand for product variety (Hill, 1994). Some disappointment resulted from these great expectations. Some commentators suggested that the problem lay in the strategic mis-use of the systems. For example, Hayes and Jaikumar (1988) suggested that managers using these new technologies in the same manner in which they used their previous conventional technologies were destined for disappointment. They stressed the need for a new mindset to experience the 'revolutionary' benefits these new flexible systems promised. Hayes and Clark (1986) observe that productivity can fall for significant periods after the introduction of new production technologies, but this can be prevented by enlightened management and reorganization. Jaikumar (1986) observed a difference in the early usage (late 1970s, early 1980s) of these technologies between certain Japanese and American manufacturers. He noticed that the flexible systems in Japan were used more for their flexible benefits than in the U.S. He also noticed that Japanese managers introduced more products every year than their American counterparts. Consequently, Japanese manufacturers also had fewer problems financially justifying these flexible technologies than American manufacturers.

Hill (1994) suggests that a great deal of disappointment resulted from managers investing in 'flexible' equipment believing that the possession of new flexible technologies would result in a 'strategic response' to competitive pressures. One lesson appears to be that flexible automation is appropriate when its capabilities (e.g., producing multiple part types in medium volumes) are aligned with the company's needs and defined manufacturing and technology strategies.

Much has been written about the economic justification of flexible technologies (see Son, 1992). There has been evidence to suggest conventional justification techniques are inappropriate for flexible automation (Kaplan, 1986) and much activity has been directed at attempting to capture the more elusive benefits of flexible technologies. Foremost among these benefits is the flexibility of making multiple products simultaneously, and many authors have attempted to capture and characterize this flexibility through mathematical programming models (e.g., Fine and Freund, 1990), real option models (Trigeorgis, 1996), and empirical studies (e.g., Upton, 1995b).

This article notes the role of **MIT**, **Sunstrand Corporation**, and the **U.S. Air Force** in the development of flexible automation.

WHAT IS FLEXIBLE AUTOMATION?

Flexible automation (FA) is a type of manufacturing automation which exhibits some form of "flexibility." Most commonly this flexibility is the capability of making different products in a short time frame. This "process flexibility" allows the production of different part types with overlapping life-cycles. Another type of flexibility that comes with flexible automation is the ability to produce a part type through many generations. Clearly, there are several other manifestations of flexibility.

Flexible automation allows the production of a variety of part types in small or unit batch sizes. Although FA consists of various combinations of technology, flexible automation most typically takes the form of machining systems, that is, manufacturing systems where material is removed from a workpiece. The flexibility comes from the programmability of the computers controlling the machines. Flexible automation is also observed in assembly systems. The most prominent form of flexible assembly is observed in the electronics industry, where flexible machines (automated surface mount technologies) are used to populate printed circuit boards with integrated circuits and other componentry. In this instance, manufacturers have found the machines' far superior accuracy and reliability to be sufficient to warrant the significant investment. Overall, however, manufacturers tend to use automation for fabrication, and leave assembly to human operators who can adapt to a greater variety of changing circumstances more rapidly and easily than machines. In this article, the discussion of flexible automation is primarily focused on machining systems.

HISTORY

Flexible automation is a form of manufacturing technology which is the culmination of a long evolution in production automation. Most of the development in industrial automation, as we know it, has largely occurred during the twentieth century.

Automation has long been the dream of engineers and scientists, whereby the simple, dirty, repetitive, and dangerous tasks traditionally done by people, could be undertaken by machines. More recently, this vision has been extended to complex physical, computational, and analytical tasks. Advances in computer technologies allowed this vision to automate several aspects of personal, administrative, industrial, and logistic activities.

Initially, automation was *fixed*, that is, it could perform a single task, or small set of tasks, efficiently and effectively but changing this task set was difficult, costly, or impossible. This feature common to earlier of fixed automation is called "rigidity." Fixed automation commonly can do a small set of well-defined tasks particularly well, but has trouble doing anything else, without significant and time-consuming intervention from human operators.

Historically, automation as a substitute for general human activity. Automation where mechanical, electronic, or computational apparatus were substituted for human activity in an organized industrial context can be traced to the 18th century.

Scale economics through fixed automation dominated the industry for much of the 20th century until competitive pressures, primarily from Japan, forced the American auto industry to change their focus to one of product variety and quick response to market needs. The emergence of the new flexible automation technologies enabled a more agile approach to cater to these pressures.

The birth of numerical control (NC), resulting from the combination of conventional machine tools and computers in the 1940s, is credited to John Parsons (Chang, Wysk, and Wang, 1998). Further development occurred at MIT, funded by the U.S. Air Force. The first full-fledged NC machine tool, which could machine complex shapes was developed in 1952 at MIT. Parson used punched cards containing programs which delivered instructions to hardwired machine tools. The hardwired controller was succeeded by an NC controller and the punched cards gave way to paper tape. These machine tools evolved into NC machine centers that could drill, bore, and mill. Developments during the 1960s included automated tool changers and indexing work tables (Viswanadham and Narahari, 1992).

Parallel developments in computing technologies resulted in much progress in the NC controller, allowing a centralized controller to issue commands to numerous numerical control machine tools. This *direct numerical control* (DNC) was appropriate when the available computing technology was bulky and expensive. However, as electronics and computers became miniaturized it became possible to place computer controllers within each machine tool, with a central controller responsible for a smaller array of operations, mostly real-time monitoring at the system level.

One of the earliest full-fledged flexible manufacturing systems was developed by the Sunstrand Corporation in 1965. It involved eight NC machine tools with a computer automated roller conveyor. Although it did not have much process flexibility, it marked the advent of flexible automation where part programs for different part types could now be loaded quickly into local microprocessors and production could switch between different part types without significant setup time. Automated movement was done using relay switches. Developments since then have been mostly refinements in the technologies and the variety of machine tools covered. Today, flexible automation technology is far more robust and cost significantly less.

FLEXIBLE AUTOMATION IN MACHINING SYSTEMS

The building block of flexible automation is the computer numerical controlled (CNC) machine tool which is typically augmented by automated materials handling systems, centralized controlling computers, automated storage and retrieval systems, and human operators. The variety of installations of flexible automation are numerous. Some typical configurations are discussed below.

A CNC machine tool is a self-contained machine, where the tool cutting movements, spindle speeds, tool exchange, and other operations are controlled by a part program executed by the computer controller based at the machine tool. The spindle is a spinning device which holds the tool used to cut into the workpiece.

Conventional machine tools (e.g., lathes, drill presses, milling machines) are not computer controlled. The operation of conventional tools is typically done by skilled craftsmen. There can be variations to dimensions on parts made on a conventional tool, whereas this variation is decreased on CNC machine tools. The elimination of this variation is one objective (benefit) of automating the discrete part production process.

Additional benefits include a reduction in required floorspace, reduced delivery and production leadtimes, higher utilization, increased quality, and smoother implementation of changes and improvements in product design. Another significant benefit is the ability to mass produce with a machine tool that is able to produce identical parts and with the ability to switch production between part types of different designs. This latter capability comes from the part programs stored in the local machine memory, or downloaded from a centralized storage device when needed. Effectively, this is a step towards gaining the benefits of both mass production and job shop customization, commonly known as mass customization. Mass customization refers to the practice of producing single parts or small batches to custom modifications to a part design. Flexible automation helps to achieve mass customization.

Flexible automation is created when the CNC machine tools are augmented by ancillary equipment such as automated materials handling systems, automated inspection, and central controllers. The materials handling systems are responsible for loading and unloading parts from the machines, transporting parts between machines in the system, and handling work-in-process inventory storage. These materials handling systems could consist of robot arms, conveyors, automated guided vehicles, and gravity feed chutes. Most commonly, a combination of these technologies are used with human operators introducing unworked parts into the system and for removing finished parts from the system.

Another aspect of flexible automation is the usage of multiple cutting tools to perform each operation on a part and the automatic changing of cutting tools at each CNC machine tool in the system. A magazine containing cutting tools is located at each machine, and cutting tools are automatically changed (i.e., without human intervention) as the part program dictates. Typical tool magazines can hold 30–90 tools. The selection of which tools should be loaded into which magazine is known as the loading problem, which is one of several production planning problems (see Stecke, 1983) associated with flexible automated machining system. Occasionally, centralized tool magazines permit the sharing of tools between various machine tools, potentially reducing the total tooling cost. However, this requires additional tool transportation devices and a more difficult coordination activity by the central computer system.

The central computer system is another feature of flexible automation systems. The central computer system differs from the local computer controller that resides at an individual machine in the system. It has an integrative role of managing the overall operation of the system. The responsibilities of the central computer can be divided into off-line and real-time activities. Typical off-line activities include effective planning and scheduling for the most productive use of the system during a given production period. Typical real-time activities include monitoring the operation of the system, adapting the schedule and production plans when problems arise, and alerting personnel when catastrophic failure occurs. The degree of 'intelligence' and automatic control in the computer system varies greatly across systems, with levels of human intervention. When the central computer system is also responsible for downloading the part programs to the workstations, the system is known as distributed numerical control.

Other ancillary equipment that is contained in flexible automation systems is some form of automated inspection system that checks the location of either the raw part, the specifications of the finished part, or some intermediate version of the part, or several of these. Video cameras and automated gauges can verify whether the part adheres to some pre-determined quality standard and alert the central computer when a part falls outside of the specifications.

Another form of flexible automation is seen primarily in the semi-conductor industry using surface mount technologies. These machines are used to 'populate' printed circuit boards (PCBs) with integrated circuits and other componentry. Typically the components in question are presented to the machine on large reels which are loaded onto the surface mount machine. As the PCBs pass alongside or though the machine, a component is extracted from the reel, and a gantry arm places the component into a specified location on the PCB. Sometimes the PCB itself is attached to the gantry arm and is moved to a location where the component is inserted into its correct position. The board is then moved along a conveyor to the next component loading position. The component locations are stored in computer memory, and the CNC gantry and inserter locations are controlled by this information. Different PCB designs can follow one another through the surface mount machine without disrupting the machine as long as there is sufficient commonality of components or capacity to load different component reels.

Due to the level of automation and presumed consistency, when a single part fails to meet specifications, it is a signal that some problem that could affect many parts could exist. This could be a faulty fixture, materials handling device, or worn or damaged tool. How the central computer deals with the situation depends upon the level of autonomy granted to it. Most systems will merely alert their human overseers. However, others may take action, checking various possible sources for the problem. The amount of artificial intelligence (AI) built into most industrial systems today is fairly low, largely limited to image processing and monitoring activities rather than the management of contingencies. These machine vision systems typically consist of a video camera, lighting, computer-based artificial intelligence to analyze and filter the image into a recognizable form, and a monitor to display the image and process status (Cohen and Apte, 1997). The use of AI is likely to expand as the adaptive control hardware technology improves. In the future, opportunities will exist for problems to be prevented before they occur. For example, tool wear can be monitored and new tools substituted before a failed tool has the opportunity to damage a part in production. Limited applications of these are beginning to appear in practice.

OTHER FORMS OF FLEXIBLE AUTOMATION

While most of the discussion so far is concerned with flexible *machining* systems because of their prevalence, there are other forms of flexible automation. Industrial robots can be used for more than part handling. If equipped with the appropriate monitoring and sensing devices, they can perform quite sophisticated and dexterous functions such as welding, inspection, and assembly. For example, a spot-welding robot arm on an automotive production line can move more quickly across an entire vehicle, performing more consistent and rapid welds than a human operator. The flexible robot can recognize the vehicle type by some sensing technology (e.g., by identifying the fixture type) or from sequencing information from the central computer, and adapt the weld location and sequence from car model to car model. Such robots are usually trained by an operator manually moving the robot arm in a learning mode, where the x-y-z Cartesian coordinates of the robot arm's position, the various arm-segment angles, and joint rotations are recorded by the robot's controller for use in real production.

Flexible automation can be configured in several different ways to achieve different production objectives. For example, a flexible cell is typically a single CNC machine tool possibly with automated materials handling system, and can make many part types at low volumes, sometimes one-off prototypes of products. Another form is where CNC machine tools can be lined serially with a conveyor for part movement in a 'transfer line' arrangement to get a high throughput of a limited number of part types. A flexible manufacturing system is typically more elaborate in design, involving several CNCs doing different sets of operations, linked together logically by computer communications and physically by materials handling devices. These systems can be operated in different ways. For example, sometimes several machines may perform identical operations for reasons of system balance and/or redundancy during periods of machine failure. This then allows for different routes for parts going through the system.

THE FUTURE

There is an obvious difference between flexible automation and the conventional equipment. Not so obvious is the change in management practice required to secure the benefits of the new technologies. This need was not fully appreciated initially, and early performance of flexible automation in America was lacklustre. Necessary changes extend to the planning processes needed to operate flexible automation. Stecke (1983) identified five production planning issues necessary for effective operation of flexible manufacturing systems. Much subsequent research into FMSs has addressed one or more of these issues, which are grouping machines, selecting part types, choosing relative mixes of products, allocating system resources to part types, and determining appropriate tool magazine loading strategies. These are unique challenges faced by production managers of flexible automation that are driven by technology.

With all the advantages that existing flexible automation offer, a legitimate question is to ask, why all manufacturing is not done on such equipment. One reason is that dedicated equipment is generally faster, operation by operation, than flexible automation, and more appropriate in high-volume environments. Another reason is that there is a cost premium in the acquisition and operation of flexible automation over dedicated systems. Also, for all the tumult about the 'agility' of flexible automation, the ability to *easily* modify the systems to accommodate entirely new part types is limited. Therefore, the next phase of flexible automation appears to be the development of reconfigurable manufacturing systems (RMSs), where the technology (Koren and Ulsoy, 1997: 1) will be "designed for rapid adjustment of production capacity and functionality, in response to new circumstances, by rearrangement or change of its components." An example of a reconfigurable machine is one that, has milling and drilling capabilities but currently has no capability for turning. But a "reconfigurable" machine can easily, quickly, and cheaply be reconfigured to acquire the new turning capability also. Although, RMS technology does not currently exist, newly constructed hardware and software tools offer the capability to produce newly introduced part types. Another example of an RMS hardware is a milling machine with room for the addition of several spindles that can be arranged in numerous configurations. The development of the hardware, the software, and the science of reconfiguration is ongoing.

Key Concepts: CAD; CAM; CIM; FMS; Flexibility in Manufacturing; Manufacturing Cell; Manufacturing Flexibility.

Related Articles: Manufacturing Flexibility; Manufacturing Strategy; Mass Customization.

REFERENCES

Chang, T.C., R.A. Wysk, and H.P. Wang (1998). *Computer-Aided Manufacturing* (2nd. Ed.), Prentice Hall, New Jersey.
Cohen, M.A. and U.M. Apte (1997). *Manufacturing Automation*. Irwin, Illinois.
Fine, C.H. and R.M. Freund (1990). "Optimal Investment in Product-Flexible Manufacturing Capacity." *Management Science*, 36 (4), 449–466.
Hayes, R.H. and K.B. Clark (1986). "Why Some Factories are More Productive Than Others." *Harvard Business Review*, September–October, 66–73.
Hayes, R.H. and R. Jaikumar (1988). "Manufacturing's Crisis: New Technologies, Obsolete Organizations." *Harvard Business Review*, January–February, 77–85.
Hayes, R.H., G.P. Pisano, and D.M. Upton (1996). *Strategic Operations*, Free Press, New York.
Hill, T. (1994). *Manufacturing Strategy* (2nd. Ed.). Irwin, Massachusetts.
Hopp, W.J. and M.L. Spearman (1996). *Factory Physics*. Irwin, Illinois.
Jaikumar, R. (1986). "Postindustrial Manufacturing." *Harvard Business Review*, November–December, 69–76.
Kaplan, R.S. (1986). "Must CIM be Justified by Faith Alone?" *Harvard Business Review*, March–April, 87–95.

Koren, Y. and G. Ulsoy (1997). "Reconfigurable Manufacturing Systems." *ERC Technical Report* #1, The University of Michigan, Ann Arbor, Michigan.

Son, Y.K. (1992). "A Comprehensive Bibliography on Justification of Advanced Manufacturing Technologies." *Engineering Economist*, 38 (1), 59–71.

Stecke, K.E. (1983). "Formulation and Solution of Nonlinear Integer Production Planning Problems for Flexible Manufacturing Systems." *Management Science*, 29 (3), 273–288.

Stecke, K.E. and I. Kim (1991). "A Flexible Approach to Part Type Selection in Flexible Flow Systems Using Part Mix Ratios." *International Journal of Production Research*, 29 (1), 53–75.

Trigeorgis, L. (1996). *Real Options: Managerial Flexibility and Strategy in Resource Allocation*, The MIT Press, Massachusetts.

Upton, D.M. (1995a). "What Really Makes Factories Flexible." *Harvard Business Review*, July–August, 74–79.

Upton, D.M. (1995b). "Flexibility as Process Mobility: The Management of Plant Capabilities for Quick Response Manufacturing." *Journal of Operations Management*, 12, 205–224.

Viswanadham, N. and Y. Narahari (1992). *Performance Modeling of Automated Manufacturing Systems*. Prentice Hall, New Jersey.

11. MANUFACTURING FLEXIBILITY

PAUL M. SWAMIDASS

Auburn University, Auburn, AL, USA

ABSTRACT

Manufacturing flexibility could refer to the capacity of a manufacturing system to adapt successfully to changing environmental conditions as well as changing product and process requirements. It could refer to the ability of the production system to cope successfully with the instability induced by the environment. Flexibility, provides the manufacturing plant the ability to maintain customer satisfaction and profitability under conditions of change and uncertainty.

For superior understanding of the term Manufacturing flexibility, we must learn to distinguish clearly between flexibility offered by a single machine as opposed to the flexibility of an entire plant. Machine level flexibility is predominantly technology based, but plant level manufacturing flexibility is a complex blend of several ingredients including, (1) hard technologies, (hardware, software, and equipment), (2) soft technologies, (i.e., know-how, procedures, organizations and techniques) (3) design, and (4) manufacturing infrastructure. Detailed description of flexibility enhancing projects at the following companies are included here: **IBM and McDonnell Douglas Aircraft Company.**

ROLE OF TECHNOLOGY

Everyday, manufacturers are investing in computer aided design (CAD), just-in-time (JIT) manufacturing, computer-aided manufacturing (CAM), and a dozen other

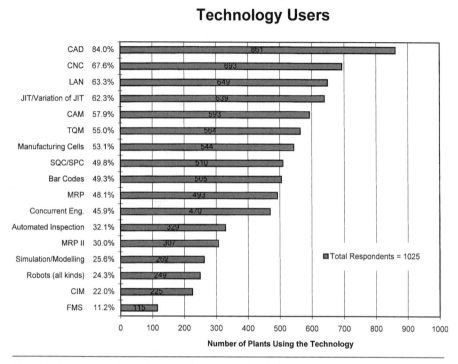

Figure 1. Percent of U.S. manufacturers using various technologies.

technologies. Although the above technologies are diverse, one thing common to all hard and soft technologies (Swamidass, 1998) is that they add flexibility to manufacturing operations. Hard technologies such as CAD, CAM, robots, and CIM are hardware and software intensive, whereas, soft technologies such as JIT, SQC and TQM are driven by know-how and techniques.

It is notable that over 70 percent of U.S. manufacturers reported that they experienced cycle time reduction due to the use of 17 different technologies (Swamidass, 1998). Reduced cycle time means improved flexibility. Figure 1 shows the use of 17 different manufacturing technologies and their use in the United States (Swamidass, 1998). The data in Figure 1 were gathered from 1025 U.S. manufacturing plants from the following industries: SIC 34: metal fabrication; SIC 35: machinery including computers; SIC 36: electrical; SIC 37: transportation; SIC 38: Instruments and photo goods. These industries employ more than 40% of all manufacturing employment in the U.S.

COMPETITIVE VALUE OF MANUFACTURING FLEXIBILITY

The environment that influences manufacturers is in a constant state of flux. However, the rate of change occurring in the environment and the extent of changes occurring may vary from one firm to another. Further, not all changes occurring in the

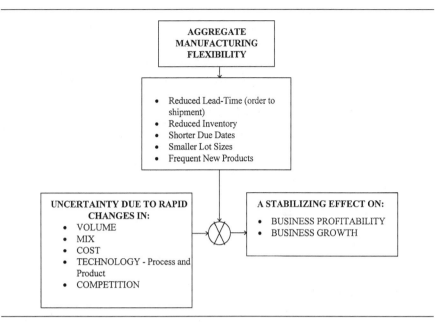

Figure 2. The strategic role of manufacturing flexibility.

environment may affect a firm's operation, its viability, and its success. But when the changing environment adversely affects a manufacturing firm, it disturbs or robs the firm of its ability to satisfy customers and operate profitably in a sustained manner. Under such circumstances, the firm needs flexibility to successfully cope with the changes in the environment.

In a changing environment, the primary challenge to manufacturing arises from the uncertainty associated with demand. The capability of manufacturing to respond appropriately to demand uncertainty will determine the stability and growth of the business unit's profitability. The competitive value of MF lies in the extend to which it can neutralize the effects of demand uncertainty. This strategic value of MF is captured in Figure 2. It shows that under conditions of demand uncertainty, MF could help maintain or stabilize the profitability of the business. It implies that the greater the demand uncertainty created by competitors, changing technology, and so on, the greater the need for MF in preserving sales growth and profitability of the business.

Manufacturing flexibility has been used strategically for offensive as well as defensive purposes as summarized in Figure 3. For example, in a given period of time, an ability to launch larger numbers of new models than one's competition could be a formidable competitive weapon in certain markets. An illustration of this is the well-documented Honda-Yamaha "war" in the early 1980's when Honda's number one position in market share for motorcycles was threatened by Yamaha. To eliminate the threat, within eighteen months, "Honda introduced 81 new (motorcycle) models and discontinued 32 models for a total of 113 changes in its product line. Yamaha retired only 3 models and introduced 34 new models for a total of 37 changes" (Abegglen & Stalk, 1985).

IN OFFENSE	IN DEFENSE	IN OFFENSE AND DEFENSE
Responding to Opportunities	Desensitizing the System to Adverse Changes	Increasing Efficiency
1. Ability to introduce large number of new models into the market (Honda vs. Yamaha motorcycles). 2. Time required to change entire product line reduced (3 years in Japan vs. 10 years in U.S. air-conditioning industry).	1. Less susceptible to changes in demand, supply and tastes due to a broader range of product mix. 2. Enables the manufacturer to cope with uncertainties caused by changes in the external environment - demand, mix, material.	1. Better utilization of capacity through wider range of product mix. 2. Reduction or elimination of setup time or change-over time. 3. Better use of capacity of the production of counter cyclical products

Figure 3. Competitive value of manufacturing flexibility.

By January 1983, about a year after Honda's successful counter attack, President Koske of Yamaha admitted defeat and declared, " . . . I would like to end the Honda-Yamaha war . . ."

MANUFACTURING FLEXIBILITY TYPES

Is manufacturing flexibility a homogeneous entity? Are there various types of flexibilities? Swamidass (1988) and Upton (1995b, 1997) support the view that manufacturing flexibility is not homogenous. Major hurdles to the understanding of manufacturing flexibility has been a homogeneous view of manufacturing flexibility, and a lack of consensus on the terms used to describe it.

Terms associated with manufacturing flexibility

Most terms used in connection with manufacturing flexibility in discrete products industries are listed in Table 1. The table illustrates some of the problems hindering our understanding of MF. First, the scope of flexibility related terms used by various authors overlap considerably. Second, some flexibility terms are aggregates of other flexibility terms used. Finally, identical flexibility related terms used by more than one writer do not necessarily mean the same thing. In addition to the terms in Table 1, Upton (1995a, 1997) defines flexibility in process industries in terms of "operational mobility" and "process range."

Process contingencies

Consider Figure 4 which has been popularized in various forms by many writers on automation (Jelinek and Goldhar, 1983) in non-process industries. Typically, many authors have used similar figures, in one form or the other, to demonstrate the appropriateness of various manufacturing processes to selected manufacturing situations. In Figure 4, the two-dimensional space represents the entire range of manufacturing possibilities or process contingencies. A comprehensive discussion on MF should be in the context of the complete range of manufacturing possibilities or process

Table 1 List of terms associated with
manufacturing flexibility in OM literature*

1. Action flexibility
2. Adaptation flexibility
3. Application flexibility
4. Assembly system flexibility
5. Demand flexibility
6. Design flexibility
7. Dispatch flexibility
8. Job flexibility
9. Machine flexibility
10. Machining flexibility
11. Material flexibility
12. Mix flexibility
13. Modification flexibility
14. Process flexibility
15. Program flexibility
16. Product flexibility
17. Production flexibility
18. Routing flexibility
19. State flexibility
20. Volume flexibility

* Adapted from Swamidass (1988).

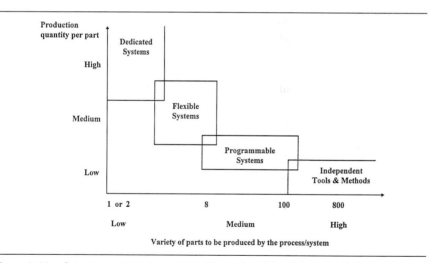

Figure 4. Manufacturing contingencies spectrum. (Figure adapted from Jelinek and Golhar, 1983.)

contingencies represented by Figure 4; after all, MF broadens a manufacturer's capability to deal with a wider spectrum of process contingencies.

Figure 5 is an extension of Figure 4 where the two-dimensional space of manufacturing contingencies is partitioned into nine regions. In Figure 5, the three regions along the diagonal represent almost all manufacturing firms: the remaining regions are

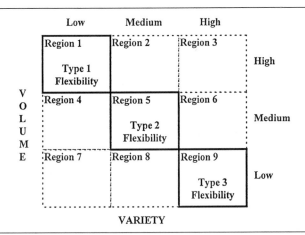

Figure 5. Three regions of the manufacturing contingency space that represent almost all manufacturing firms. (Figure adapted from Swamidass, 1988.)

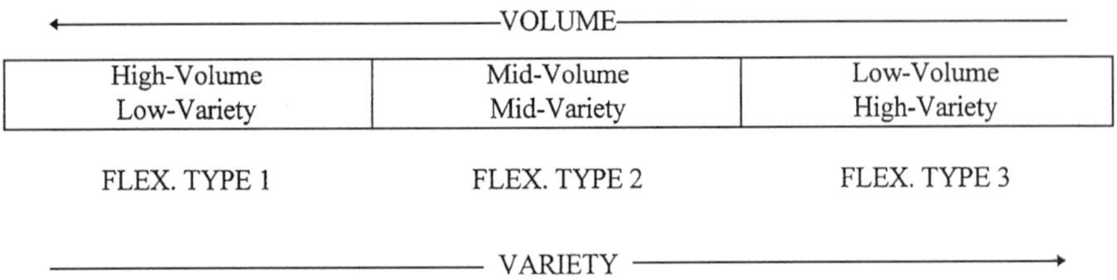

High-Volume Low-Variety	Mid-Volume Mid-Variety	Low-Volume High-Variety

| FLEX. TYPE 1 | FLEX. TYPE 2 | FLEX. TYPE 3 |

VARIETY

Figure 6. Volume-variety continuum and flexibility types.

redundant. If all the redundant regions are dropped, Figure 5 reduces to the manufacturing flexibility continuum shown in Figure 6.

Three manufacturing flexibility types

The manufacturing flexibility continuum in Figure 6 shows three regions with very distinct flexibility needs. For the sake of simplicity and ease of identification, the manufacturing flexibility associated with the three regions are labeled Type 1, Type 2, and Type 3 flexibilities in the context of discrete products industries.

The first step towards an understanding of MF requires our accepting multiple types of manufacturing flexibilities to cover all possible manufacturing contingencies. Different types of manufacturing flexibilities are needed to cope with the process contingencies that arise within the diverse regions in Figure 6. Since the process contingencies within the regions are different, MF to deal with the contingencies in these regions are also different. Failure to recognize the differences in process contingencies leads to erroneous assumptions about the nature and usefulness of MF. A major consequence of such erroneous assumptions is the difficulty faced by practitioners when justifying investments in flexibility enhancing systems.

Examples of flexibility types: The three types of flexibilities introduced here are made more obvious to the reader through the following examples.

Type 1 Flexibility: An example of Type 1 flexibility is the Honda Motorcycle Division, whose head-to-head competition with Yamaha, was described earlier. In this example, a mass producer of motorcycles was able to introduce 81 new models and discontinue 32 models for a total of 113 changes in its product line in about 18 months using Type 1 flexibility.

Another good example of Type 1 flexibility is then IBM's (now Lexmark) automated Selectric typewriter/printer plant at Lexington, Kentucky. During the period 1981–1986, this plant underwent a major 350 million dollar renovation to implement flexible automation which enabled the plant to manufacture almost any product that can be assembled "from above" (to permit assembly by robots) within an envelop of size $18'' \times 22'' \times 28''$. This case is described in greater detail in a later section.

Type 2 Flexibility: In 1983, General Electric's Series 8 locomotive plant in Erie, Pennsylvania needed to machine a family of motor frames and gear boxes. The plant installed a Gidding and Lewis FMS with a capacity of 5,000 motor frames of sizes up to $4' \times 4' \times 5'$ with over 100 machining surfaces. The system included two vertical milling machines, three horizontal machining center, three heavy horizontal boring mills and one medium horizontal mill. The machines included robotic or automatic tool changers with over 500 cutting tools. Machining time was reduced from 16 days to 16 hours per frame with the installation of the FMS.

Type 3 Flexibility: Ingersoll Milling Machine Co., a special machinery producer, uses a CIM system which includes CAD/CAM. Orders for special machines come in lots of one and the firm seldom builds a duplicate of a machine, thus each order individually goes through the various stages if design, engineering, and manufacturing.

The foregoing examples illustrating the three types of manufacturing flexibilities are summarized in Table 2.

AGENTS OF MANUFACTURING FLEXIBILITY

Manufacturing flexibility is the result of several contributing factors. In order to make substantial gains in MF there has to be a coordinated effort to address most of the contributing factors of flexibility. According to Figure 7, process, product design, and infrastructure induce MF.

In recent years, technological developments such as Flexible Manufacturing Systems (FMS), Computer Integrated Manufacturing (CIM), Computer Aided Manufacturing (CAD), and other technologies come to a manager's mind when considering

Table 2 Examples of the three types of flexibilities

Type 1: Flexibility in automation	Type 2: Flexibility in manufacturing	Type 3: Flexibility in design and manufacturing
IMB Lexington, Kentucky: Selectric System 200 typewriters and printers plant. Annual production about 1,000,000 units. Capability to produce an infinite number of models a low cost. Lead time for new product introduction reduced to 18 months. (as of 1985)	GE's series 8 locomotive plant, Erie, Pennsylvania. Gidding and Lewis FMS for machining a family of motor frames and gear boxes. Yearly capacity 5000 motor frames of sizes up to $4' \times 4' \times 5'$ with over 100 machining surfaces. The system includes two vertical milling machines, three horizontal machining centers, three heavy horizontal machining centers, three heavy horizontal boring mills and on medium horizontal mill. The machines include robot or automated tool changers with over 500 cutting tools. Machining time reduced from 16 days to 16 hours per frame. (as of 1983)	Ingersoll Milling Machine Co., a very special machinery producer with a CIM system which includes CAD/CAM. A typical lot is one or two pieces; seldom builds a duplicate. (as of 1984)

manufacturing flexibility. It may turn out to be a serious error to consider manufacturing flexibility only in terms of process hardware. Only by blending flexibility in manufacturing hardware, product design, and infrastructive can one assure plant level flexibility and ensure the effective use of investments in flexibility.

Infrastructure is vital to integrating the benefits provided by process technology and product design. Manufacturing infrastructure is the framework made up of people, procedures, information systems, and decision making systems (Skinner, 1978) which enables process flexibility and product design flexibility to prosper.

Figure 8 is a matrix that presents four flexibility types in the columns and five agents of flexibility in rows. In the figure, flexibility types 1 to 3 are found in non-process industries whereas type 4 flexibility is found in process industries. The cells in Figure 8 (i.e., intersections of columns and rows) contain selected examples of agents appropriate to the attainment of the type of flexibility unique to the column using a method that is unique to the row.

Figure 8 shows that while *five agents* are used in attaining all three types of flexibility, their use differs across flexibility types. For example, whereas product design is an agent of flexibility, the contribution of design to flexibility depends upon the type of flexibility sought. While Type 1 flexibility requires a few, stable designs for its success, extreme variability in design, from one order to the other, and an ability to rapidly accommodate customer initiated changes to design are essential to the success of Type 3 flexibility. The reader is encouraged to extend to other cells the above example of how Figure 8 is to be interpreted.

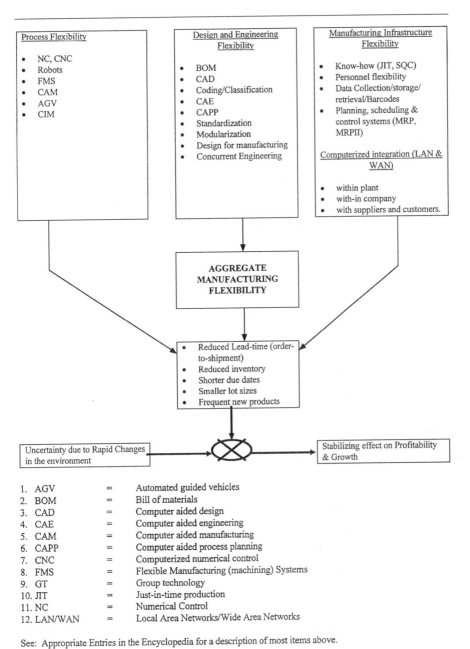

1. AGV	=	Automated guided vehicles
2. BOM	=	Bill of materials
3. CAD	=	Computer aided design
4. CAE	=	Computer aided engineering
5. CAM	=	Computer aided manufacturing
6. CAPP	=	Computer aided process planning
7. CNC	=	Computerized numerical control
8. FMS	=	Flexible Manufacturing (machining) Systems
9. GT	=	Group technology
10. JIT	=	Just-in-time production
11. NC	=	Numerical Control
12. LAN/WAN	=	Local Area Networks/Wide Area Networks

See: Appropriate Entries in the Encyclopedia for a description of most items above.

Figure 7. Ingredients of aggregate manufacturing flexibility.

		Flexibility Context			
		A. Flexibility In Automated Line (Type 1)	B. Flexibility In Manufacturing (Type 2)	C. Flexibility in Design and Manufacturing (Type 3)	D. Process Industry (Type 4)
	Strategic Choices →	1. Low-cost high-volume 2. New product introduction 3. Multiple models	1. High variety, mid volume 2. Different configurations 3. Different routings	1. Custom design 2. Very low volume 3. Design change frequent	1. Low Cost Change over 2. Range of Process 3. Process Mobility
Flexibility Agents	1. Design	A few stable frozen designs	Variety of moderately stable designs	Extremely variable custom designs	not relevant
	2. Process/ Technology/Flow	• Flexible automation, Robotics, AS/RS • Continuous flow	• FMS/AGV, CAD, CAM • Automated flow, AGV	FMS, NC/CNC. CAD/CAM, CIM Intermittent flow	Fixed at Installation
	3. Infrastructure	• JIT • A few dependable vendors • Flexible employees	• GT, Cells • MRP/JIT • Flexible multi-skilled employees	• GT • Flexible PPC • Alternate schedules and routings	• Flexible employers • Computer Control
	4. Computerized Integration among Design, Process, and Infrastructure	• Enhances planning, scheduling flexibility. • Reduces inventory • Reduces lead-time	• Enhances mix, schedule and routing flexibility • Reduces lead-time	• Enables Concurrent engineering, • Reduces design time • reduces design changes • Enables easier design changes.	• Improves scheduling • Decision making at the plant floor
	5. Computerized Integration with vendors and suppliers	• Smoothes production scheduling • Reduces Inventory • Reduces lead-time	• Reduces inventory • Reduces lead-time	• Improves concurrent engineering • Reduces lead-time	• Reduces inventory • Reduces lead-time • Customer Responsiveness • Better planning and forecasting

Figure 8. A manufacturing flexibility typology.

IMPLICATIONS OF MANUFACTURING FLEXIBILITY TYPES

There are several powerful implications of the three types of flexibilities proposed in Figure 8. These implications are discussed under the following headings:

• The strategic goals associated with flexibility types are different.
• Certain aspects of manufacturing flexibility types are mutually conflicting.
• Product design is integral to the attainment of MF in non-process industries.
• Justification procedures for investment in MF should account for the differences in flexibility types.
• An advanced information system is essential to the attainment of manufacturing flexibility.

Strategic choice and flexibility types: In Figure 8, each flexibility type is defined by a unique set of strategic choices or goals. Each set of goals are made of a unique *mix* of the following: (a) variety of models produced, (b) stability of design, (c) number of units produced per model, (d) capability for custom design and production, (e) routing flexibility, and (f) lot size.

The goal of Type 1 flexibility is: low cost, high volume production of very few models of stable design on a common line with ease of new product introduction. While low cost, high volume production has always been the result of automated lines, modern technologies such as robots provide the additional capability to introduce new models on the line with ease and more rapidly—i.e., greater flexibility. Such lines have mixed model production capability and can handle lot sizes as small as one.

The goal of Type 2 flexibility is: moderate variety, mid-volume production of different configurations in lot sizes as small as one. In Type 2 environments, although product design change capabilities exist, the goal is to minimize disruptions due to design changes by concentrating on the production of relatively stable designs. This type of flexibility may be found most frequently in the manufacture of components or subassemblies requiring several machining operations. There is a moderate level of routing flexibility in Type 2 flexibility.

The goal of Type 3 flexibility is: custom design and manufacturing of an infinite variety of very low volume (often one-of-a-kind) product design and redesign. The design, engineering, and manufacturing systems are properly matched to cope with frequent minor as well as major changes to design. There is a high level of routing flexibility in Type 3 environments.

The goal of Type 4 flexibility is: increased process mobility and process range, and lower change over cost and time.

Flexibility types can be mutually conflicting: As can be inferred from the discussions above, MF is heterogeneous. The heterogeneity of MF results in some mutual incompatibilities among the three types of flexibilities. For example, the methods used for attaining Type 3 flexibility may be at cross purpose to the methods used for attaining Type 1 flexibility. For example, while it is extremely important for the hardware and infrastructure in Type 3 environments to have the capability to frequently deal with minor and major changes in the design and manufacture of components, such changes are very disruptive and incompatible with the low cost manufacturing objectives that invariably go with Type 1 environments—see IBM's example described later in this paper.

The important role of design in flexibility: The typology proposed here recognizes that, regardless of the type of flexibility, design plays an important role in the attainment of the three types of flexibilities, although as shown in Figure 8, its role varies substantially from one type of flexibility to the other. The message of Figure 8 is clear and simple—product design should be taken into account when developing manufacturing flexibility.

Economic justification of investments in flexibility: The thorny issue of justifying investments in automation technologies has occupied many researchers and practitioners. One reason it confounds those working on the topic is the difficulty associated with the evaluation of flexibility for use in investment analyses. Any progress in evaluating the benefits of MF must be preceded by an improved understanding of the concept of MF.

In three recent instances of actual investments in automated manufacturing by manufacturers, the author found that the firms made their investments with either inadequate, or no justification. Kaplan (1986) calls this practice of justification of investment in manufacturing technology, " . . . justification by faith." Kaplan also reminds his readers that the practice of justification of investment in manufacturing technologies is difficult because there exist only extremely weak methods for analyzing the value of flexibility to the manufacturer.

The heterogeneous view of MF proposed here should improve investment analyses because it encourages the consideration of different as well as narrow and specific types of flexibilities in justifying investments in manufacturing technology. For example, increasing flexibility in Type 1 environments may mean increasing the number of models that can be produced on an existing line or the ease with which new models can be introduced on an existing line—none of which may be relevant to Type 2 and Type 3 flexibilities. Therefore, analyses of investments in manufacturing flexibility with explicit recognition of the three flexibilities can bring greater focus to the analyses and thus reduce the difficulties associated with their justification.

The need for computerized integration: As shown in Figure 8, all three types of flexibilities require integration of design, technology, and infrastructure. Further, for the manufacturing system to be highly responsive to changing needs, setup time between changes in design, technology, and infrastructure must tend towards zero. This would make massive data and information processing an imperative. Furthermore, given that today's manufacturing technologies are invariably computer controlled and need computerized data-bases and knowledge based systems, the need for real-time, on-line processing of huge centralized or distributed data-bases becomes inevitable. For example, real-time computer integrated manufacturing today integrates activities all the way from customers, suppliers, production planning to real-time control/monitoring of thousands of operations on the factory floor. The instantaneous flow of manufacturing relevant information to relevant users reduces manufacturing response time to changes and increases flexibility.

ILLUSTRATIVE CASES

Beginning around 1980, Manufacturing flexibility became a new goal aggressively pursued by manufacturers. In the last two decades, while manufacturers have explored and exploited several new techniques for increasing manufacturing flexibilities, the basic principles of attaining manufacturing flexibility have not changed since early 1980's. This section, details the experience of two different manufacturers in the U.S. who took very different, yet successful, approaches to attaining manufacturing flexibility. These classic examples of enhancing manufacturing flexibility provide generalizable lessons that are valid today just as they were more than 15 years ago (Swamidass, 1988).

Case 1: IBM's former Lexington Plant (Now Lexmark)

An example of systematically developing a plant for increased manufacturing flexibility for competitive advantage is the investment at the former facility of IBM's Information Products Division at Lexington, Kentucky now operating under Lexmark (Swamidass, 1988). This factory, which is one of the largest automated factories in the world, manufactured typewriters and printers in a million square feet of floor space.

The most noteworthy aspect of 350 million dollar project was the extent of flexibility that was planned and achieved along with a high degree of automation. The top down planning and execution of this project (lasting almost seven years)

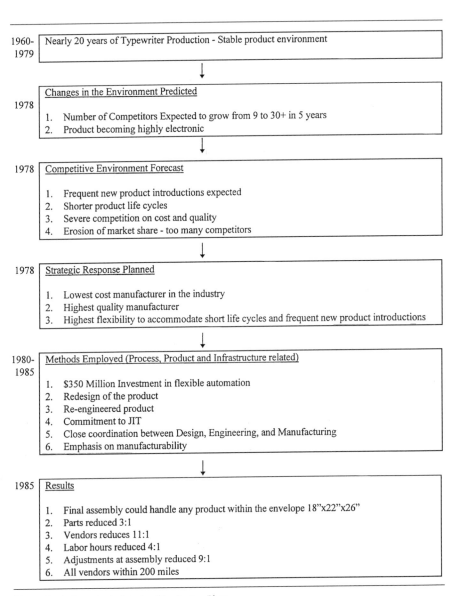

| 1960-1979 | Nearly 20 years of Typewriter Production - Stable product environment |

| 1978 | Changes in the Environment Predicted

1. Number of Competitors Expected to grow from 9 to 30+ in 5 years
2. Product becoming highly electronic |

| 1978 | Competitive Environment Forecast

1. Frequent new product introductions expected
2. Shorter product life cycles
3. Severe competition on cost and quality
4. Erosion of market share - too many competitors |

| 1978 | Strategic Response Planned

1. Lowest cost manufacturer in the industry
2. Highest quality manufacturer
3. Highest flexibility to accommodate short life cycles and frequent new product introductions |

| 1980-1985 | Methods Employed (Process, Product and Infrastructure related)

1. $350 Million Investment in flexible automation
2. Redesign of the product
3. Re-engineered product
4. Commitment to JIT
5. Close coordination between Design, Engineering, and Manufacturing
6. Emphasis on manufacturability |

| 1985 | Results

1. Final assembly could handle any product within the envelope 18"x22"x26"
2. Parts reduced 3:1
3. Vendors reduces 11:1
4. Labor hours reduced 4:1
5. Adjustments at assembly reduced 9:1
6. All vendors within 200 miles |

Figure 9. A chronological study of Lexington Plant.

exemplified two things. First, it illustrated the strategic value of flexibility in the changed competitive environment, and second, it demonstrated the comprehensive approach needed to achieve meaningful manufacturing flexibility throughout the plant. Figure 9 summarizes chronologically the steps taken at Lexington leading to the planning and execution of a manufacturing system very highly biased towards flexibility and automation at the same time.

Changing competitive environment: Beginning around 1960, the plant manufactured typewriters for nearly 20 years in a relatively stable environment characterized by very slow changes to the product and very few competitors. In 1978, IBM's planners recognized the signs of major change in the competitive environment. Primarily, market forces were shaping typewriters into electronic products as opposed to the traditional electro-mechanical machines they used to be. The transformation of typewriters into electronic products led IBM to expect several giant manufacturers of electronics products to enter this product market. IBM estimated that the number of competitors in the market would grow from nine in 1978 to over thirty by 1985.

Adoption of new goals: For the IBM plant, the full impact of the ongoing changes in the market place was not difficult to see. The history of electronic product markets showed that the consequence of making an electronic product with over thirty competitors was to compete in a market with a very high rate of new product introduction and very short product life cycles. Additionally, under these conditions, intense competition on the basis of cost and quality was to be expected. The above scenario called for a strategic response on the part of IBM if the firm was serious about protecting its large market share. IBM's strategic response was very specific. It required the conversion of the facility at Lexington to manufacture typewriters and related products at the lowest cost and highest quality with maximum flexibility for introducing new products. These goals were instrumental in shaping the plant in the years that followed.

Implementation: What followed was a determined commitment to implement the above mentioned strategy at a cost of 350 million dollars over a period of five years beginning in 1980. In attaining maximum flexibility, all three sources of flexibility—process, product design, and infrastructure—were fully exploited. Lexington plant's actions implicitly recognized that major gains in flexibility, quality, and cost reduction could not be attained without coordinated changes in process, product, and infrastructure. According to Reichenback (*Modern Materials Handling*, 1985), Lexington's Automation Manager, this recognition is demonstrated in the following ten-point guideline for automation used at the plant:

1. Ensure early involvement of manufacturing engineering in new product designs to keep products compatible with automation.
2. Set up modular manufacturing areas.
3. Make products for shipment without finished goods storage.
4. Closely schedule and monitor manufacturing operations. Provide just–in–time parts delivery from vendors.
5. Minimize the work-in-process inventory.
6. Maximize automation of inplant manufacturing operations.
7. Require zero defects in manufacturing and vendor parts supply.
8. Minimize engineering changes.
9. Streamline information handling in all areas.
10. Train people to take on new responsibilities as "owner-operators" of equipment areas within production lines.

The use of the owner-operator concept at Lexington is another example of how investment in equipment alone was not sufficient to achieve higher levels of MF. According to Lexington plant's Reichenbach, an owner-operator,

... monitors equipment, operations, and parts supplies, and will refill hoppers when necessary. The worker can also make corrective adjustments to equipment, and perform minor maintenance as required to keep the equipment operating.

Targets and results at lexington: The concerted efforts on all fronts by the plant yielded some significant results in manufacturing flexibility. By the end of 1985 the assembly line was so flexible as to be able to assemble any product that can fit within an envelope size $18'' \times 22'' \times 28''$ as long as the assembly process was performed from above to accommodate the robots on the line.

While the above description of the assembly process capability demonstrates enormous gains in flexibility, similar gains in flexibility were also made in product design and manufacturing infrastructure. In the pursuit of flexibility, the product was modularized and redesigned to reduce the number of components by a ratio of about 3:1. Further, component varieties were drastically reduced. For example, a single type of ball bearing replaced over 60 sizes or varieties.

In pursuit of infrastructure flexibility, Lexington focused on ever increasing attainment of just-in-time production. This involved targeting component and parts inventory requirements for no more than 2.4 shifts usage. Additionally, purchasing cut down the number of vendors by a ratio of 11:1 and set targets to restrict the purchase of materials from vendors situated within a radius of 60 miles from the plant.

The plant at Lexington which used to manufacture a single model prior to 1978, by 1985 was making more than seven different models with almost unlimited capability for making many more models. The lead time for introducing a new product dropped from about four years to less than 18 months.

Trade-offs in flexibility: In recent years, in the pursuit of manufacturing flexibility by trial and error, manufacturers have come to realize that certain flexibilities have to be surrendered to gain other more important flexibilities.

This trade-off between various forms of flexibility is evident in Lexington plant's actions. For example, the new system at Lexington consciously compromised the flexibility of indiscriminate design change and the flexibility to use multiple vendors. Before the new system, minor design changes were implemented by quick and informal agreement between the designer and manufacturing. In the new system, the procedure for even minor design changes involved careful study of the technical and cost implications of the change on engineering, purchasing, manufacturing, and production planning and control.

If the freedom to use many vendors is a form of flexibility, then IBM chose to sacrifice this form of flexibility too. The number of vendors was reduced from over 700 to about 60 in a period of five years from 1980–85.

Thus, some flexibilities were sacrificed to gain greater flexibility in the rate of new product introduction, in the variety of models that can be simultaneously manufactured, and in the ability to deal with shorter product life cycles without sacrificing the ability to be one of the lowest cost manufacturers. At the Lexington plant, the preeminent flexibility among the various competing flexibilities was never in doubt. A prioritization of flexibilities facilitated the resolution of trade-offs between conflicting flexibilities.

Case 2: McDonnell Aircraft Company (MAC)

This case pertains to the Automated Wing Drilling System installed at the firm's St. Louis plant. The system described here went into operation in November 1986 (Swamidass, 1988).

Process description: In the manufacture of the advanced version Harrier II jet combat aircraft, the manufacture of the wing is a major task. Drilling about 6,500 holes in each wing structure for mounting its skin is an important and substantial part of wing manufacturing.

Since the surface of the wing is highly contoured, unique set ups of the drill in three dimensional space are mandated for each of the 6,500 holes to ensure absolutely perpendicular drilling. Further, different versions of the wing used in the British and the U.S. Marine models of the aircraft add to the variety of drilling set ups. Additionally, modifications to the wing or occasional rework may cause variations in drilling. Finally, the drilling operation also accommodates normal variability that occurs during the fabrication of the wing before it reaches the drilling station.

Corporate goals: Corporate or divisional goals called for 90 percent improvement in quality as measured by rework and rejects, 40 percent reduction in cost, and 25 percent reduction in cycle time over a five year period starting in 1985.

The need for a flexible automated drilling system: A major cost component of wing drilling is associated with the three dimensional set up of the drill thousands of times on the highly contoured wing surface. McDonnell realized that substantial cost reductions could be made by automating the slow and difficult set up operation, and that major reductions in set up time would require flexible technology for wing drilling. Thus, the implication of corporate goals for the wing drilling operation was flexibility in addition to cost and quality.

Project initiation: The acquisition of a flexible automated drilling systems (ADS) at MAC was the result of a bottom up process in response to the corporate goals described above. The project to acquire ADS was initiated by the Manufacturing Equipment and Process Engineering Group. In contrast, at Lexington the project that resulted in great increases in flexibility at the plant was the result of top down planning. The scope of the project at MAC was much smaller than that at the IBM; one covered an entire plant, while the other covered one major operation.

Benefits of flexible ADS: The investment in ADS which amounted to 4.5 million dollars required careful justification. The benefits of the system could be categorized into three groups. First, cost related benefits were very obvious. Labor was reduced by 250 hours per wing. The wing drilling operation which formerly took nine days for completion, was reduced to just three days.

Second, quality related benefits include a 50 percent reduction in rework and rejection in hole drilling. Positioning accuracy increased from $+/-1/64''$ (about .016) to a range between $+/-.003$ to $+/-.007''$.

Third, the flexible ADS permits the drilling of any configuration, which can fit within an envelope that is 108′ long, 14′ high, and 11′ wide. The benefits of flexibility include the random order in which different wing configurations can be drilled with little effect on set up time. The cost of retooling for new wings yet to be designed will be substantially lower than the system replaced by the new ADS. Further, the lead time required for retooling and start up for an entirely new wing will be much shorter than the older system.

Justification of the investment: The strictly cost based justification of the equipment revealed three-year pay back period. The anticipated economic life of the equipment was estimated to be about 10 years due to its ability to accommodate wings (or fuselages) of all configurations that do not exceed the envelope's dimensions.

Benefits due to reduced rejection and rework were not included in the analysis. Nor was reduction in work-in-process due to the reduction in cycle time included in the justification analysis. The investment was justifiable on labor cost savings alone.

Supporting infrastructure: Unlike the experience of the Lexington plant described above, MAC did not acquire flexibility in wing drilling as part of a master plan for enhancing plant-wide flexibility. While Lexington altered the entire manufacturing system, infrastructure and all, MAC took a more localized approach in its acquisition of the flexible ADS.

Relatively speaking, the supporting infrastructure at MAC was little disturbed by the introduction of flexible ADS. If anything, the ADS had to meet existing infrastructure demands. The design function was already flexible through the use of a CAD system. Therefore, an important requirement for the ADS was its ability to interact with the existing CAD, and its ability to access design parameters directly from the CAD system and automatically convert them into drilling instructions.

Production planning and control, and inventory management, which are part of the infrastructure, were treated as though there would be minimal disturbance to them due to the acquisition of the flexible ADS. Actually, ADS fits into the existing system with little change to the infrastructure. The most significant interface involved software development to tie together a supervisory computer, CAD system, drilling system, a vision system, and parts program. A total of 12 months was required for the development of this interface at a cost of about six to seven man-years.

IMPLEMENTING MANUFACTURING FLEXIBILITY PROJECTS

Comparing the experiences of Lexington and MAC

The two cases described here are both successful yet very dissimilar in nature. The experiences of the two firms are compared in Table 3. There are some notable items in this comparison. First, Lexington needed Type 1 flexibility and MAC needed Type 2 flexibility. Second, Lexington employed several agents to gain flexibility while MAC essentially employed hardware (i.e. process technology) to gain flexibility. The various strategies one could use in implementing flexibility are explored below.

Strategies for implementation

The examples from Lexington and MAC and the discussions throughout this chapter should be of use to those interested in taking on projects intended to increase

Table 3 Comparisons between the Lexington and MAC projects

Item	Lexington	MAC
1. Scope of the flexibility project	Entire plant, used technology, design and infrastructure	Major assembly, used only technology
2. Flexibility Type	Type 1	Type 2
3. Corporate goals	Lowest cost, highest quality, most flexible plant	90% improvement in quality, 40% less cost in 5 years
4. Design	New products, CAD Introduced	No change, CAD already in use
5. Process	Flexible lines	Massive drilling station
6. Volume (units)	About 1 million/year	About 100/year
7. Investment	$350 million	$4.5 million
8. Infrastructure	Total change JIT, etc.	Little change
9. Effect	Low cost, high quality very flexible, competitive plant	60% cost reduction, 15% quality gains, very flexible process, cycle time 60% less
10. New product Introduction	Increased frequency, faster	Faster
11. Project Initiation	Top-down	Bottom-up

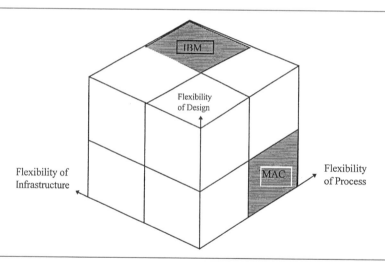

Figure 10. Strategies for implementing manufacturing flexibility. [Comparison between International Business Machines (IBM) and McDonnel Aircraft Corp. (MAC) Cases.]

manufacturing flexibility. In prior discussions, we saw that flexibility can be enhanced through technology, design, and infrastructure dimensions. Flexibility enhancing projects may emphasize one or more of these three dimensions to attain flexibility. If we scale the emphasis placed on these dimensions to gain flexibility as either high, or low, then, as shown in Figure 10, there are eight possible implementation strategies for manufacturing flexibility projects.

Lexington's implementation strategy shown in Figure 10 is an all-out strategy, which is high on technology, high on design, and high on infrastructure changes. It is a

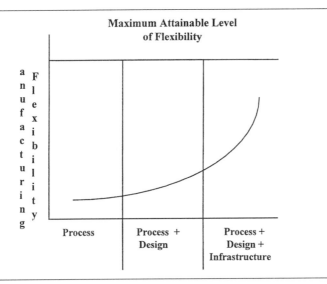

Figure 11. Importance of process, design and infrastructure flexibilities in achieving manufacturing flexibility.

strategy that calls for large investments, a long implementation time table, and maximum disruption during implementation—but it results in the greatest benefits.

The hardware only approach used by MAC shown in Figure 9 is a good example of the implementation of localized flexibility projects with limited investments and equally limited results. Localized projects make sense if they provide logical progression from one project to the other. For example, in the MAC case, the addition of ADS to existing CAD system exploited the flexibility of the drilling system fully while expanding the benefits of the CAD system already in place. While MAC's implementation of ADS was preceded by a CAD system implementation, the two implementations were the result of two separate decisions.

Figure 11 shows that total MF results from a buildup of the results from the employment of flexible technology, design, and infrastructure. Several conclusions can be made based on Figure 11 including the following: only so much flexibility can be gained in a manufacturing system without infrastructure flexibility. The same can be said about technology and design.

Key Concepts: Design and Engineering Flexibility; Economic justification of Flexibility; Flexibility in Automated Line; Flexibility in Design and Engineering; Flexibility in Manufacturing; Flexibility Types; Infrastructure Flexibility; Process Flexibility; Trade-Offs in Flexibility.

Related Articles: Flexible Automation; Just-in-Time Manufacturing; Manufacturing Technology Use in the U.S. and Benefits.

REFERENCES

Abegglen, J.C. and G. Stalk (1985). *The Japanese Corporation*, New York: Basic Books, Inc.

Gerwin, D. (1993). Manufacturing flexibility: A Strategic Perspective. *Management Science*, 39 (4), 395–409.

Gerwin, D. and H. Kolodny (1992). *Management of Advanced Manufacturing Technology*. New York: Wiley-Interscience.

"IBM's Automated Factory—A Giant Step Forward" (1985). *Modern Materials Handling*, March.

Jelinek, M. and J. Goldhar (1983). "The Interface Between Strategy and Manufacturing Technology." *Columbia Journal of World Business*, Spring.

Kaplan, Robert S. (1986). "Must CIM be Justified by Faith Alone?" *Harvard Business Review*, 86 (2) March–April, 87–95.

Skinner, W. (1978). *Manufacturing in Corporate Strategy*, New York: John Wiley and Sons.

Stalk, G., Jr. (1988). Time—The Next Source of Competitive Advantage. *Harvard Business Review*, July–August, 41–51.

Swamidass, Paul M. (1986). *Manufacturing Strategy*: Its Assessment and Practice, *Journal of Operations Management*, 6 (4) (combined issue) August.

Swamidass, Paul M. (1988). *Manufacturing Flexibility*. Monograph #2, Operations Management Association, Plano, TX.

Swamidass, Paul M. (1996). "Benchmarking Manufacturing Technology use in the United States." In *Handbook of Technology Management* edited by Gus Gaynor. New York: McGraw-Hill.

Swamidass, Paul M. (1998). *Technology on the Factory Floor III: Technology use and Training in the United States*. Washington, D.C.: Manufacturing Institute of the National Association of Manufacturers.

Upton, David M. (1995a). "Flexibility as process mobility: The management of plant capabilities for quick response manufacturing." *Journal of Operations Management*, 12 (3&4) June, 205–224.

Upton, David M. (1995b). What really makes factories flexible? *Harvard Business Review*, July–August, 74–84.

Upton, David M. (1997). Process range in manufacturing: An empirical study of flexibility. *Management Science*, 4 (8) August, 1079–1092.

U.S. Dept. of Commerce, Washington, D.C. (1987). *A Competitive Assessment of the U.S. Flexible Manufacturing System Industry*, July.

Venketesan, R. (1990). Cummins flexes its Factory. *Harvard Business Review*, 90 (2) March–April, 120–127.

Voss, C.A. (1986). *Managing New Manufacturing Technologies*, Monograph No. 1, Operations Management Association, September.

12. MANUFACTURING TECHNOLOGY USE IN THE U.S. AND BENEFITS

PAUL M. SWAMIDASS

Auburn University, Auburn, AL, USA

ABSTRACT

Industrial revolution over the last few centuries was made possible by the progressive use of more and more manufacturing technologies for the purpose of replacing manual labor, and for improving quality and quantity of output. The process of new manufacturing technology introduction goes on even today. Today, manufacturing technologies are more diverse and complex than they used to be.

Manufacturing technologies may be classified as either hard or soft technologies. Table 1 is a list of hard and soft technologies included in a recent study by Swamidass (1998). Hard technologies are hardware intensive with associated software. Soft technologies are manufacturing principles, practices, techniques and know-how that may be enhanced by hardware and software.

The annual growth rate of U.S. manufacturing productivity (output per work hour) is at about three percent since early 1980s (Bureau of Labor Statistics; The Manufacturing Institute). U.S. manufacturing productivity has increased 285 percent since 1960. Due to higher productivity growth in manufacturing, inflation rates in manufacturing have been lower than the whole U.S. economy—the inflation rate in manufacturing between 1995–1997 was 1.2 percent per year while the overall inflation rate was 2 percent (Department of Commerce; Manufacturing Institute). Manufacturing

Table 1 A list of hard and soft technologies

The technology	Explanation
Hard technologies	
1. Automated inspection	
2. CAD ...	Computer aided design
3. CAM ...	Computer aided manufacturing including programmable automation of single or multi-machine systems
4. CIM ...	Computer integrated manufacturing
5. CNC ...	Machines with computerized numerical control
6. LAN ...	Local area networks
7. FMS ...	Flexible manufacturing systems; automated multi-machine systems linked by an automated material handling system
8. Robots	All kinds of robots
Soft technologies	
1. Bar Codes	
2. Concurrent Engineering	
3. JIT ...	Just-in-time manufacturing
4. Manufacturing cells	
5. MRP ...	Material requirements planning
6. MRP II ...	Manufacturing resource planning
7. SQC ...	Statistical quality control
8. Simulation and Modeling	
9. TQM ...	Total quality management

productivity in the U.S. exceeds that of other major industrial nations including Japan and Germany; 1996 Japan's productivity was 78 percent of the U.S., and Germany's productivity was at 82 percent of the U.S. (Manufacturing Institute; Van Ark and Pilat, 1993). The productivity of the U.S. manufacturing sector enables it to hold its share of world exports; U.S. share of world merchandise exports has risen from 11.4 percent in 1986 to 12.9 percent in 1997.

This article describes the results of a survey of manufacturing technology use in the U.S. and its findings.

A SURVEY OF TECHNOLOGY USE IN THE U.S.

Given the impressive performance of U.S. manufacturing in recent years, the use of manufacturing technology in the U.S. can be revealing. Figure 1 is based on responses from 1025 manufacturing plants to a survey on manufacturing technology use in the United States; the study was sponsored by the National Association of Manufacturers (NAM) and the National Science Foundation, both of the USA (Swamidass, 1998).

Figure 1 shows the extent of manufacturing technology use in the United States. According to the figure, 84 percent of U.S. manufacturers use CAD; 67.6 percent use CNC and so on. Several benefits of technology use are also reported by U.S.

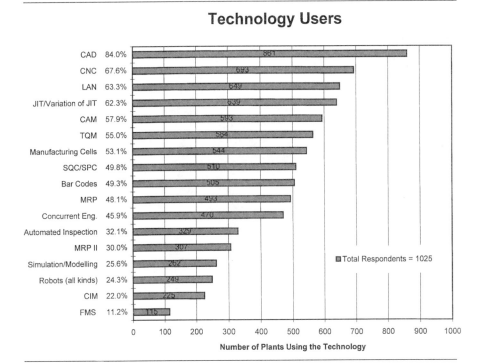

Figure 1. Technology users.

manufacturers. More than 2/3, or more than 67 percent of the manufacturers report that the benefits of technology use included the following (Swamidass, 1998):

1. Decreased manufacturing cycle-time
2. Decreased manufacturing cost
3. Increased product line
4. Increased return on investment

The benefits listed above are more than mere cost reduction from the use of technology; decreased cycle-time provides strategic and competitive advantages. This study of technology use (Swamidass, 1998) found that the average three-year return on investment was at 16.85 percent in 1997 compared to 12.99 percent in 1993 (Swamidass, 1996). One explanation for this performance is the judicious use of the various manufacturing technologies in Table 1.

Additionally, the evidence from the 1998 study leads one to infer that plants are making substantial gains in manufacturing flexibility and agility through the increased use of computerized integration, and manufacturing cells. These benefits provide competitive advantage to manufacturers.

Table 2 A comparison of selected 1993 and 1997 data

	1993 sample	1997 sample
1. Sample	1042	1025
2. Sales ($ million)	47.2	34.5
3. Employment	228	168
4. Sales per employee ($000)	133	147
5. Inventory turns	8.0	9.7
6. Direct labor	18.3%	19.8%
7. Rejection and rework	4.0%	3.5%
8. Lead time (weeks)	7.2	7.4
9. ISO certified plants	4%	20%
10. Foreign-owned (foreign ownership >50%)	3.4%	6.3%
11. No sales to defense department	54%	64.8%
12. Percent reporting cycle time reduction	66%	76%
13. Return on investment	13.0%	16.8%

IMPROVED OPERATIONAL EXCELLENCE AND TECHNOLOGY USE

Table 2 compares key operational measures for U.S. Manufacturers participating in the 1993 and 1997 studies (Swamidass, 1996; 1998). The following are notable trends in the table.

Inventory turns increase. Inventory turns are an index of manufacturing health; larger the better. Inventory turns have increased to 9.7 since 1993 when it was 8.0—i.e., the average plant in 1997 had 1.23 months of inventory while it was 1.5 months of inventory in 1993.

Sales per employee increases. Sales per employee are an index of productivity. Sales per employee in 1997 were $147K as opposed to $133K in 1993.

Rejection and rework rate decreases. Rejection and rework rate is an index of the ability of the manufacturing process to produce quality products; lower the rate, the better the quality. Rejection and rework rate has decreased from 4 percent in 1993 to 3.5 percent in 1997.

Cycle time and manufacturing cost reduction. In 1997, 76 percent of all manufacturers reported reduction in cycle time as a result of technology use and 75 percent reported reduction in manufacturing costs. Other benefits of technology use and the number of manufacturers reporting these benefits are recorded by Swamidass (1998).

Extremely skilled use of technology has superior pay off. Not all manufacturers use manufacturing technologies with extreme skill. Extremely skilled users of JIT and its variations report the best inventory turns (19.4), lowest manufacturing lead times (5.7 weeks), and one of the highest return on investment (20.8%). Extremely skilled users of manufacturing cells report the highest return on investment (21.8%). Extremely skilled use of technologies requires a thorough understanding of the technology being used as well as constant training and retraining of employees.

NETWORK TECHNOLOGY USE INCREASES

The users of local area network (LAN) have grown more than the users of any other technology since 1993; from 49 percent of respondents to 63.3 percent. LAN enables the integration of various computerized technologies in the factory, and it is essential to the integration between factories and their customers or vendors. LAN also permits the expanded and better use of existing technologies on the factory floor through the synergy generated by the integration of several hard and soft technologies.

The 1997 study found that 58 percent of all transactions between the shop floor and production/materials planning is now computerized, and 41 percent of all transactions between design and customers is now computerized. For the extent of computerized integration among other units internal and external to the factory, see Swamidass (1998).

WORKFORCE TRAINING

Modern manufacturing technologies need educated and trained workforce. Training and retraining of the workforce and management ensures the best use of investment in technologies. The average training budget in 1997 was five percent of payroll (Swamidass, 1998). The training and retraining needs of manufacturing industries using technologies will remain high in the U.S. because (1) the training needs of manufacturing cell users increases because of the need to train operators to handle multiple machines in cells; (2) manufacturers experiencing down-sizing and reengineering face the need to train and retrain employees that are retained; (3) manufacturers hiring employees leaving a downsized facility need to retrain incoming employees; (4) the extremely skilled use of technologies requires frequent training and retraining of the employee; and (5) flexibility and agility in operations requires more frequent retraining of employees.

On-the-job training (OJT) is the most commonly used training technique for training operators in the U.S., although it takes more time to train an employee (8.3 months). Training by vendors is the quickest training method (5.3 months). Manufacturers report that the availability of trained operators delays the skilled use of technologies by as much as 4.9 Months.

CONCLUSIONS

Increased LAN use means increased computerized integration

The secret to understanding one of the trends in technology use in manufacturing plants lies in our ability to interpret the increased use of LAN technology. Clearly, by 1997, the use of some technologies had reached saturation levels—this is made evident by the data gathered over time by Swamidass (1996, 1998). Manufacturers normally begin to invest in several different technologies in the form of "islands of automation." LAN technology enables the integration of these "islands" and taps the synergistic benefits that flow from the integration of several technologies. Thus, one trend in technology use in the nineties is the computerized *integration of the various factory floor technologies*.

Swamidass (1998) also found computerized integration among internal and external units of a plant. For example, in 1997, 48 percent of all transactions between design and shop floor was computerized, and 41 percent of all transactions between design and customers was computerized. While the growth in the use of individual technologies may taper off, growth is to be expected in the integration of technologies through the computerization of transactions between internal and external units of manufacturing plants.

Increased agility

Computerized integration of units inside and outside a factory contributes to agile manufacturing by enhancing the speed of information flow and the ability of manufacturing systems to respond to changes. Swamidass (1998) found that more plants used manufacturing cells in 1997 than in 1993. Manufacturing cells require operators to be skilled in the use of multiple tasks, which adds to the agility of the plant. Through the increased use of LAN and manufacturing cells in the U.S., the flexibility of the average plant is being enhanced.

Systemic changes brought on by soft technology use

Since 1993, slightly fewer plants are reporting the use of JIT, TQM and SQC techniques; however, the benefits associated with the use of these soft technologies are on the rise. For example, inventory turns and rejection and rework rates have improved to 9.7 and 3.5 percent, respectively (Table 2). This is interpreted to mean that, since mid 1980s, essential features of JIT, TQM and SQC are becoming generic and ingrained manufacturing practices in the U.S. without being associated with any specific technique. These systemic changes bring permanence to manufacturing improvements, and the continuous improvement theme underlying these practices should continue to improve manufacturing performance in the U.S.

Transportation industry is leader in inventory turns

A remarkable finding of the 1997 survey is the inventory turns of 19.4 reported by the transportation industry (SIC 37), which includes the auto industry; this is twice that of the national average at 9.7. It appears that the transportation industry has become a mature user of lean manufacturing principles and the industry is reaping the benefits of long-term and consistent use of such practices.

The auto industry, which forms a sizable portion of the transportation industry, has been one of the earliest industry groups to adopt soft technologies such as JIT, TQM and SQC in the mid-1980's because of severe competition from Japan and the rapid erosion of their market share. The big three automakers instituted standards and certification for their suppliers based on world-class manufacturing principles. In the process, the industry has proved to be a good training ground for world-class manufacturing techniques. Individuals moving from this industry to other industries may have contributed to the spread of such techniques to other industries.

Training methods

On-the-job training (OJT) is the most commonly used training technique but it is also the most time consuming method. Vendors provide the quickest training; if vendors can provide training, small plants have more to gain by using vendors than larger plants- small plants can save more than four months while larger plants can save less than two months by going to technology vendors for training. Each plant should evaluate if the added cost of vendor-provided training justifies the savings in training time.

More larger plants use technologies than small plants

Swamidass (1998) found the unmistakable effect of size on technology use. He categorized the plants into three groups to study the effect of plant size (employees); the sample had 570 small plants with less than 100 employees, 428 larger plants with more than 99 employees, and a subset of 69 very large plants with more than 499 employees. See Table 3 for extensive comparisons across size.

The following were found to increase with size: (1) sales per employee; (2) inventory turns, (3) LAN users; (4) cell users; (5) percent of transactions computerized between shops and production planning; (6) percent of plants with 90%+ computerization between shops and production planning; and (7) percent of plants

Table 3 The average plant in the study by size

	Small plants*	Larger plants*	Very large plants*
Sample	570	428	69
Average employment	45	329	995
Average sales ($ million)	5.9	59.6	188.7
Sales per employee	$135K	$163K	$190.5K
Rejection and rework (% of manufacturing costs)	3.4%	3.6%	3.2%
Direct labor cost (% of sales)	23%	15.3%	11.8%
Average manufacturing lead time	6.2 weeks	8.3 weeks	6.9 weeks
Inventory turns	9.5	10.1	12.9
LAN users (% of all plants)	51.9%	79.2%	97.1%
Cell users (% of all plants)	38.4%	72.9%	88.4%
Age of the plant	28 years	30 years	33 years
Exporters (percent of all manufacturers)	56.8%	82.2%	78.3%
On-the-job training time for skilled employees	9.4 months	7.1 months	5.1months
Vendors' training time for skilled employees	5.3 months	5.2 months	4.6 months
Training budget (% of payroll)	5.6%	4.3%	3.5%
Extremely skilled operators (% of all operators)	43%	36.7%	34.5%
Percent of all transactions computerized between shops and production/materials planning	49%	71%	78%
Plants with 90%+ computerized transactions between shops and production/materials planning (% of all plants)	20%	42.1%	55.1%
Delay in skilled technology use due to the non-availability of skilled labor (months)	5.5 months	4.4 months	3.7 months
Plants reporting cycle time reduction as a result of technology use (% of all plants)	73%	85%	84.1%
Return on investment	15.6%	18.3%	23.3%

* Small plants: employees <100; Larger plants: employees >99; Very large plants: employees >499 (included in column 2).

reporting cycle time reduction as a result of technology use; and (3) return on investment.

The following decrease with size: (1) on-the-job training time; (2) training budget as a percent of sales; (3) percent of extremely skilled operators in the plant; and (4) the delay in the skilled use of technologies for want of skilled workers.

The needs of small plants

The data raises the questions, why small companies do not use technologies as often as larger plants? While not all technologies may be appropriate for small plants, the following are some of the reasons:

- They may need assistance to understand the use and benefits of several new technologies. Large companies have in-house experts to investigate and evaluate investments in new technologies.
- The training expense associated with the use of technologies may be holding back small manufacturers from using technologies more aggressively.
- Funds for investments may be more difficult to obtain for smaller manufacturers.

Key Concepts: Bar Codes; CAD; CAM; CIM; CNC; FMS; JIT; LAN; Lean Manufacturing; Manufacturing Cell; MRP II; Robots; Simulation and Modeling; SQC; TQM.

Related Articles: Concurrent Engineering; Flexible Automation; Manufacturing Flexibility.

REFERENCES

Swamidass, P.M. (1996). Benchmarking Manufacturing Technology use in the United States, in *The Handbook of Technology Management*, edited by Gus Gaynor, New York, NY: McGraw-Hill.

Swamidass, P.M. (1998). *Technology on the Factory Floor III: Training and Technology Use in the U.S.*, Manufacturing Institute of the National Association of Manufacturers, Washington, DC. The Manufacturing Institute. *The Facts About Modern Manufacturing*, fourth edition. Washington, D.C.

Van Ark, B. and D. Pilat (1993). "Productivity levels in Germany, Japan and the United States: Differences and Causes." in *Brookings Papers on Economic Activity, Microeconomics, #2, 1–49*.

13. AGILE MANUFACTURING

PRATAP S.S. CHINNAIAH
SAGAR V. KAMARTHI
Northeastern University, Boston, USA

ABSTRACT

Businesses are witnessing unprecedented changes today. New products, new processes, new technologies, new markets, and even new competitors are appearing and disappearing within short periods of time. Historically, mass production has evolved into lean production. Now lean production is evolving into agile manufacturing. Until the 1950s companies focused on productivity improvement and in the 60s and 70s they concentrated on quality enhancement. In the eighties, while companies worked hard to achieve flexibility, in the 90s they are challenged by the need to increase agility. To review current issues in agility, this article describes the market forces that demand agility, the elements that constitute agility, agility enablers, and agility implementation.

The National Science Foundation (NSF) and the Defense Advanced Research Projects Agency (DARPA) have jointly established the Agile Manufacturing Research Institutes (AMRI) at three US Universities. The purpose of the AMRIs is to enhance the understanding of agile manufacturing enterprises, develop system performance measures based on quantitative data, structure a program of research to meet industry-defined needs, and move emerging agile technologies into the next stage where functional prototyping or proof-of-concepts test can take place (DeVor et al., 1997). Agile manufacturing practices in the following companies are included in this article: **AT & T; Bally Engineered Structures; Chrysler; Ford; GE Fanuc; GM; Honeywell; John Deere; Mars Company; Matsushita; Panasonic; USCAR.**

ELEMENTS OF AGILITY

The central idea in agile manufacturing is that an enterprise should be built on the competitive foundations of continuous improvement, rapid response, quality improvement, social responsibility, and total customer focus.

Agility is the ability of producers of goods and services to thrive in rapidly changing, fragmented markets. Agility is a comprehensive response to the challenges posed by a business environment dominated by changes and uncertainty. For a company, to be agile is to be capable of operating profitably in a competitive environment of continually and unpredictably changing customer opportunities. An agile enterprise must have broad change capability that is in balance across multiple dimensions. Would the enterprise be considered agile if a short-notice change was completed in time, but at a cost that eventually bankrupted the company? Would it be agile if the changed environment thereafter required the specialized wizardry and constant attention of one specific employee to keep it operational? Is it agile if change is virtually free and painless but out of synch with market opportunity timing? Is the enterprise agile if it can readily accommodate a broad category of changes that are no longer needed, or too narrow for the latest requirements? These questions lead to four principal change proficiency metrics: to define agility time, cost, robustness, and scope.

Completing a change in a timely manner is the only effective way to respond, while the *time* of the change alone does not provide a sufficient metric for agility. Virtually anything can be changed if *cost* is not a constraint, but change at any cost is not a viable solution. If the response to the change costs too much relative to a competitor's cost, it will limit shareholder profits. A change, quick and economical, is still not a sufficient profile of agility. If after the change, the modified system requires continual attention to remain functional, the change accommodation is insufficiently *robust*. A company may end up with a fragile result if standard procedures are bypassed in the process of changing quickly and economically. Finally, a system is considered to be agile precisely because it is able to thrive on change, but how much change can the system handle? The metric of *scope* addresses this question. Scope is the principal difference between flexibility and agility. Flexibility is the capability the deal with planned responses to anticipated contingencies. On the other hand, being agile means the system can operate within a defined range of changes, and it can be deconstructed and reconstructed to handle unplanned changes. Thus, for an enterprise to be agile, it must have a balanced response-to-change capability across the four change proficiency metrics: time, cost, robustness, and scope.

MARKET FORCES THAT DRIVE AGILITY

As the context of commercial competition changes dramatically, the conduct of business changes correspondingly. Today, the following market forces demand changes in the conduct of business.

Market fragmentation: Markets of all kinds are being fragmented at an accelerating pace. Companies are segmenting customer groups and pricing the same goods or services differently depending on the circumstances of the transaction. For example, with virtually unlimited packaging possibilities, Mars company prices their M&M candies differently when they are sold in different packages of the same weight. Companies are segmenting markets according to function and are exploiting economies of scope which enables high variety and low-volume production. (Stipp, 1996).

Production in smaller lot sizes: Many companies are already capable of making several products, that are different from each other, on a high-volume line with little or no increase in production costs. Lot sizes can be as low as one. This type of production capability has revolutionized marketing. (Kaplan, 1993).

Information capacity to treat masses of customers as individuals: More and more companies are discovering that they can produce customer-configured products "to order" instead of "to forecast," and in doing so they can generate benefits far beyond elimination of inventories. The consumer initiates the interactive relationship with the company through which the product is jointly defined (Pine II, 1993). Massive customer databases and production equipment innovations permit effective production of smaller and smaller orders at significantly lower costs (Lau, 1995).

Shrinking product cycle times: The decreasing product development cycle times, increasing proliferation of models, and accelerating pace of the introduction of new or improved models are among the most aggressive challenges of contemporary competition. Today, Panasonic's consumer electronics product development cycle time is three months. That is, the cycle time of any given model of CD player, TV, VCR, cassette deck, or stereo receiver is just 90 days. During that time its successor is being designed, tested, and put into production. The design, development, production, distribution, and marketing processes are continuous and overlapping.

Convergence of physical products and services: The traditional distinction between goods and services and companies is vanishing. A direct result of this convergence is that manufacturing companies are acquiring the capability to create information and provide additional service to customers.

Global production networks: No market is domestic anymore, and no producer is restricted to domestic production only. The addition of high-capacity information and communication systems to the existing global transportation systems opens foreign markets to many producers.

Simultaneous intercompany cooperation and competition: To a degree unprecedented in history, companies are entering into partnerships, joint ventures, and collaborations of every imaginable kind including the formation of virtual companies. The three U.S. Companies, GM, Ford, and Chrysler were unwilling to develop catalytic converters cooperatively in the 1970s. As a result, each company spent hundreds of millions of dollars developing essentially the same product. Today, the three firms have joined a multifaceted consortium, USCAR, which will permit the joint development of technologies, materials, and components ranging from structural plastics to electric vehicle-control systems.

Distribution infrastructures for mass customization: Agile competition rests on individualized products and interactive customer relationships. Inevitably, new product distribution mechanisms are emerging as part of agile manufacturing. For example, direct marketing by producers of goods require customer-centered production.

Corporate reorganization frenzy: US companies have been implementing a wide range of initiatives such as just-in-time, total quality management, flexible manufacturing systems, business process reengineering, ISO 9000 in order to improve their competitiveness and profitability.

ENABLING SYSTEMS FOR AGILITY

The change in a manufacturing system's behavior and structure needed for agility is enhanced and enabled by a number of items. Special skills are required of the people employed in agile systems. Because it is neither possible nor desirable to automate all activities in these systems, people are required to be highly skilled, multiskilled craftsmen. Continuous improvement of the skills of the personnel will be an essential part of an agile manufacturing facility. The teams working on individual orders must operate in a frictionless manner and must stress the *process* rather than traditional *functions*. For example, at Bally Engineered Structures, Inc. factory floor workers are now responsible for planning and supervising their own work. The concept of virtual teams consisting of product designers, tool planners, and production engineers located at far-flung geographical locations is another example. The objective in agile manufacturing is to combine the organization, technology, and people into an integrated whole.

The implementation and operation of agile systems are possible with both low and high technologies. The more the complexity of the technology involved in an agile manufacturing system, the higher the cost of implementation and operation of the production system. Generally, highly flexible general-purpose technologies are preferred in agile manufacturing systems.

Several technologies such as computer-aided design (CAD), virtual reality, customer database, electronic kiosks, and multimedia communications may be used at the marketing stage for identifying individual customer requirements. Some other technologies specifically required for meeting the customized orders in arbitrary lot sizes are configurators, stereolithography, internet, multimedia workstations, personalized smart cards, electronic data interchange (EDI), etc. A configurator is an expert system that assists in developing valid product and process descriptions to meet a customer's specific needs. Configurators create accurate drawings and designs, speed up proposal creation, and increase sales efficiency for mass customization. Many software vendors are now offering configurator modules integrated with Enterprise Resource Planning (ERP). These software systems permit integration of sales, order processing, and engineering. For example, a software known as PROSE (Product Offerings Expertise) is a knowledge-based engineering and order processing platform that supports sales and order processing at AT&T Network Systems. A sophisticated information management system called CDIN (Computer-Driven Intelligence Network) at Bally

Engineered Structures, Inc. connects everyone in the company as well as independent sales representatives, suppliers, and customers together. Similarly, John Deere Harvester Works uses a configurator software called OptiFlex to configure the harvester features such as row count, or the nature of fertilizer it can handle. It is reported that literally thousands of configurations can be produced using OptiFlex.

Virtual reality can play an important role in the process of translating a customer's requirements into design specifications. For example, customers buying their kitchen units from Matsushita Appliances can experience their kitchen environment with the help of data gloves and 3-D vision systems before placing an order. Such virtual reality tools are known as experience simulators. Similarly, close communication between manufacturers and customers can be realized with multimedia communications technology. Several geographically separated product development teams such as a customer group, a manufacturing group, and a design group can work together with the help of multimedia communication technology. Various teams can exchange documents, see each other on a tiled video display in a window on a computer screen, talk, exchange ideas, modify product specifications and solve manufacturing problems. Such facilities are labeled as virtual meeting rooms.

A new line of factory automation software called CIMPLICITY that facilitates mass customization is available from GE Fanuc. With the help of an FMS in its production shop, Honeywell Microswitch Corporation is able to drive product customization. Personalized smart cards that have details of personal features such as physical body dimensions and medical information are being visualized to make mass customization work faster. At the development stage, technologies such as personalized smart cards, electronic kiosks, internet, CAD, CAE, configurators, and concurrent engineering are useful in reducing design lead times. Some technologies such as rapid prototyping, web-aided design (or internet-aided design), virtual machining, virtual assembly & testing, cost & risk analysis models, and product realization process are being developed by TEAM (Technologies Enabling Agile Manufacturing) program established by the US Government (Herrin, 1996). At the production stage, in addition to flexible integrated manufacturing facilities, skilled people play a very important role. The technology should augment peoples' skills and abilities. However, technology is being used for automating certain tasks where productivity and safety can be improved considerably. Achieving seamlessness among the different phases of the value chain must be the goal of employing technology and skilled people together. The emerging production concept—agile manufacturing—offers strategies to integrate organization, people, and technology into a coordinated interdependent system.

DEVELOPING AGILE CAPABILITIES

The journey to agility is a never-ending quest to do better than the competition, even as the competitive environment constantly changes. The agile paradigm is concerned principally with an unpredictable change. To understand the kinds of change impacting an enterprise and analyze the enterprise's ability to respond, the change is decomposed into various domains. Building a model of *change domains* gives makes it easier to analyze potential agile characteristics. Dove et al. (Dove, 1996, Dove et al., 1996) identified the following eight domains for analyzing and constructing agile

Table 1 Change domains of agility

Domain	Production	Organizational structure	Information automation	Human resources
Creation/Deletion: Build something new or remove something completely	Build new production plant	Build new team with new people	Build information access and e-mail infrastructure	Hire all new people for the new facility
Expansion/ Contraction: Increase or decrease existing resource mix	Add similar production equipment	Add more people with similar skills to a team	Add acquired company to the network	Increase or decrease employee head count
Addition/ Subtraction: Add or delete resource types	Add different production equipment	Add more people with different skills to a team	Add access to a new database	Add people with new and different skills
Reconfiguration: Change relationships among modules	Convert a production line to different purpose	Abolish old teams and reform new teams	Change the network structure	Adjust dental vs. medical benefit mix
Migration: Event-based change of fundamental concepts	Convert to bid-based cellular scheduling	Institute self-direction in work teams	Full access to outside databases and e-mail	Institute on-the-job continuous learning
Variation: Ability to cope up with real-time operating surprise	Setup/changeover for unscheduled part	Function when team members are absent	Video traffic swamps the network	Deal with a union wildcat work shutdown
Augmentation: Continuous incremental upgradation	Daily control system upgrades	Continuous learning of team-work skills	Personal agents get smarter	Strat monthly company communication sessions
Correction: Reincorporate corrected failures or alternatives	Return broken station to service	Fix dysfunction in a team structure	Route around a bad network node	Return to EEOC compliance

capabilities: creation/deletion, expansion/contraction, addition/subtraction, re-configuration, migration, variation, augmentation, and correction.

The last three change domains are common to both lean and agile systems. Table 1 shows the eight change domains and simple examples of how they might manifest themselves in four different areas of an enterprise. These eight change domains when combined with the four proficiency metrics of agility (time, cost, robustness, and scope) explained earlier gives an analytical tool for prioritizing problems and opportunities. For example, when considering alternatives, the creation/deletion change domain can be assessed on the four metrics of agility by the use of ratings on a scale of 1 to 10. Based on the rating the best alternatives can be chosen for increasing the agile capabilities to an enterprise.

IMPLEMENTATION

Agile business practices may become core requirement for conducting business in future. Concepts of lean production, flexible manufacturing, mass customization, just-in-time are all operating strategies that a firm can implement to become more agile. Additionally, concepts such as reengineering, total quality management, statistical process control are the transformation tools that may be used to transform a company into a more competitive organization. There are five key issues that directly influence a firm's agility: communication connectedness, interorganizational participation, management involvement, production flexibility, and employee empowerment (DeVor et al., 1997).

Communication connectedness: The level of communication connectedness in the firm refers to: (1) the technical aspects of network linkages, computer systems compatibility, and hardware interface flexibility; and (2) the behavioral aspects of communication. These two aspects of communication connectedness refer to external connectedness with customer and supplier computer systems, as well as internal connectedness.

Interorganizational participation: This refers to the level of business activities that cross boundaries of a company. The involvement of suppliers and customers in the product design and process improvements is an example. In some cases, much tighter linkages are employed, which eliminate the need for quotations based on price, quality, delivery, etc. Then, purchasing processes can be made quicker. Sourcing and sales decisions involving personnel from product development, quality, and manufacturing functions increases agility. Such tightly-knit relationship will require much trust and dependence.

Production flexibility: This issue refers to the ease and timeliness with which a product line or the entire factory can be re-configured to produce a new product or process. It also relates to the automation of the processes of a firm and their reconfigurability.

Management involvement: This refers to the level of active participation of the management in the implementation of agile practices. Management involvement is a critical element in increasing the level of agility in a firm. Management provides physical and organizational resources in support of the creativity and initiative of the workforce. Management must be actively committed to continuous workforce education in the pursuit of the quality and capability of the workforce. It is management's role to implement business processes that lead to reconfigurability and flexibility.

EXAMPLES OF AGILE MANUFACTURING

Practice of agility is highly context-dependent. Therefore what a company must do in order to become agile is based on its own understanding of its customers, markets, competitors, products, competencies, and resources. Agility is a continual process of managing change, and a constant adaptation of internal practices and external relationships to meet new opportunities.

Many companies including Boeing, Bellcore, Chrysler, Caterpillar, Citicorp, Ford, Honda, Hitachi, IBM, Hyundai, Intel have realized the benefits of agility by applying the concepts developed at the AMRIs (Goldman et al., 1995). The following case illustrates the agile competitive behavior (Goldman et al., 1995).

The case of agility in the apparel industry: $\{TC^2\}$, the Textile/Clothing Technology Corporation is a consortium of 200 companies. This consortium demonstrated the evolving agility potential of the US apparel industry at the September 1994 Bobbin show, held in Atlanta, Georgia. For the first time at a major trade show, a fully operational, integrated virtual apparel manufacturing capability was demonstrated. A blouse and coordinated skirt were designed, cut to customer order, printed, sewn, and distributed daily to customers at the show and at three other remote retail sites.

A designer located at the Fashion Institute of Technology (FIT) in New York City linked by two-way audio and video to $\{TC^2\}$'s booth at the Bobbin show in Atlanta designed the blouse print pattern interactively for customers at the following locations (1) at the show; (2) at a retail store in Cincinnati, Ohio; (3) at EDS headquarters in Dallas, Texas; and (4) at the National Center for Manufacturing Sciences in Ann Arbor, Michigan. Approval from the customers for the print pattern, sizes, and colors was obtained by 3 p.m. on each day of the show. The FIT designer then transmitted the designs electronically to the Atlanta show booth of Computer Design, Inc. (CDI). CDI translated the designs into digital screen print color separation data and digital garment pattern design data. The former were transmitted to a screen-making facility in Atlanta, and the latter to computer-controlled laser cutting machines at the show. Concurrently, full-color posters displaying the designs being produced on the show floor were printed, as were custom garment tags that accompanied each order to fulfillment.

The screen print manufacturer in Atlanta produced the screens overnight and delivered them by 8 o'clock the next morning to the show booth of Precision Screening Machine (PSM) for off-site printing of that day's designs. Concurrently, laser cutting machines at $\{TC^2\}$'s booth used the digitized point-of-sale garment pattern and size data to cut garment pieces out of white fabric and deliver them to PSM for printing during the show hours. The printed pieces, collated by the customer orders for size, style, and design, were then shipped to two separate locations on the show floor for assembly by flexible, cross-trained manufacturing teams: blouses for assembly by Sunbrand, and skirts for assembly by Juki. When sewing was completed, the garments were then delivered to customers at the show and shipped overnight to the customers at the remote locations.

This demonstration which involved a partnership among 35 companies and the electronic coordination of four remote sites and five different locations at the Bobbin show, reveals how the apparel industry is to realizing agile production, marketing, and distribution by producing customized products, "to order," for traditionally for mass-market customers. In this case, the firms leveraged resources to cut product cycle time by means of intensive interactive cooperation among companies, utilized people and information, and achieved flexibility, empowerment, and mutually advantageous sharing.

Furthermore, the limited Bobbin show demonstration is only a piece of $\{TC^2\}$'s larger agile manufacturing effort called "Apparel-on-Demand" demonstration project. In this project, $\{TC^2\}$ members have participated in the development of three-dimensional body scanning hardware and software, programs to translate this data into two-dimensional design data for printing fabric and cutting garment pieces using computer-controlled laser cutters, and high-speed computerized garment assembly machinery. Together with the software linking point-of-sale data to fiber, fabric, and garment manufacturers true agile clothing production moves to imminent reality.

POTENTIAL PROBLEMS AND BENEFITS

There are many potential problems that act as barriers to the implementation of agility. The concepts of agility have yet to be incorporated into commercial systems, measures, laws, customs, and habits. The barriers to assimilating agility can be grouped into four categories: technological barriers, financial barriers, internal barriers, and external barriers. Many technologies such as intranets, internets, and rapid prototyping are still slow and costly. The commonly used financial accounting system is another barrier because it is reasonably accurate for classical mass production but is unsuitable for agile commerce. Internal barriers include performance measurement systems, budgeting procedures, dysfunctional organizations and information systems, lack of trust, and lack of knowledge of successful users of agility. External barriers include legal systems that are suitable for mass-production environment, tax laws, trade agreements and enterprise zone definitions, static and rigid requirements of public bodies (Goldman et al., 1995). Despite these hurdles, some users experience benefits of agility such as increased customer satisfaction through customization, shortened cycle times for product development, and customer involvement in product specification and design. These outcomes of agility in conjunction with high quality and reliability promise improved competitive position, high market share, and profitability to the users.

CONCLUSIONS

Manufacturing evolved from mass production to lean production, and finally to agile manufacturing. Lean production practices that began in Japan at Toyota, Inc., in the 1950's deserve full credit to Japan's ascendancy in the automotive world. The lessons of lean production are extremely important for understanding agility (Kidd, 1994; Preiss, 1995a). *Lean* is a response to competitive pressures with limited resources. Whereas, *agile* is the response to complexity brought about by constant change. *Lean* is a collection of operational techniques focused on productive use of resources; *agile* is an overall strategy focused on thriving in an unpredictable environment. A very discernible difference surfaces when we look at the architectural roots of manufacturing paradigms. Craft production is based upon the comprehensive single unit: one person builds the entire rifle, or one team builds an entire car. Mass production introduced specialized work modules and sequential work flow. Lean production brought flexibility. Now agile manufacturing brings reconfigurable work modules and work environments. *Lean* is interested in those things that can be controlled; *agile* is interested in dealing with those things that cannot be controlled. Agile manufacturing may not solve all the problems, nor it is the correct approach for all things at all times. Agile manufacturing is a new option that needs to be understood and applied when relevant.

Key Concepts: Agility Change Domains; Agility Proficiency Metrics; Employee Empowerment; Reconfigurability; Virtual Reality.

Related Articles: Business Process Reengineering; Lean Manufacturing Implementation; Virtual Manufacturing.

REFERENCES

DeVor, R., R. Graves, and J.J. Mills (1997). "Agile Manufacturing Research: Accomplishments and Opportunities." *IIE Transactions*, 29, 813–823.

Dove, R. (1996) *Tools for Analyzing and Constructing Agile Capabilities*. Agility Forum, Bethlehem, PA.

Dove, R., S. Hartman, and S. Benson (1996). *An Agile Enterprise Reference Model with a Case Study of Remmele Engineering*. Agile Forum, Bethlehem, PA.

Goldman, S.L., R.N. Nagel, and K. Presiss (1995). *Agile Competitors and Virtual Organizations*. Van Nostrand Reinhold, New York, NY.

Herrin, G.E. (1996). "Industry thrust areas of TEAM." *Modern Machine Shop*, 69, 146–147.

Kaplan, G. (1993). "Manufacturing S La Carte—Agile Assembly Lines, Faster Development Cycles." *IEEE Spectrum*, 30, 24–34.

Kidd, P.T. (1994). *Agile Manufacturing-Forging New Frontiers*. Addison-Wesley Publishing Company, Wokingham, England.

Lau, R.S.M. (1995). "Mass Customization: The Next Industrial Revolution." *Industrial Management*, 37, 18–19.

Pine II, J.B. (1993). *Mass Customization—The New Frontier in Business Competition*. Harvard Business School Press, Boston, MA.

Preiss, K. (1995a). *Mass, Lean, and Agile as Static and Dynamic Systems*. Agility Forum, Bethlehem, PA.

Preiss, K. (1995b). *Models of the Agile Competitive Environment*. Agility Forum, Bethlehem, PA.

Richards, C.W. (1996). "Agile Manufacturing: Beyond Lean?" *Production and Inventory Management Journal*, 37, 60–64.

Roos, D. (1995). *Agile/Lean: A Common Strategy for Success*. Agility Forum, Bethlehem, PA.

Sheridan, J.H. (1993). "Agile Manufacturing: Beyond Lean Production." *Industry Week*, 242, 34–36.

Stipp, D. (1996). "The Birth of Digital Commerce." *Fortune*, 134, 159–164.

14. VIRTUAL MANUFACTURING

PRATAP S.S. CHINNAIAH
SAGAR V. KAMARTHI
Northeastern University, Boston, MA, USA

ABSTRACT

Business is becoming more global and complex every day. Technologies are changing so fast that no single company can meet every market opportunity all alone any more. Customers are new participating in the product specification process helping designers to create and develop new products and services. Even competitors are embracing one another to develop new technologies, enter new markets, and make products that they can't produce on their own. Because of these reasons more and more business partnerships are emerging among companies and entrepreneurs.

Today's joint ventures, strategic alliances, and outsourcing activities represent only a very small beginning of a bigger future for spontaneous partnerships. These partnerships span both manufacturing and service industries. Virtual organization (VO), virtual enterprise, virtual factory, cooperative manufacturing, and networked manufacturing are synonymous terms for virtual manufacturing (Buzzacott, 1995; Camarinha-Matos et al., 1997; Davidow and Malone, 1992). Such partnerships among companies will be less permanent, less formal, and more opportunistic. In VO, companies band together to meet a specific market opportunity but may go apart once the need evaporates. These partnerships are facilitated by high-speed communication networks, information databases, and common standards for swapping design drawings. Information networks enable far-flung companies and entrepreneurs link up and work together from start to finish of a project. These relationships make companies far more reliant on one another and require far more trust than ever before. They share a sense of "co-destiny," in the sense that the fate of each partner is dependent on the others. This new corporate model redefines the traditional boundaries of the company. The closer cooperation among competitors, suppliers, and customers makes it harder to determine where one company ends and another begins.

Read here about the various ways of practicing virtual manufacturing in the following companies: **Aerotech Services Corp; Apple; Applied Materials, Inc.; Applied Services Corp; AVO Inc.; Chrysler; C-TAD Systems; Ford; General Motors; Hewlett-Packard; IBM; Intel; John Deere Company; Matsushita Appliance Company; McDonnell Douglas Aerospace; Motorola; Sony; Telepad Inc; UCAR Composites; U.S. Air Force; Wal-Mart Stores.**

REASONS FOR VIRTUAL MANUFACTURING

The VO model could become the most important organizational innovation since the 1920s when Pierre Du Pont and Alfred Sloan developed the principles of decentralization to organize giant and complex corporations. There are six strategic reasons, which are listed below, that motivate companies to use a virtual organization model (Goldman et al., 1995).

1. *Shared infrastructure, R&D efforts, risk, and costs:* The first strategic consideration is the value of sharing risk, infrastructure, R&D, and the cost of resources (human or technological). For a small company, access to specialized manufacturing equipment might be justification enough to join a virtual organization. For a large company, it might be the R&D need that encourages the company seek virtual partnerships. For example, Chrysler, Ford, and GM, together with various government agencies formed the USCAR consortium for cooperatively developing a wide range of generic automotive technologies such as catalytic converters.

2. *Linking complementary core competencies:* The second strategic reason to use the virtual organization concept is to unite complementary core competencies of different companies in order to serve customers, whom the separate companies could not serve on their own. Each member of the virtual organization brings something unique that is needed to meet a customer need. The joint development of Hewlett-Packard HP-100 palmtop computer with a built-in Motorola pager is an example of this strategy.

3. *Reduced concept-to-cash time through concurrency:* The possibility for multiple companies to operate in parallel and perform many tasks concurrently is the third reason for forming a virtual organization. In this process, the speed of development is increased and concept-to-cash duration is reduced. In the commercial world, concept-to-cash time is an important measure of a company's agility. In a virtual organization, this time is reduced, in part by not having to build new facilities, or find, hire, and train new people, and in part by the concurrent operations of multiple partners. For example,

Motorola, Apple, and IBM collaborated to bring out PowerPC chip in the quickest possible time. Another example is Apple and Sony working together to bring out the line of PowerBook products in a short time.

4. *Increased facilities and apparent size:* The virtual organization is a way of leveraging a company's ability to satisfy and enrich its customers. For small organizations, serving the customers through a virtual organization allows the company to show depth of backup, breadth of capability, and the financial ability to deal with what the customers may view as a large company. The need to increase apparent size, facilities, and scope is not limited to small companies competing for larger opportunities. As a truly global economy emerges, with customers scattered all over the world, the ability of a virtual organization to serve the customers anywhere in the world becomes an important competency in its own right. Agile Web in Pennsylvania is an example of cooperative efforts of small companies, who need, on occasion, a critical mass of capability to secure business. This web can help form focused virtual organizations quickly out of an amorphous pool of resources.

5. *Gaining access to markets and sharing market or customer loyalty:* Market access and product loyalty are two very valuable core competencies of an organization. Even these values may be shared in a virtual organization. For example, Intel successfully made the public aware of its name by having its customers label their computers "Intel Inside." This idea of the end user being aware of contributing organizations will become prevalent as virtual organizations become more common.

6. *Migrating from selling products to selling solutions:* A product may offer different value to different customers at the same time. The value of a solution is contextually dependent on the significance of what the customer is able to do or avoid doing. Virtual organization is an enabling mechanism for the enrichment of the customer. It is sometimes necessary for skill-based product or service suppliers to join in a virtual organization with other skill-based providers under what one calls the general contractor principle. A general contractor synthesizes the skills of a group of skill-based providers into a solution-based product with known value to the customer. For example, home builders play the role of a general contractor.

CHARACTERISTICS OF VIRTUAL ORGANIZATION

A virtual organization is a temporary network of independent companies, suppliers, customers, even rivals linked by information technology to share skills, resources, costs, and access to one another's markets. It may not have neither a central office nor an organization chart. It may have no hierarchy, and no vertical integration. A virtual manufacturing organization is fluid and flexible, created by a group of collaborators who quickly unite to exploit a specific opportunity. The venture may disband once the opportunity is met. In businesses as diverse as movie making and construction, companies have come together for executing specific projects. Opportunities may last for many years as in the case of building Boeing 777 airplanes, or they can be as short as only a few weeks or months. A virtual organization can be defined as a new organizational model that uses technology to dynamically link people, assets, and ideas.

The following are the five key characteristics of a virtual organization (Goldman et al., 1995).

1. *Opportunism:* A virtual organization has high degree of adaptability. It is agile in its internal organizational structure, rules, and regulations. A virtual organization is an opportunity-pulled and opportunity-defined integration of core competencies distributed among a number of real organizations.

2. *Excellence:* A virtual organization is assumed to be world-class and excellent in its core competencies. A member of the virtual organization contributes only what is regarded as its *core competencies.* A VO will mix and match what it does best with the best of other member companies and entrepreneurs in the virtual organization. In the virtual organization, a capable manufacturer will do manufacturing, an expert design firm will produce product-designs, and a successful marketing company will sell the products.

3. *Technology:* A virtual organization is assumed to offer world-class and state-of-the-art technology in its product as well as service solutions.

4. *Borderlessness:* A virtual organization's competencies can remain dispersed physically and still be synthesized into a coherent productive resource to meet a customer requirement.

5. *Trust:* The members of a virtual organization team must act responsibly toward each other in a trusting and trustworthy manner. If mutual trust does not exist, they cannot succeed in attracting customers to do business with them.

The common benefits of a virtual organization are the reduction in time, costs, and risks; the increase in product service capabilities; and the expansion of resources throughout the concept-to-cash cycle. Rapidly formed virtual corporations composed of the best of several firms have competitive advantage. For instance, the success of Applied Materials, Inc. is based on a collaborative web of suppliers and customers.

Many large corporations are using the virtual concepts to broaden their offerings to their customers or produce sophisticated products less expensively. The virtual concept can also provide muscle and reach for some smaller companies and entrepreneurs. For example, TelePad, Inc. which produces a hand-held pen based computer, is a small company that has limited in-house design talent, a handful of engineers, and no manufacturing plants. The product was designed and co-developed with GVO, Inc.; an Intel Corporation's swat team was brought in to resolve some engineering problems; several other companies developed software for the product; a battery maker developed a portable power supply; and finally spare manufacturing capacity at an IBM plant was used to manufacture the product (Goldman et al., 1995).

PROBLEMS AND PROSPECTS

While power and flexibility are the obvious benefits of a virtual organization, the virtual model has some real risks too. For starters, a company joining such a network loses control of its functions it cedes to its partners. Proprietary information or technology may be tapped unethically by the other members of the virtual organization. The structure may pose new stiff challenges to managers, who must learn to build trust with outsiders and learn to manage beyond the walls of their company.

One of the big drawbacks of the virtual organization is that it results in loss of total control over technology or operations. For example, to get into the personal computer market quickly, IBM relied on a couple of outsiders for the key technologies: Intel for micro processors and Microsoft for the operating system software. At first, IBM won widespread praise for its unprecedented decision to develop a major product by forming partnerships with teams outside of its corporate walls. Ironically, the approach also meant that IBM's computer system wasn't proprietary, and IBM soon found that it had created a market it could not control. Hundreds of clone makers emerged with lower prices and better products (Goldman et. al., 1995).

The three main technologies that companies have employed to create a virtual factory—electronic data interchange (EDI), proprietary groupware such as Lotus Notes, and dedicated wide-area networks—are not in themselves complete solutions. If a large-scale virtual factory has to succeed, it must fulfill the following three demands (Upton and McAfee, 1996).

Sophistication: A virtual factory must be able to accommodate the network of members whose information technology sophistication varies widely—from a small machine shop with a single PC to a large site that operates an array of engineering workstations and mainframes.

Security: While maintaining a high level of security, a virtual factory must be able to cope with a constantly churning pool of suppliers and customers whose relationships fluctuate in closeness and scope.

Functionality: A virtual organization must give its members the capacity to transfer files between computers, the power to access common pools of information, and the capability to access and utilize all the programs on a computer located at a distant site.

EDI, groupware, and wide-area networks can each deal with some of these demands, but none can deal with all of them, nor can combinations of the three technologies (Upton and McAfee, 1996). Nevertheless, the standards and technologies are improving to make the virtual factories possible. For example, AeroTech, a small and young information-services company has built a virtual factory for McDonnell Douglas Aerospace. This networked manufacturing community is open and friendly to even the most unsophisticated users; it provides a very high degree of functionality, and works eventhough the community's membership is constantly changing. Some technologies such as rapid prototyping, web-aided design (or internet-aided design), virtual machining, virtual assembly & testing, cost & risk analysis models, and product realization process are being developed by TEAM (Technologies Enabling Agile Manufacturing) program established by the US Government (Herrin, 1996); Mo et al., 1996; Steven, 1997).

Apart from technological problems, networked partners face new management challenges. Before companies can more routinely to engage in collaboration, they must build a high level of trust in each other. Virtual organizations demand a different set of skills from their managers. They will have to build relationships, negotiate *win-win* deals, find the right partners with compatible goals and values, and provide the temporary organizations with right balance of freedom and control.

MECHANISMS FOR IMPLEMENTATION

Technologies for VM include all the modules of a manufacturing enterprise such as computer-aided design (CAD) work stations, computer numerical control (CNC) machines, material handling systems, which are linked by either an enterprise network, or the Internet (Arnst, 1996; Cortese, 1996). A futuristic confederation of National Information Infrastructure that can link computers and machine tools across the US is being envisaged (Byrne, 1993; Stipp, 1996). This communication superhighway would permit far-flung units of different companies to quickly locate suppliers, designers, and manufacturers through an information clearing house. Once connected, they would sign *electronic contracts* to link up with less legal headaches. Teams of people in different companies would routinely work together, concurrently rather than sequentially, via computer networks in real time. Artificial intelligence systems and sensing devices would connect engineers directly to the production line. The key is to act like a single-source provider, with full accountability to the customer.

A virtual organization uses one or more of the existing organizational mechanisms such as partnership, joint venture, strategic alliance, new cooperation, supplier-subcontractor, cooperative agreement, royalty or license, outsourcing contract, or the web. Of these mechanisms, the web is a new concept. A web is an open-ended collection of prequalified partners that agree to form a pool of potential members of a virtual organization. To meet individual customer needs, a unique combination of companies is pulled into a virtual relationship because of the distinctive requirements of a customer or a group of customers.

In the world of virtual manufacturing, the types of relationships between supplier-companies (denoted as supplier) and customer-companies (denoted as customer) expand to cover a wide range of possibilities (Preiss et al., 1996; Preiss and Wadsworth, 1995). The scope of the relationship is shown in Figure 1. In the graph, X-axis indicates the value-added by the supplier to the customer's product; Y-axis indicates the financial arrangement between the customer and the supplier; and Z-axis indicates the degree of business process integration. On the X-axis is the total value-added conferred on the customer by the supplier. Simple predefined parts with no value-added service are near the origin of the X-axis. As we move to the right of the X-axis, the customer receives much more before-sale and after-sale information, knowledge, and value-added service.

The Y-axis represents the financial arrangement employed to control the flow of revenue from customer to supplier. Near the origin of the axis, the customer pays the supplier a fixed unit price. Further up on the Y-axis, the customer pays the supplier a fixed price per unit quantity of products, plus some additional variable price depending on the level of service and knowledge provided in conjunction with the products. At the top of the Y-axis, the customer shares the risks and revenues with the supplier.

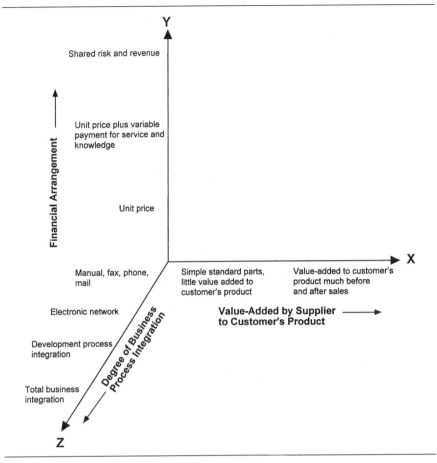

Figure 1. Agile customer-supplier relationships.

The Z-axis represents the continuum of order-process linkages, from passively taking orders to increasingly active and interactive supplier-customer linkages, culminating in full integration of the process. Combinations of the characteristics represented on X, Y, and Z axes reveals the facets of customer-supplier relationships, and also the scope for improvement. For example, a supplier at a low rank on the Z-axis (manual fax, phone, mail) providing complex products and services (high on the X-axis) would be operating slowly. To speed up the business process, the supplier would have to move to electronic network or beyond on the Z-axis.

When the customer-supplier relationship moves to the right on the X-axis to transfer value-added knowledge and services, companies may utilize some type of EDI methods. The use of electronic transfer of design drawings and specifications is necessary in order to operate quickly and competitively. More sophisticated design tools utilizing expert systems, digital interactive video, etc. may be needed in order to provide leading-edge designs quickly.

The basis of payment far up the Y-axis labeled "shared risk and revenue" is undefined in a classical mass production environment in which the products being supplied are predefined by the supplier. If the supplier and customer do not have a relationship whereby revenue and risk are shared, then using a more integrated process (shown along Z-axis) for exchange of know-how and services will not be possible.

Other combinations of relationships on the X-Y-Z coordinate system can be interpreted. The specific relationships identified along the axes X, Y, and Z in Figure 1 are just a few examples of the possible combinations in the total spectrum of customer-supplier relationships. Wal-Mart Stores, Inc. implemented EDI to integrate manufacturers, wholesalers, retailers, and customers into a single customer-focused process. Many analysts believe that Wal-Mart has outperformed the information systems (IS) of other retail chains because it was the first to realize that IS is important to success (Preiss and Wadsworth, 1995).

VIRTUAL MANUFACTURING IN PRACTICE

Virtual manufacturing is a relatively new concept. However, many companies already have been practicing the concepts of VM. The US Air Force MANTECH group along with Defense Advanced Research Projects Agency (DARPA) and Defense Modeling and Simulation Office (DMSO) has been promoting the development of VM technology (Herrin, 1994; Scott, 1996). Their objective is to develop VM simulation and control models for manufacturing. These models will have the capability to evaluate if a product is manufacturable. They can analyze various production and supplier scenarios. The end result is intended to enhance manufacturing operations by providing timely answers to such questions as: Can we make the product? How much will it cost? What is the best way to produce the product? And, what are the alternatives?

Applications of VM encompass the entire life cycle of a product, from design, specification, production, testing, distribution, and to delivery to the customer. Matsushita Appliances Company used VR technology to develop a virtual kitchen space decision support system. This system allows customers experience a kitchen environment, evaluate alternatives, and select the best combination of kitchen appliances. The customer choices are stored as drawings and subsequently transmitted to the company's production facilities for manufacturing and delivery (Shukla et al., 1996).

Ford worked with C-TAD Systems, Inc. to develop a virtual CAD tool, which helps any designer to work on any car design or to interact with production engineers located anywhere in the world (Owen, 1997b). John Deere Company used VM software to install a robotic arc welding production facility. This project involved a virtual 3-D environment for design, evaluation, and testing of the robotic production system. The VM approach shortened the design-to-manufacturing lead time (Owen, 1997a).

Boeing 777 plane was initially developed as a digital plane, on which virtual machining, virtual assembly, and virtual testing were carried out.

THE CASE OF McDONNELL DOUGLAS AEROSPACE

AeroTech Services Group, which is based in St. Louis, Missouri, has built a highly effective virtual factory for McDonnell Douglas Aerospace (Upton and McAfee, 1996). There are around 400 companies in this manufacturing network. The open and flexible nature of this network accommodates users, whose information technology sophistication and relationships with one another vary greatly. Besides, the network permits the members carry out a wide variety of collaborative tasks very securely.

Consider the relationship between McDonnell Douglas and UCAR Composites, a $12 million manufacturer of tooling for high-performance composite components based in Irvine, California. To build a new part, CAD files are translated at McDonnell Douglas into the NC machine code needed to operate UCAR's metal cutting machines. Using standard internet protocols over a dedicated high-speed link, McDonnell then translates the CAD file and the metal cutting program to AeroTech's safe network node. AeroTech's system subsequently forwards them to UCAR on normal phone lines. Once information on the job arrives in California, UCAR engineers view it on their own CAD/CAM systems to make last-minute checks on the program. They then run the cutting program on their machines to manufacture parts.

This method of transferring NC programs is also being used by hundreds of small machine shops, many of whose information management systems and expertise are much less sophisticated than UCAR's. Many of these companies, including five- or six-person machine shops, can dial AeroTech using a regular modem, download an NC program or a drawing onto their PCs, and use the data to manufacture the parts.

The virtual factory arrangement also helps its members find the best suppliers much more quickly than before. In the past, McDonnell Douglas would invite representatives of qualified suppliers to come to St. Louis to participate in the bidding process that often took several days. Using the virtual factory's electronic bidding system, both the suppliers and McDonnell Douglas save time and money. AeroTech also helps the manufacturing community coordinate their schedules better by allowing remote members access scheduling software on one another's machines. AeroTech has made it possible for both longtime and casual partners to collaborate easily, securely, and cheaply without having to invest in new and proprietary information technology.

CONCLUSIONS

The existence of a global market, with increasing demands for customized products, is promoting the creation of virtual organizations as dynamic mechanisms to focus on time-based market opportunities. New technologies are enabling simulation of all steps from design to delivery without actually producing a product physically, or without requiring the participating members to be physically near each other. Virtual manufacturing utilizes the enhanced competitive capabilities resulting from cooperation between members of a virtual organization. The success of the virtual manufacturing organization is tied to the ability of the real companies to form the virtual organization rapidly to meet an emerging time-based opportunity. The ability to work intensively with other companies and to be able to trust them from the start of a project is enhanced by prequalification agreements based on company attributes and contractual commitments. The use and role of virtual organizations are subtle and evolving. Once the benefits of virtual manufacturing are understood, many more companies may take advantage of virtual manufacturing.

Key Concepts: Agile Manufacturing; Lean Manufacturing; Mass Customization; Supply Chain Management.

Related Articles: Agile Manufacturing; Lean Manufacturing Implementation; Supplier Partnership as Strategy; Supply Chain Management: Competing Through Integration.

REFERENCES

Arnst, C. (1996). "Wiring Small Business." *Business Week*, Nov. 25, 164–172.

Buzzacott, J.A. (1995). "A Perspective on New Paradigms in Manufacturing." *Journal of Manufacturing Systems*, 14, 118–125.

Byrne, J.A. (1993). "The Virtual Corporation." *Business Week*, Feb. 8, 98–103.

Camarinha-Matos, L.M., H. Afsarmanesh, C. Garita, and C. Lima (1997). "Towards an Architecture for Virtual Enterprises." *Proceedings of the Second World Congress on Intelligent Manufacturing Processes & Systems*, Budapest, Hungary, 531–541.

Cortese, A. (1996). "Here comes the Intranet." *Business Week*, Feb. 26, 76–84.

Davidow, W.H. and M.S. Malone (1992). *The Virtual Corporation*. Harper Collins, New York, NY.

Goldman, S.L., R.N. Nagel, and K. Preiss (1995). *Agile Competitors and Virtual Organizations*. Van Nostrand Reinhold, New York, NY.

Herrin, G.E. (1994). "Virtual Manufacturing." *Modern Machine Shop*, 66, 158–159.

Herrin, G.E. (1996). "Industry thrust areas of TEAM." *Modern Machine Shop*, 69, 146–147.

Irani, S.A., T.M. Cavalier, and P.H. Cohen (1993). "Virtual Manufacturing Cells: Exploiting Layout Design and Intercell Flows for the Machine Sharing Problem." *International Journal of Production Research*, 31, 791–810.

Kannan, V.J. and S. Ghosh (1996). "A Virtual Cellular Manufacturing Approach to Batch Production." *Decision Sciences*, 27, 519–539.

Mo, J.P.T., Y. Wang, and C.K. Tang (1996). "The use of the Virtual Manufacturing Device in the Manufacturing Message Specification Protocol for Robot Task Control." *Computers in Industry*, 28, 123–136.

Owen, J.V. (1197a). "Virtual Manufacturing." *Manufacturing Engineering*, 119, 78–83.

Owen, J.V. (1997b). "Virtual Manufacturing." *Manufacturing Engineering*, 119, 84–90.

Preiss, K., S.L. Goldman, and R.N. Nagel (1996). *Cooperate to Compete-Building Agile Business Relationships*. Van Nostrand Reinhold, New York.

Preiss, K. and B. Wadsworth (1995). *Agile Customer-Supplier Relations*. Agility Forum, Bethlehem, PA.

Scott, W.B. (1996). "Integrated Software to cut JSF, F-22 Costs." *Aviation Week and Space Technology*, 144, 64–65.

Shukla, C., M. Vazquez, and F. Chen (1996). "Virtual Manufacturing: An Overview." *Computer and Industrial Engineering*, 31, 79–82.

Stevens, T. (1997). "Internet-aided Design." *Industry Week*, 246, 50–55.

Stipp, D. (1996). "The Birth of Digital Commerce." *Fortune*, 134, 159–164.

Upton, D.M. and A. McAfee (1996). "The Real Virtual Factory." *Harvard Business Review*, 74, 123–133.

VIII. LEAN MANUFACTURING

15. JUST-IN-TIME MANUFACTURING

GREGORY P. WHITE

Southern Illinois University at Carbondale, Illinois, USA

ABSTRACT

Just-in-Time (JIT) is a philosophy of operation that seeks to utilize all resources in the most efficient manner by eliminating anything that does not
contribute value for the customer. In this philosophy, resources include—
but are not limited to—equipment, facilities, inventory, time, and human
resources. Because of this broad definition of resources, JIT includes many
different components. Some of those components emphasize the efficient
use of material resources; other components of JIT focus on the efficient
use of human resources. In fact, experts don't even agree completely on
the factors that should be included as components of JIT. To make matters
worse, some components of JIT are also important components of other
techniques, such as Total Quality Management (TQM).

One possible way to define JIT that avoids the problem of identifying
specific components is to define JIT as "flow manufacturing." The model
for this definition is a continuous flow type of process such as an oil refinery or
a paper mill. Such processes generally operate very efficiently, with material
flowing rapidly from raw material to finished product. The objective of
JIT is to make manufacturing processes resemble such flow processes in as
many ways as possible. Companies mentioned in this article include **Harley-
Davidson; Kawasaki; NUMMI**

HISTORY

Although Henry Ford purportedly used some JIT concepts to produce the Model T (Womack et al., 1990), Just-in-Time as a complete philosophy was developed by the Japanese company Toyota and was used only in Japan until the late 1970s, when some U.S. companies began to notice that certain Japanese competitors were producing better products for lower cost. Subsequently, delegations from U.S. manufacturers visited Japanese plants. What most impressed those delegations was the drastically reduced inventory level with which most Japanese plants functioned. Other, more important results of JIT, that were not as noticeable, did not originally receive much attention. As a result, JIT was known early on in the U.S. under such names as "zero inventory" (Hall, 1983) or "stockless production", or even as "kanban", which, as will be described later, is just one means of controlling production and inventory levels under the JIT system of manufacturing.

As more individuals studied JIT, they eventually realized that it involves more than just an inventory control system (Schonberger, 1982; Monden, 1981; Monden, 1983). In fact, many organizations today believe that other components of JIT are even more important than inventory reduction, and that inventory reduction is really just a by-product of other changes brought about by JIT. Keller and Kazazi (1993) provide a good overview of the JIT literature.

STRATEGIC PERSPECTIVES

(Krajewski et al., 1987) used a computer to simulate various approaches to manufacturing, including the use of MRP and JIT. Surprisingly, they found that the use of a particular methodology was not nearly as important as the implementation of certain procedures for improving manufacturing performance. Such procedures include reduced setup times, small lot sizes, worker flexibility, and preventive maintenance, all of which are components of JIT. Thus, it does not appear that a company must necessarily adopt the entire JIT philosophy to obtain performance improvements; the use of key components that enable flow manufacturing may be sufficient. However, giving lip service to JIT by requesting smaller lot sizes from suppliers without reducing one's own lot sizes or setup times, as some companies have reportedly done, probably will generate only marginal, if any, performance improvement.

IMPLEMENTATION

Sakakibara et al. (1993) identified sixteen "core JIT components" and divided them into six categories. Those categories and the components of each are described briefly below.

(1) Production Floor Management: Production floor management refers those to components of JIT that are directly applied to control production activities at the shop floor level with the objective of maintaining a smooth flow of materials. Those components are:

- Setup-time reduction.
- Small lot sizes.

- Preventive maintenance.
- Kanban (if applicable).
- Pull system support (if applicable).

Small lot sizes and setup-time reduction: Most companies find that significant changes must be made in shop floor procedures as they move toward the flow manufacturing goal of JIT. One way to achieve smooth, continuous flow is by using small lot sizes with the objective of eventually achieving a lot size of one. Of course, if such small lot sizes are to be feasible, then setup times must be reduced close to zero, or to the minimum level possible. A considerable body of knowledge has been developed concerning setup time reduction (Shingo, 1985), and some companies have achieved dramatic results, often going from setups that required hours to than ten minutes (single-minute setups; see SMED).

Preventive maintenance: While reduced lot sizes and setup times are important first steps toward achieving flow manufacturing, companies implementing JIT must also take steps to ensure that their manufacturing equipment will be able to maintain the requisite smooth flow of material. Thus, preventive maintenance is used extensively to avoid unplanned down time. Often, a predetermined portion of each day, or even each shift, is set aside for preventive maintenance. Gupta and Al-Turki (1998) describe approaches to preventive maintenance in a JIT system.

Kanban and pull system support: Batch-oriented manufacturing relies on an extensive paperwork system—including shop orders and dispatch lists—to control and monitor production. However, when small lot sizes are used, the usual paperwork system would become unmanageable. Thus, kanban is often used as a system for controlling the production and movement of material through a JIT manufacturing process. Kanban is a "pull" system in which material movement, and often production, is based on *actual need* for more materials at downstream work centers, not on *planned need* as in an MRP system, which is a "push" system. The kanban system operates by using a signaling device to indicate the need for materials. Although the signaling device is often a card, it can be any visible or audible signal. For example, Harley–Davidson uses empty containers for the pull signal while one Kawasaki plant has used colored golf balls. The important point is that the number of such devices are strictly limited so that excess inventory cannot be built up. Further, the number of units that can go into a container is predetermined. Thus, two rules that apply when using a kanban system are: (1) produce or move material only when authorized by a kanban; (2) fill containers only with the predetermined number of parts.

Figure 1 depicts the basic flow in a two-card kanban system, in which one kanban controls conveyance (or withdrawal) of material from storage to the work center using that material while another kanban controls production of the material. Often, only the *conveyance kanban* is used, when production is based on a predetermined schedule. A conveyance kanban (*C- kanban*) can be returned for more materials only when the level of materials at the using work center has dropped to a predetermined level. Likewise, *a production kanban (P-kanban)* can be returned to the producing work center to authorize more production only when a container has been removed by a downstream work

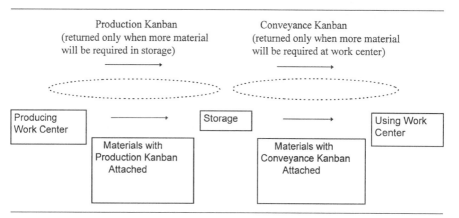

Figure 1. Flow of kanban cards.

center from storage using a C-kanban. In this way, inventory build up is strictly limited by the number of kanban cards.

(2) Scheduling: Experience tends to indicate that JIT works best in an environment in which the production schedule remains fairly stable. Thus, two components of JIT related to scheduling are:

- Repetitive master schedule.
- Daily schedule adherence.

Companies using JIT often develop a master schedule that is repeated each day over an extended period of time, usually for at least several months. Each day's production of a given product is based on the average daily demand for that product, thus minimizing the amount of excess inventory at the end of the day. Ideally, this "rate-based" production schedule should produce each product in the smallest possible lot sizes. As a result, instead of producing each product in large batches that may extend over several days or weeks, small batches of each product are produced each day. This is referred to as "uniform load scheduling" or "mixed model sequencing." Producing in this manner not only minimizes inventory buildup of any product, but also levels the requirements for all component parts and for the resources used to produce those components. For example, if a company manufactures three products (A, B, and C) and the average daily demand is 5 units for product A, 10 units for product B, and 15 units for product C, then an "ideal" mixed model sequence could be CBCBCA, which would be repeated five times per day. Of course, setup times or other considerations might require that larger lot sizes be used for certain products, but the basic goal is to produce every product every day, if possible.

(3) Process and Product Design: As companies implement JIT, they often find that flow manufacturing is facilitated through modifications to both the process and the product. Thus, the following are two components of JIT that influence design of processes and products:

- Equipment layout.
- Product design simplicity.

Equipment layout: With the JIT focus on smooth material flow and minimal inventory, equipment layouts are often changed under JIT so that work centers are close together. In fact, manufacturing cells are frequently used to produce a family of parts or products. These manufacturing cells group machines together in close proximity so that material handling times and workforce requirements are minimized. By working only on a family of parts or products that have similar machining requirements, setup times in these manufacturing cells can be reduced or eliminated.

Product design simplicity: By simplifying product design to reduce the number of different parts, and even the number of parts required, inventory is reduced. Products are designed so that they are easy to manufacture and materials flow smoothly through production.

(4) Work Force Management: One significant change under JIT, which is also being used widely by companies that do not use JIT, is a different approach to managing human resources. With JIT the focus is on fully utilizing a person's capabilities, including mental as well as physical capabilities, through the following components:

- Multi-functional workers.
- Small group problem solving.
- Training.

Employees are trained to perform several different tasks or jobs so they can move around as needed, and can help other employees if problems occur. Further, workers' problem-solving capabilities are fully utilized by allowing them to work in self-directed groups to solve production problems. Finally, to ensure that employees can work effectively in groups, possess the know-how to solve problems, and can perform various jobs without errors, a significant amount of time is devoted to training.

(5) Supplier Management: The emphasis on reducing inventory and achieving a smooth flow of materials involves suppliers. However, this step should usually be delayed until the customer company has its JIT house in order and has gained substantial experience working with JIT. At that time, a company may consider including the following JIT components:

- JIT delivery.
- Supplier quality level.

After a company has implemented JIT in its own operations, it should have a level production schedule that can be transmitted to suppliers, who can then work to provide JIT deliveries. However, if deliveries occur just as material is needed, then there is no time for inspection and no room for quality problems. Consequently, the supplier must also be able to achieve a consistently high level of quality so that inspection of incoming material is not necessary because very few, if any, incoming components are defective.

(6) Information System: Because JIT brings about significant changes in manufacturing processes, it also produces major alterations to the information systems associated with those processes. Two major areas of change are:

- MRP adaptation to JIT.
- Accounting adaptation to JIT.

Although it may appear that MRP and JIT are mutually exclusive, that is not true. In fact, many companies are now combining MRP and JIT, although doing so does require adaptations to the MRP system. These adaptations are prompted generally by the smaller lot sizes and shorter lead times of JIT. Thus, companies may find that *bucketless MRP* systems are more appropriate for JIT use. While this modification can help in adapting to reduced lead times and increased rate of material flow, it does not necessarily alleviate the paperwork problems. MRP systems are based on the usual approach of shop orders, allocation, and picking tickets. Applying this approach to JIT can quickly bog the system down in paperwork. To avoid that outcome, many companies utilize "backflushing" to enter material changes into the system. With backflushing, material usage is calculated from the production of final products. If a certain number of units of final product were produced, then it can be assumed that an appropriate number of units of each component part were also used. This same backflushing approach can also be used to adapt cost accounting systems to a JIT environment.

Supporting JIT components

The preceding six categories have been identified as "core JIT components." However, there is also a large set of "supporting JIT components." This latter group is discussed briefly in the following paragraphs.

Top Management Support: JIT, because it requires some rather drastic organizational changes, must have top management support to motivate those changes and keep everyone on track through the periods of turmoil that are nearly inevitable with change. Further, JIT often involves functions other than just manufacturing and, thus, needs top management support for cross-functional cooperation.

Quality Management: The need for a high level of quality from suppliers has already been mentioned. However, the same—or even higher—quality level will also be needed from internal manufacturing processes. Achieving such high quality often requires strong management leadership in the area of quality, implementation of methodologies such as statistical process control, and an orientation toward meeting customer needs and expectations.

Human Resource Management: The use of human resources in JIT frequently require changes throughout the entire organization. For example, the need for employees who can work in groups, make their own decisions, and perform multiple jobs will influence the new-employee recruitment and selection process. With employees work-ing in groups the employee reward and compensation programs may need to be mod-ified. Likewise, the role of managerial personnel may change from that of a decision maker to a facilitator.

Technology Management: The need for smooth material flow and consistently high qual-ity under JIT can often lead to changes in manufacturing technology such as the use of flexible manufacturing systems and robotics. Similarly, simplified product designs and the utilization of common parts are increased through the use of computer aided design systems.

Manufacturing Strategy: Changes that occur with JIT will often lead to changes in a com-pany's manufacturing strategy. For instance, JIT companies often develop distinctive competencies in terms of product quality, delivery speed, and delivery reliability. These competencies can be used advantageously as part of the company's competitive strategy. Further, the JIT philosophy helps companies to develop a plant-wide manufacturing philosophy and a long-range orientation, both of which are useful characteristics for developing and implementing competitive strategies.

Company-Wide Continuous Improvement and Problem-Solving Activities by Workers, Engi-neers, and Management: An important part of the JIT philosophy is that waste, in some form, can always be found, and that problem-solving efforts can be used to find ways to eliminate that waste. Thus, continuous improvement is both a component and a result of JIT. Because JIT spreads across functional boundaries, the continuous improvement and problem-solving activities eventually become company-wide across all levels of the organization.

TECHNOLOGY ISSUES

JIT does not require any particular technology. In fact, the simplification that results from using JIT can often lead to a reduction in the use of technology. For example, companies frequently find that sophisticated material handling systems are no longer needed as workers are moved closer together so they can easily pass parts, by hand, from one to the other. However, this is not to say that technology is not used with JIT. Many companies find that as they simplify and integrate processes, a point is reached at which those processes can be automated. Automation is not seen merely as a way of eliminating people, but as a way to make a process more consistent and reliable. The Japanese word "jidoka" is often used to describe the use of technology to monitor automated processes so that the process will be stopped automatically if a problem occurs (Hirano, 1988).

Computer technology has also been applied to planning and control in JIT. This can involve either the integration of JIT with computerized MRP or actual stand-alone computer software for JIT (Vollmann et al., 1997).

RESULTS

Plossl (1985) reports the results of a group of companies that implemented JIT. Within a five-year period, those companies experienced reductions in manufacturing cycle times in the 80 to 90 percent range, inventory reductions of 35 to 90 percent, labor cost reductions of 10 to 60 percent, and quality cost reductions of 25 to 60 percent.

A common analogy used for the inventory reduction that results from JIT implementation uses a stream of water in a channel strewn with rocks. Water represents the inventory level and rocks are problems which are uncovered as the inventory is reduced. This means that, as inventory is reduced, problems will be uncovered and must be solved before smooth operations can resume. Unfortunately, some companies tend to avoid the problems by "canoeing around the rocks." Doing so goes against the basic concept of JIT, which is continuous improvement. This means that companies implementing JIT will not always have smooth sailing, but will often be faced with new problems that must be solved. However, solving them will eventually lead to further improvements in performance.

PREREQUISITES

Although MRP and JIT are sometimes thought of as being at different ends of a spectrum, they are both actually similar in many ways. One similarity is that both are formal systems, and both require discipline for success. Thus, companies have found quite often that, once they have established the necessary discipline to use MRP effectively, it is not so difficult to implement JIT. On the other hand, a company that has no discipline and is operating with informal systems will often have a much more difficult time of implementing JIT. Thus, the implementation of JIT appears to proceed most smoothly when a company an has established a formal manufacturing system such as MRP.

Most of the early successes with JIT were experienced by automobile companies or other repetitive manufacturers. Consequently, it was believed for a while that JIT was only applicable in such industries. Today, we have learned that many of the JIT components can be applied in nearly every manufacturing environment, and even in service industries. It is important to realize that individual components of JIT can be implemented without adopting the entire JIT philosophy. Some companies may focus only on the worker flexibility aspects of JIT while others may implement only setup and lot size reduction. Thus, JIT, in at least some form, is applicable in any manufacturing situation.

EXAMPLES

One of the best known JIT success stories is Harley-Davidson, the U.S. motorcycle manufacturer. In the early 1980s Harley was the last remaining company of its kind in the U.S., and from all indications its days were numbered as Japanese-made motorcycles with higher quality and a lower price tag were rapidly stealing Harley's market share. Harley-Davidson management decided that, to fight back, it would have to learn its competitors' secret weapon, which turned out to be JIT. Over several years Harley successfully implemented JIT and was able to drastically reduce its inventory levels and production costs while significantly increasing product quality. Harley-Davidson today is a strong proponent of JIT as demand for its motorcycles outstrips production capacity and the product sells throughout the world, even in Japan.

Harley-Davidson and other success stories are presented in an excellent collection of cases developed by (Sepehri, 1986). Those cases describe the use of JIT by companies in the United States, including Toyota USA and NUMMI, which is a joint venture between Toyota and General Motors, located in Fremont, California. Schonberger (1987) has developed a similar collection that includes companies using JIT, TQM, or both.

LESSONS

Implementation JIT is a complex, lengthy task. Fortunately, all components need not be implemented simultaneously, so companies have generally found it best to begin with a few and add others progressively. Hallihan et al. (1997) and Sohal et al. (1993) describe models for implementing JIT based on surveys of industry experience. The following points summarize some wisdom gained from experience:

- Top-management commitment and support are essential for successful implementation.
- Everyone in the organization must be educated about the importance of JIT and what they can expect from JIT.
- Begin with a small pilot project and get it operating smoothly before expanding the use of JIT throughout the organization.
- Focus on an application with a high probability of success as the first step. Rearranging an operation for improved product flow can often produce visible results quickly.
- Do not involve suppliers until your own operation is working smoothly and a stable production schedule has been achieved.

Key Concepts: C-Kanban; Flow Manufacturing; JIT Philosophy; Lean Manufacturing; Mixed Model Sequencing; P-Kanban; Rate-Based Production Schedule; Repetitive Production; Signaling Device; Stockless Production; Total Productive Maintenance; Uniform Load Scheduling.

Related Articles: Activity-Based Costing; Agile Manufacturing; Lean Manufacturing Implementation; Total Productive Maintenance; Transition to Cell Manufacturing: The Case of Duriron Inc., Cookeville Valve Division.

REFERENCES

Berkley, B. (1992). "A review of the kanban production control research literature." *Production and Operations Management*, 1 (4), 393–411.
Gupta, S.M. and Y.A.Y. Al-Turki (1998). "Adapting just-in-time manufacturing systems to preventive maintenance interruptions." *Production Planning and Control*, 9 (4).
Hall, R.W. (1983). *Zero Inventories*. Dow Jones-Irwin, Illinois.
Hallihan, A.P. Sackett, and G.M. Williams (1997). "JIT manufacturing: The evolution to an implementation model founded in current practice." *International Journal of Production Research*, 35 (4), 901–920.
Hirano, H. (1988). *JIT Factory Revolution*. Productivity Press, Massachusetts.
Keller, A.Z. and A. Kazazi (1993). "Just-in-time manufacturing systems: A literature review." *Industrial Management and Data Systems*, 93 (7), 1–32.
Krajewski, L.J., B.E. King, L.P. Ritzman, and D.S. Wong (1987). "Kanban, MRP, and shaping the manufacturing environment." *Management Science*, 33, 39–57.

Monden, Y. (1981). "What makes the Toyota production system really tick?" *Industrial Engineering*, 13, 36–46.

Monden, Y. (1983). *Toyota Production System*. Industrial Engineering and Management Press, Georgia.

Plossl, G.W. (1985). *Just-in-Time: A Special Roundtable*. George Plossl Educational Services, Georgia.

Sakakibara, S., B.B. Flynn, and R.G. Schroeder (1993). "A framework and measurement instrument for just-in-time manufacturing." *Production and Operations Management*, 2, 177–194.

Schonberger, R.J. (1982). *Japanese Manufacturing Techniques: Nine Hidden Lessons in Simplicity*. The Free Press, New York.

Schonberger, R.J. (1987). *World Class Manufacturing Casebook*. The Free Press, New York.

Sepehri, M. (1986). *Just-in-Time, Not Just in Japan*. American Production and Inventory Control Society, Virginia.

Shingo, S. (1985). *A Revolution in Manufacturing: The SMED [Single-Minute Exchange of Die] System*. The Productivity Press, Massachusetts.

Sohal, A.S., L. Ramsay, and D. Samson (1993). "JIT manufacturing: Industry analysis and a methodology for implementation." *International Journal of Operations and Production Management*, 13 (7), 22–56.

Vollmann, T.E., W.L. Berry, and D.C. Whybark (1997). *Manufacturing Planning and Control Systems* (4th ed.). Irwin/McGraw-Hill, New York.

Womack, J.P., D.T. Jones, and D. Roos (1990). *The Machine That Changed the World*. R.A. Rawston Associates, New York.

16. LEAN MANUFACTURING IMPLEMENTATION

JT. BLACK

Auburn University, AL

ABSTRACT

Many companies have implemented lean production in the last decade. Lean production describes a manufacturing system that uses less resources to make a company's products. The term lean production was first used in the book, *The Machine that Charged the World* (Womack, Jones, and Roos, 1990), to describe the Toyota Production System. Lean manufacturing systems are called by various names in the United States. Most commonly used names for lean manufacturing are listed in Table 1. Some of the practices of Toyota are included in this article.

CONVERTING TO LEAN PRODUCTION

Figure 1 outlines the ten key steps to converting a factory from a job shop/flow shop manufacturing system to a linked-cell manufacturing system in achieving a true lean production system. The requirements for the successful implementation of lean manufacturing are:

1. All levels in the plant, from the production worker (the internal customer) to the president must be educated in lean production philosophy and concepts.
2. Top management must be totally committed to this venture and provide necessary leadership. Everyone must be involved in the change, and the internal customer must be empowered to play a vital role in this evolutional process.

Table 1 Various names for lean manufacturing system

System name	Name used by
LEAN PRODUCTION	MIT group of authors, researchers
TOYOTA PRODUCTION SYSTEM—(TPS)	Toyota Motor Company
OHNO SYSTEM	Taiichi Ohno, Inventor of TPS
INTEGRATED PULL MANUFACTURING SYSTEM	AT&T
MINIMUM INVENTORY PRODUCTION SYSTEM—(MIPS)	Westinghouse
MATERIAL NEEDED—(MAN)	Harley Davidson
JUST-IN-TIME/TOTAL QUALITY CONTROL (JIT/TQC)	Schonberger
WORLD CLASS MANUFACTURING—(WCM)	Schonberger
ZERO INVENTORY PRODUCTION SYSTEMS—(ZIPS)	Omark and Bob Hall
QUICK RESPONSE—OR MODULAR MANUFACTURING	Apparel Industry
STOCKLESS PRODUCTION	Hewlett Packard
KANBAN SYSTEM	Many companies in Japan
THE NEW PRODUCTION SYSTEM	Suzaki
ONE PIECE FLOW	Sekine
CONTINUOUS FLOW MANUFACTURING—(CFM)	Chrysler
INTEGRATED MANUFACTURING PRODUCTION SYSTEMS—(IMPS)	Black

3. Everyone in the plant must understand that cost, not price determines profit. The customer determines price, the plant determines the cost.
4. Everyone must be committed to the elimination of waste. This is fundamental for becoming lean.
5. The concept of standardization must be taught to everyone and applied to documentation, methods, processes as well as system metrics.

The steps for implementing lean manufacturing identified in Figure 1 are elaborated step by step.

Step 1: Form U-Shaped Cells—Restructure the Factory Floor. In the linked-cell manufacturing system (L-CMS), cells replace the job shop. The first task is to restructure and reorganize the basic manufacturing system into manufacturing cells that manufacture families of parts. This prepares the way for a systematically created linked-cell system designed for one-piece movement of parts within cells and for small-lot movement between cells. Creating cells is the first step in designing a manufacturing system in which production control, inventory control, quality control, and machine tool maintenance are integrated.

Step 2: Rapid Exchange of Tooling and Dies. Everyone on the plant floor must be taught how to reduce setup time using SMED (single-minute exchange of dies) principles. A setup reduction team acts to facilitate the SMED process for production workers and supervisors. The key is that everyone must get involved in setup reduction. The team may attack the plant's worst setup problem as a demonstration project. Reducing setup time is critical to reducing lot size.

Step 3: Integrate Quality Control. A *multiprocess* worker can run more than one kind of manufacturing process. A *multifunctional* worker can do more than operate machines;

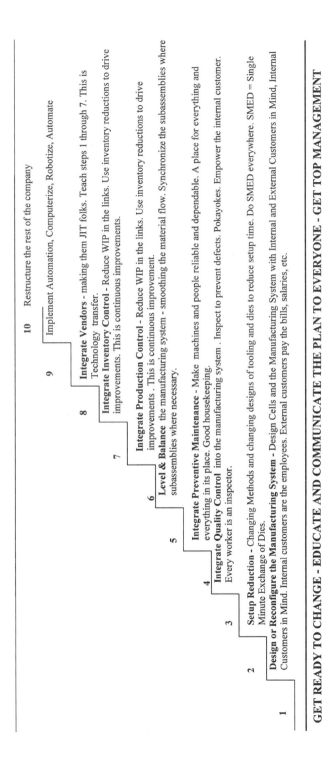

10   Restructure the rest of the company

Implement Automation, Computerize, Robotize, Automate

9

Integrate Vendors - making them JIT folks. Teach steps 1 through 7. This is
Technology transfer.

8   **Integrate Inventory Control** - Reduce WIP in the links. Use inventory reductions to drive
improvements. This is continuous improvements.

Integrate Production Control - Reduce WIP in the links. Use inventory reductions to drive
7   improvements . This is continuous improvement.

Level & Balance the manufacturing system - smoothing the material flow. Synchronize the subassemblies where
subassemblies where necessary.

6

Integrate Preventive Maintenance - Make machines and people reliable and dependable. A place for everything and
5   everything in its place. Good housekeeping.

Integrate Quality Control into the manufacturing system . Inspect to prevent defects. Pokayokes. Empower the internal customer.
Every worker is an inspector.

4

Setup Reduction - Changing Methods and changing designs of tooling and dies to reduce setup time. Do SMED everywhere. SMED = Single
3   Minute Exchange of Dies.

Design or Reconfigure the Manufacturing System - Design Cells and the Manufacturing System with Internal and External Customers in Mind, Internal
Customers in Mind. Internal customers are the employees. External customers pay the bills, salaries, etc.

2

GET READY TO CHANGE - EDUCATE AND COMMUNICATE THE PLAN TO EVERYONE - GET TOP MANAGEMENT

1

Figure 1. Black's ten steps to lean production, or how to design a linked-cell manufacturing system.

he/she is also an inspector, who understands process capability, quality control, and process improvement. In lean production, every worker has the responsibility to make the product right the first time and every time, and has the authority to stop the process when something goes wrong. This integration of quality control into the manufacturing system markedly reduces defects while eliminating roaming inspectors and associated costs. Cells provide a natural environment for the integration of quality control. The fundamental idea here is to use inspection to *prevent defects from occurring*. The cells do not release defective parts to downstream subassembly areas.

Step 4: Integrate Preventive Maintenance. To make machines operate reliably, begin with the installation of an integrated preventive maintenance (IPM) program by giving workers the training and tools to maintain equipment properly. The extra or additional processing capacity obtained by reducing setup time allows operators to reduce equipment speeds or feeds and to run processes at less than full capacity, thus reducing the pressure on workers and the wear and tear on tools. Reducing pressure on workers and processes fosters in workers a drive to produce near perfect quality. A key to IPM is housekeeping. Simply put, there is a place for everything and everything is put back in its place. In addition, each worker is responsible for cleanliness of workplace and equipment.

Step 5: Level, Balance and Synchronize. Level the entire manufacturing system by producing a mix of final assembly products in small lots. This is called smoothing of production in the JIT literature. The objective is to reduce the effect of lumpiness of demand for component parts and subassemblies. Standardize cycle time in the system. Cycle time is the reciprocal of the production rate of the final assembly line. Use a simplified and synchronized system to produce the proper number of components everyday, as needed. Some companies begin at mixed-model final assembly, and work backward through subassembly and manufacturing cells. Each process, cell, and subassembly tries to match the daily build quantity and cycle time of final assembly. This is called *balancing*. If the time for the manufacture of the subassembly is matched to that of final assembly, then the system is synchronized. Only through the effective use of JIT manufacturers can accomplish this.

The cycle time for the final assembly line is determined by the demand rate for parts according to the following calculations. Let CT = cycle time; PR = production rate.

$$\text{Daily Demand} = \frac{\text{monthly demand (forecast plus customer orders)}}{\text{number of days in month}}$$

$$\text{CT} = \frac{1}{\text{PR}} \quad \text{where PR} = \frac{\text{Daily Demand (in units)}}{\text{available hours in day (hrs)}}$$

This incredibly simple approach highlights the method used by JIT companies to calculate cycle time. Manufacturing becomes simpler when a linked-cell system is installed.

Step 6: Use Linked Cells. The simplification and integration of production control is materialized by linking the cells, subassemblies, and final assembly elements by utilizing

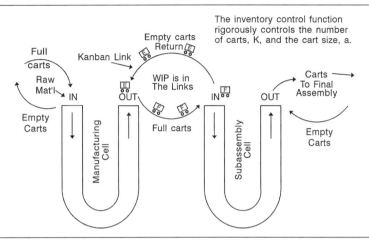

Figure 2. Cells are linked by a kanban system. The arrival of an empty container at the manufacturing cell is the signal to produce more parts.

the kanban subsystem. By connecting process elements with kanban links, the need for route sheets is eliminated. All the cells, processes, subassemblies, and final assemblies are connected by kanban links that pull materials to final assembly. This achieves the integration of production control into the manufacturing system, forming a linked-cell manufacturing system (see Figures 2 and 3).

Step 7: Integrate Inventory Control. People on the plant floor can directly control the inventory levels in their areas through the utilization of kanban control. This integrates the inventory control system with the manufacturing system while systematically reducing work-in-process (WIP). Reduction of WIP exposes problems that must be solved before inventory can be further reduced. The minimum level of WIP (which is actually the inventory *between* the cells, subassemblies, and assembly) is determined by the quality, the reliability of the equipment, setup time, and the transport distances to the downstream cell or assembly line.

Step 8: Integrate the Suppliers. Educate and encourage suppliers (vendors) to develop their own lean production system for superior quality, low cost, and rapid on-time delivery. They must be able to deliver parts to the customer, when needed, and where needed without incoming inspection. The linked-cell network ultimately should include every supplier. Thus, suppliers become remote cells in L-CMS.

After these eight steps have been completed, and the manufacturing system has been redesigned and infused (integrated) with the critical production functions of quality control, inventory control, production control, and machine tool maintenance, the autonomation of the integrated system is the next step.

Step 9: Institute Autonomation. Converting manned cells to unmanned cells is an evolutionary process initiated by the need to solve problems in quality, reliability, or capacity

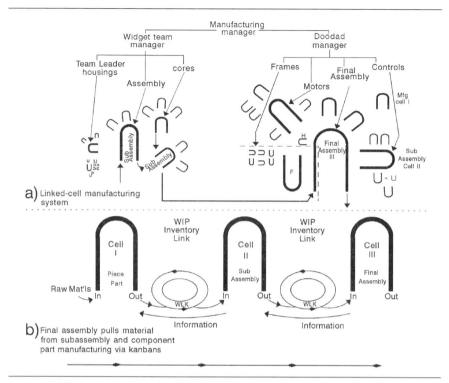

Figure 3. In the linked-cell manufacturing system, the cells are linked with controllable inventory buffers call kanban links or loops.

(i.e., bottlenecks). It begins with mechanization of operations such as load, unload, inspect, and clamp, and logically moves toward more advanced mechanization in the form of automatic detection and correction of problems and defects. This is called autonomation, the automatic control of quality and quantity.

Step 10: Restructure the Production System. Once the factory (the manufacturing system) has been restructured into a JIT manufacturing system, and the critical control functions are integrated and matured, the company will find it expedient to restructure the rest of the company. Production system will require removing or reducing the functional orientation of the various departments and forming cross-functional teams, often along product lines. The implementation of concurrent engineering teams decrease the time needed to bring new products to market. This movement is gaining ground in many companies and is often called *Business Process Reengineering* (BPR); it is basically the restructuring of the plant operation to be as waste free and efficient as the manufacturing system.

RESTRUCTURING THE REST OF THE BUSINESS

Changing to a Lean Manufacturing system will affect product design, tool design and engineering, production planning (scheduling) and control, inventory control, purchasing, quality control and inspection, the production worker, supervisors, middle managers and top management. Such a conversion cannot take place overnight and must be viewed as a *long-term transformation* from one type of *production system* to another. This effort often begins with building product realization teams designed to bring new products to the marketplace faster. In the automotive industry, one example of this is called, platform teams, which enable concurrent engineering. They are composed of personnel from design, engineering, manufacturing, marketing, sales, finance, and so on. As the notion of team building spreads and the lean manufacturing system gets implemented, a leaner factory/company takes shape. Unfortunately, many companies are restructuring the their non-manufacturing side without first completing the initial necessary steps 1 through 8 to get the manufacturing system lean and productive.

BENEFITS OF LEAN PRODUCTION

The conversion to lean manufacturing results in significant cost savings over a two- to three-year period. Specifically, manufacturing companies report significant reduction in raw materials, in-process inventories, setup costs, throughput times, direct labor costs, indirect labor costs, staff, overdue orders, tooling costs, quality costs, and the cost of bringing new designs on line.

However, this reorganization has a greater and immeasurable benefit. It prepares the way for Computer Integrated Manufacturing (CIM). The progression from the functional shop to the factory with linked cells and, perhaps, ultimately to robotic cells under computer control gradually turns the entire system into a CIM system, which is accomplished in several logical steps.

OBSTACLES TO LEAN PRODUCTION

A major effort on the part of a business is required to undertake the conversion to lean manufacturing. The hurdles to implementation are:

1. The top management person (or the real leader) does not totally buy into the conversion.
2. System changes are inherently difficult to implement. Changing the entire manufacturing production system is a huge task.
3. Companies spend freely for new manufacturing processes (i.e., machines) but not for the conversation to a new manufacturing system. It is easier to justify new hardware than to implement a new manufacturing system (i.e., linked cells).
4. Fear of the unknown. Decision making is often choosing among alternatives in the face of uncertainty; the greater the uncertainty, the more likely that the "do-nothing" option will prevail.
5. Faulty criteria for investment decisions. Manufacturing decisions should seek to enhance the ability of the company to compete (on items such as quality, reliability, delivery time, flexibility for product change or volume change) rather than merely minimize the cost of the investment.
6. Lack of blue-collar involvement in the decision-making process of the company. Getting the production workers involved in the decision-making process is, in

itself, a significant change. Some managers may have problems adjusting to this change.

7. The conversion to lean manufacturing represents a real threat to middle managers. In lean manufacturing systems, some of the functional tasks that middle managers have been responsible for will be integrated into the manufacturing system.

Clearly, education and careful planning at all levels of the enterprise is needed to overcome these constraints. The attitudes of management and workers must change. By the next century, we may see significantly fewer workers on the plant floor. These workers will be far better educated and more productive than today's workers. They will be involved in: (1) solving daily production problems, (2) working to improve the entire system, and (3) making decisions to improve their job, the processes, and the manufacturing system. Table 2, obtained from a first-tier supplier to Toyota, summarizes some of the key managerial aspects of the Lean Production System.

Table 2 mentions the use of the seven tools of quality in managing the lean production system. The seven tools of quality are: flow diagram; histogram; pareto chart or

Table 2 Managing the lean production system (LPS)

The LPS includes the linked-cell manufacturing system (cells linked by a kanban pull system), the 5 S's listed below, standard operation, the seven tools of QC and other key elements listed here.

Kanban Pull System

It is the production process that uses a card system, standard container sizes, and pull versus push production to accomplish just-in-time production.

5 S's (Seiri, Seiton, Seiketsu, Seiso, Shitsuke)

The five S's in Japanese language are proper arrangement, orderliness, cleanliness, cleanup, and discipline.

Standard Operation in Manufacturing Cells

The components of standard operation include: cycle time, work sequence, and standard stock on hand in the cells.

Morning Meeting

A daily meeting is held for the purpose of sharing production and safety information, quite often by a quality circle.

Key Points in Process Sheets

The process sheets, which are visually posted at each work station detail the work sequence and most critical points for performing the tasks.

Change-Over and Setup

The machine setup that takes place when an assembly line changes products.

Seven Tools of Quality

The seven tools of quality are: Pareto diagrams, check sheets, histograms, cause and effect diagrams, run charts for individuals, control charts for samples, and scatter diagrams. (See Figure 4)

Production Behavior

Rules which include information on personal safety, safety equipment, clothing, restricted areas, vehicle safety, equipment safety, and housekeeping.

Visual Management

Each production line in the plant has a complete set of charts, graphs, or other devices like *Andons* (a Japanese word for an electronic line stop indication board) for reporting the status and progress in the area.

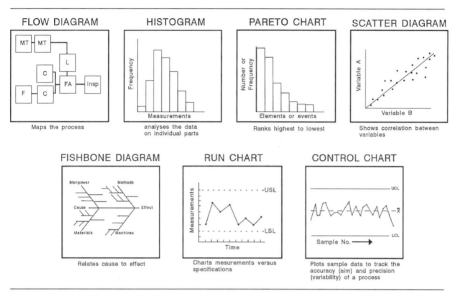

Figure 4. The seven tools of quality control.

ABC analysis; scatter diagram; fishbone diagram; run chart; and control chart. These tools are displayed in Figure 4. In the figure, Histogram, Pareto Chart, Scatter Diagram, Run Chart and Control Chart use quantitative or numerical data to understand, and/or identify and explain a problem.

A Histogram identifies the more frequently occurring events or problems. A scatter Plot visually shows the nature of relationship between two variables of interest. The shape of Histograms may reveal if the process is skewed to the left or right.

Run Chart and Control Chart use data for statistical Process control. A Run chart is used with small batches to detect process shift. Each item is inspected and plotted on the Run Chart. A sequence of plots with upward or downward trend suggest a process that is shifting and may go out of control, if not corrected. Run charts are bounded by USL and LSL, which are upper and lower specification limits. Plots outside the limits indicate the failure to meet specifications by the product it should be reworked or discarded and the process should be adjusted.

Fishbone Diagram is effective for use in a team for the identification of root cause to a problem. It is also called "cause effect" diagram, which is a tool for systematically identifying and investigating all causes, subcauses, and so on.

Key Concepts: Autonomation; Five S's; Integrated Quality Control; Islands of Automation; Just-in-Time Manufacturing; Linked-Cell Manufacturing System (L-CMS); Platform Teams; Production Systems.

Related Articles: Agile Manufacturing; Just-in-Time Manufacturing; Total Productive Maintenance.

REFERENCES AND BIBLIOGRAPHY

Black, J.T. (1991). *The Design of The Factory With A Future*. McGraw Hill.

Hall, Robert W. (1983). *Zero Inventories*. Dow Jones-Irwin.

Harmon, Roy L. and Leroy D. Peterson (1990). *Reinventing the Factory: Productivity Breakthroughs In Manufacturing Today*. The Free Press.

Ishikawa, Kaouru (1972). *Guide to Quality Control*. Asian Productivity Organization.

Monden, Yasuhiro (1983). *Toyota Production System*. Industrial Engineering and Management Press, IIE.

Nakajima, Seiichi (1988). *TPM, Introduction to TPM: Total Productive Maintenance*. Productivity Press.

Ohno, Taiichi (1988). *Toyota Production System: Beyond Large-Scale Production*. Productivity Press.

Schonberger, Richard J. (1986). *World Class Manufacturing*. The Free Press.

Sekine, Kenichi (1990). *One-Piece Flow: Cell Design for Transforming the Production Process*. Productivity Press.

Shconberger, Richard J. (1982). *Japanese Manufacturing Techniques: Nine Hidden Lessons in Simplicity*. The Free Press.

Shingo, Shigeo (1985). *A Revolution in Manufacturing: The SMED System*. Productivity Press.

Shingo, Shigeo (1986). *Zero Quality Control: Source Inspection and the Poka-Yoke System*. Productivity Press.

Shingo, Shigeo (1988). *Non-Stock Production: The Shingo System for Continuous Improvement*. Productivity Press.

Suzaki, Kioyoski (1987). *The New Manufacturing Challenge*. The Free Press.

Womack, J.P., D.T. Jones, and D. Roos (1990). *The Machine that Changed the World*. New York: Harper Perenial.

17. TOTAL PRODUCTIVE MAINTENANCE (TPM)

KATHLEEN E. McKONE

University of Minnesota, Minneapolis, MN, USA

ELLIOTT N. WEISS

University of Virginia, Charlottesville, VA, USA

ABSTRACT

Substantial capital investments are required for manufacturing almost all goods of economic significance. The productivity of these investments enable companies and nations to compete. The maintenance of capital investments involves significant recurring expenses. For example, in 1991, DuPont's expenditure on company-wide maintenance was roughly equal to its net income. Maintenance expenses vary depending on the type of industry; however, they are typically 15–40% of production costs. Companies attempt to control maintenance costs by keeping them at a specified budget level, a level often based on the previous year's expenses.

During the last decade, manufacturers found this approach to be insufficient. As companies have invested in programs such as JIT and TQM in an effort to increase organizational capabilities, the benefits from these programs have often been limited by unreliable or inflexible equipment. In the context of JIT & TQM use, rather than being seen simply as an expense that must be controlled, maintenance is now regarded as a strategic competitive tool. Total Productive Maintenance (TPM) has evolved as an effective program for improving equipment performance and increasing organizational capabilities.

TPM has resulted in significant improvements in plant performance. McKone, Schoreder, and Cua (1998) evaluated the impact of TPM practices on manufacturing performance and found that TPM has a positive and significant relationship with low cost (as measured by high inventory turns), high levels of product quality (as measured by higher levels of conformance to specifications), and strong delivery performance (as measured by higher percentage of on-time deliveries and by faster speeds of delivery). Their research indicates that TPM plays a significant role in improving manufacturing performance.

BACKGROUND

TPM originated from the fields of reliability and maintenance—a pair of closely related disciplines that have become standard engineering functions in many industries. The primary objective of these functions is to increase equipment availability and overall effectiveness.

There have been four major periods in the history of maintenance management:

1. The period prior to 1950 was characterized by *reactive maintenance*. During this phase little attention was placed on defining reliability requirements or preventing equipment failures. Typically, equipment specifications included requirements for individual parts without the consideration of the reliability or availability of the entire system.
2. The second period, which saw the growth of *preventive maintenance*, involved an analysis of current equipment to determine the best methods to prevent failure and to reduce repair time. This period resulted from the emergence of the military equipment industry during World War II. Emphasis was placed on the economic efficiency of equipment replacements and repairs as well as on improving equipment reliability to reduce the mean time between failures.
3. The third period, called *productive maintenance*, became well established during the 1960s when the importance of reliability, maintenance, and economic efficiency in plant design was recognized. Productive maintenance has three key elements: maintenance prevention, which is introduced during the equipment-design stages; maintainability improvement, which modifies equipment to prevent breakdowns and facilitate ease of maintenance; and preventive maintenance, which includes periodic inspections and repairs of the equipment. General Electric Corporation is typically credited for initiating productive maintenance in the 1950s but the approach did not gain popularity until the 1960s (Hartmann, 1992). In the late 1950s the concepts of productive maintenance were also promoted in Japan.
4. The most recent period is represented by *Total Productive Maintenance (TPM)*. TPM officially began in the 1970s in Japan. Seiichi Nakajima, vice-chairman of the Japanese Institute of Plant Engineers (JIPE), the predecessor of the Japan Institute of Plant Maintenance (JIPM), promoted TPM throughout Japan and has become known as the father of TPM. In 1971, TPM was described by JIPE as follows:

TPM is designed to maximize equipment effectiveness (improving overall efficiency) by establishing a comprehensive productive-maintenance system covering the entire life of the equipment, spanning all equipment-related fields (planning, use, maintenance, etc.) and, with the participation of all employees from top management down to shop-floor workers, to promote productive maintenance through motivation management or voluntary small-group activities (Tsuchiya, 1992, 4).

TPM provides a comprehensive company-wide approach to maintenance management.

In the 1980s, TPM was introduced in the United States. There are several reasons why American companies are utilizing TPM. Dominant, among other reasons, is that many organizations face competitive pressures from firms that have improved plant productivity, often through successful TPM implementation. One particular group of competitors, the Japanese, have demonstrated in many industries that maintenance is a critical component of their success. JIPM has emphasized the importance of maintenance and has awarded over 830 preventive-maintenance prizes to companies that have achieved a high level of success with TPM implementation.

It would, however, be a gross oversimplification to suggest that the Japanese influence is the only factor that has brought about the popularity of TPM. There are also several other changes in the manufacturing environment that have increased the importance of maintenance. JIT, TQM, and Employee Involvement (EI) programs have become more commonplace in industry to address the increasing demands of customers, companies have attempted to reduce costs, shorten production lead-time, and improve quality. These improvement programs require reliable and consistent equipment throughout the entire plant. Moreover, as customers become more demanding and processes become more interrelated, the need for an effective maintenance strategy also increases. Figure 1 shows the interrelationships between TPM, JIT, TQM, and EI. From this diagram, it is clear that proper implementation of TPM will enhance the effectiveness of the other improvement programs. Schonberger (1986) also highlights the importance of JIT, TQM, EI *and* TPM to World Class Manufacturing.

As the technology of production equipment becomes more sophisticated, it is essential that operators and maintenance personnel are provided with the tools and training to support the new demands of the equipment. It is also important that production personnel take advantage of the new maintenance technologies, such as vibration analysis and infrared thermography, that have become useful means for *predicting* and diagnosing equipment problems.

The trends in manufacturing techniques and processes make it important for organizations to maintain manufacturing equipment in working condition to meet the higher performance criteria, produce maximum returns, and compete aggressively worldwide. Thus, TPM has emerged as a result of heightened corporate focus on making better use of available resources.

KEY FEATURES OF TPM

Figure 2 provides a framework for considering TPM activities. The elements of the framework have been developed based upon popular TPM literature (Hartmann, 1992;

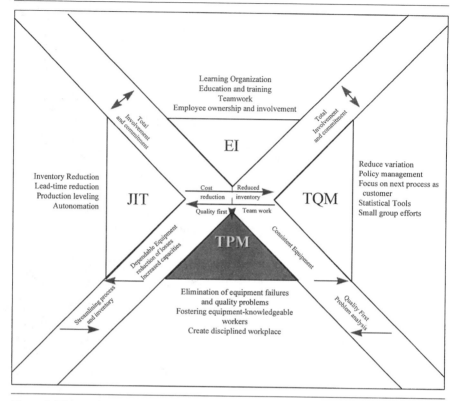

Figure 1. The interdependence of JIT, TQM, TPM and EI.

Nakajima, 1988; Shimbun, 1995; Suzuki, 1992; and Tsuchiya, 1992; Gotoh, 1991), and a number of site visits and interviews with practicing managers. While maintenance activities primarily focus on cost reduction and equipment effectiveness, TPM, which emphasizes a company-wide approach to maintenance, also plays a vital role in improving other manufacturing performance measures. Therefore, improved cost, quality, delivery, flexibility, and innovativeness are important goals of a TPM program and are represented in the framework.

TPM Values: Critical to successful TPM implementation are its core operating values. Although these values should be tailored to any given organization, they offer broad guidelines for the management of equipment. The following represent the general values of TPM:

- Process and product quality are a key part of every person's performance.
- Equipment failures and off-quality product can and will be prevented.
- If it ain't broke, fix it anyway.
- Equipment performance can be managed.

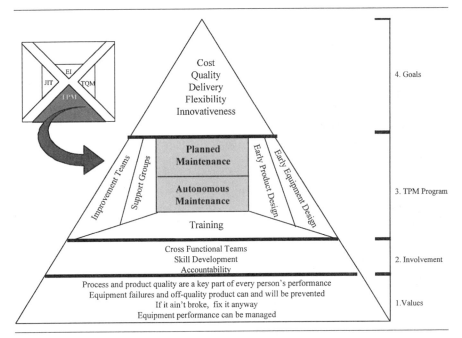

Figure 2. A model of TPM.

Employee Involvement: Every employee should be involved in the TPM process. *Accountability* for equipment performance is important to the success of the TPM program. However, employees cannot be accountable for quality and predictable production if they cannot impact equipment performance. Employees' *skills* must be developed to meet the needs of their expanded roles under TPM.

The core values and skills of the organization provide the foundation for TPM implementation. Cross-functional teams also encourage employee involvement. Teams help to break down the barriers that are inherent in the traditional approach to maintenance. Teams also help to identify problems and suggest new approaches for elimination of the problems, introduce new skills, initiate training programs, and define TPM processes.

The TPM Program: While TPM is commonly associated with autonomous and planned maintenance activities, the program also includes other activities that help to improve equipment effectiveness over the entire life of the equipment. These activities include: Training, Early Equipment Design, Early Product Design, Focused Improvement Teams, and Support Group Activities (Tajiri and Gotoh 1992; Nakajima, 1988; Gotoh, 1991). The six major TPM activities are reviewed below.

TPM training should be given to every employee. Although internal capabilities may not be sufficient to achieve the goals of TPM at the start of the program, the capabilities for continuation of TPM must be developed. Also, as needs develop, training should support new TPM activities such as inspection, periodic restoration, and prediction analysis.

Early equipment management activities are typically driven by the research and development or engineering functions within the organization. Early equipment management considers the trade-offs between equipment attributes encompassing reliability, maintainability, operability, and safety. The engineering efforts involve the consideration of the life-cycle costing for equipment purchases as well as comprehensive commissioning periods prior to full production.

Early product design involves efforts to simplify the manufacturing requirements and improve quality assurance through product design. By considering these factors in the product-design stage, it is easier to meet the diversified needs of consumers in terms of product features, design, quality, and price. The shop-floor employees can focus on maintaining the process and equipment rather than on working out the logistics of manufacturing the product.

Focused improvement team efforts help to eliminate the major equipment-related losses including breakdown losses, setup and adjustment losses, minor stoppage losses, speed losses, quality defects and rework losses, and start-up/yield losses. Typically, selecting a process-improvement team to resolve a particular problem eliminates the major losses. The team then identifies and analyzes the cause of the problem, plans and implements a solution, checks the results, and if improvement occurs, develops standards to ensure that the improved conditions will remain. The improvement team also documents its work so that others can learn from its improvement efforts.

Support group activities ensure that the production department does not produce useless or wasteful products, and that orders are filled on time, at the quality and costs that the development and engineering departments prescribe. This is not the sole responsibility of the production department; it requires a TPM program that embraces the entire company, including the administrative and support departments.

Much of the day-to-day maintenance planning and execution is performed by the production and maintenance personnel. These efforts are represented by the *Autonomous and Planned Maintenance* activities in Figure 2. Together, production and maintenance groups help to improve the effectiveness of the maintenance program. As operators are trained, they begin to inspect and maintain the equipment and perform basic maintenance tasks. This assignment of maintenance tasks to production operators frees up time for maintenance personnel to perform long-term improvement efforts and plan maintenance interventions.

FIVE PHASE IMPLEMENTATION

The autonomous and planned maintenance efforts are typically divided into four phases that correspond to the four stages of maintenance development proposed by Nakajima (1988); they are also frequently included in subsequent books (Tajiri and Gotoh, 1992; Nachi-Fujikoski, 1990; Suzuki, 1992). We also include a fifth stage that foresees the future direction of TPM efforts. This phase incorporates the many TPM activities that are not directly part of autonomous or planned maintenance roles. Figure 3 shows a five-phase framework for autonomous and planned maintenance development under TPM

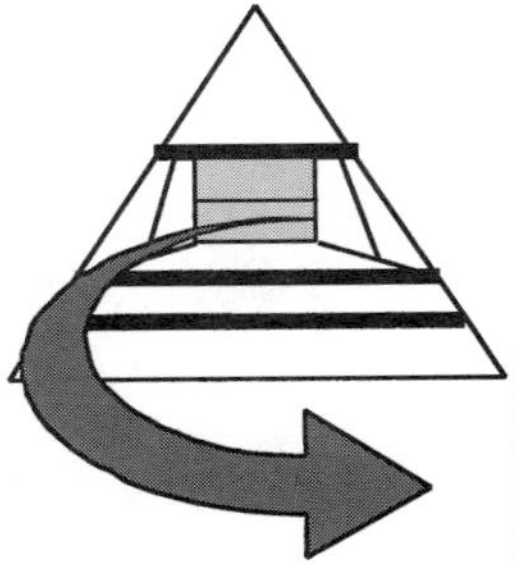

	Phase I	Phase II	Phase III	Phase IV	Phase V
	Reduce Life Span Variability	Lengthen Average Life Span	Estimate Life Span	Predict Life Span	Design Life Span
Planned Maintenance	Step 1: Evaluate Equipment Step 2: Restore Deterioration	Step 3: Correct Design Weaknesses Step 4: Eliminate Unexpected Failures	Step 5: Build Periodic Maintenance System & Identify Symptoms of Deterioration	Step 6: Build Predictive Maintenance System Step 7: Prevent Quality Defects	Step 8: Technical Analysis of Any Deterioration & Design Step 9: Implement in All Support Areas
Autonomous Maintenance	Step 1: Basic Cleaning Step 2: Eliminate Source of Problems Step 3: Set Standards	Step 4: General Inspection of Equipment	Step 5: Autonomous Inspection	Step 6: Maintenance for Quality Step 7: Autonomous Maintenance	Step 8: Process Improvement & Design Team Step 9: Implement in all Support Areas

Figure 3. Five phases of TPM development.

and includes typical steps for implementation. The five phases are described in detail below.

A practitioner's framework for *autonomous and planned maintenance* is shown in Figure 3. It is accepted by practitioners, easy to understand, and closely depicts the managerial decisions made for maintenance on a day-to-day basis within the plant.

(1) Phases I and II—Equipment Improvement: The first phase of TPM involves efforts to reduce the variability in equipment life span. This phase involves the removal of assignable causes that reduce equipment life span. The efforts include restoration and cleaning of the equipment. At this stage, practitioners must decide how much time and money to invest in restoring the equipment to a base condition.

The main objective of Phase II is to lengthen the equipment's life. In this stage, the cleaning and lubrication procedures become standardized and operators are educated to conduct detailed inspection of all equipment. Companies must decide how much time and money to invest in training operators and technicians. Cross-training personnel can make an organization more flexible and more responsive to maintenance needs; however, some tasks will still be too difficult or unsafe for operators to perform. Another Phase II decision is to make investments in eliminating the root causes of contamination and failure, and to simplify the inspection and maintenance tasks. Finally, practitioners must decide how much time to invest in order to sustain the equipment at its base condition.

Both Phase I and II assume that it is possible to change the distribution of equipment failure occurrences. This assumption is central to the TPM philosophy. In traditional maintenance, the assumption is made that equipment conditions deteriorate over time, leading to failure or the need for replacement. With TPM, investments are made to reduce equipment problems; the assumption is that a goal of zero failures and defects is achievable.

(2) Phase III: After implementing the first two phases, equipment conditions should be dependable and operating conditions should be consistent. Equipment life span can be accurately estimated, and mechanics and operators can plan periodic inspections and renovations. Phase III of TPM development begins to determine the best type and interval of inspections and repairs.

It is interesting that much of the traditional maintenance literature supports the decisions involved in Phase III of TPM development. Essentially, the failure rate is assumed to be a function of the equipment, and little attention is given to the possibility of improving the equipment or restoring the equipment to a better condition (Phase I and II decisions).

See McCall (1965), Pierskalla and Voelker (1976), and Valdez-Flores and Feldman (1989) for a comprehensive reviews of literature relevant to this phase of maintenance management.

Most of the Phase III literature compares the cost benefits of various maintenance policies: preparedness or preventive maintenance, and periodic- or sequential-replacement policies. More recent work extends the base models to incorporate multi-component equipment and multi-state equipment conditions, or relax some of the assumptions of earlier models.

Phase III models have been extremely effective in the development of a philosophy of preventive versus reactive maintenance. Most organizations realize that it can be beneficial to replace equipment prior to failure, when the costs of breakdowns are high. However, in practice, the replacement interval is often chosen based on the equipment manufacturer's recommendations. Therefore, intervals are often established without considering the actual failure distribution (based on the environment in which it operates) or maintenance costs (based on the downtime costs associated with a particular plant). Research is needed to develop better estimates of the actual failure distribution and the associated maintenance costs.

(3) Phase IV: Phase III allows operators and technicians to gain a deeper understanding of the equipment and process. Phase IV permits personnel to use this new knowledge about equipment deterioration, together with diagnostic techniques, to predict failures of equipment, and eliminate equipment-related quality problems. Condition-based maintenance helps get additional time out of the equipment as well as eliminate unexpected failures. The practitioner must thus decide where, when, and how to use prediction tools. In addition, maintenance policies must be coordinated with quality-control policies in order to reduce product quality problems.

Traditionally, equipment maintenance was treated as a method for increasing equipment availability. The goal of maintenance was to keep the equipment running. With the advent of quality-management efforts, however, the condition of the equipment became important to control the quality of the product. Several recent papers address the relationship between maintenance and quality; some examples are presented here. Tapiero (1986) considers the problem of continuous-quality production and machine maintenance. In his model, quality is assumed to be a known function of the machine-degradation state. He considers open-loop (based on quantity or age of equipment) and closed-loop (based on the equipment condition) stochastic control-maintenance problems. Rahim (1994) presents a model for jointly determining the economic production quantity, inspection schedule, and control-chart parameters for an imperfect production process.

Predictive maintenance is commonly discussed in trade journals; however, very little academic research focuses on this area. McKone and Weiss (1999) consider the joint use of continuously monitoring prediction tools as well as periodic maintenance policies in order to minimize maintenance costs. They recommend that periodic tools *not* be abandoned for the use of predictive tools, and provide decision rules for selecting the appropriate periodic policy, when predictive tools are available.

The product quality literature clearly provides some support for this stage of TPM development. Tools and techniques that help to define, identify, and eliminate known and/or potential failures, problems, and errors in the system are important to equipment improvement efforts. For example, Failure Model and Effect Analysis (FMEA) (Stamatis, 1995) helps to identify plant equipment and process problems that result in product quality problems. FMEA can help to link both maintenance and quality improvement efforts.

(4) Phase V: The fifth phase involves an organization-wide focus on plant productivity. First, design teams made up of engineers, maintenance personnel, and operators

prepare equipment cleaning and inspection standards, and personnel are trained to produce efficiently and effectively. Phase V decisions also consider non-maintenance systems, such as spare parts, raw materials, and production scheduling, that impact the equipment productivity and quality. Finally, efforts are made to eliminate losses in labor, energy, and materials in addition to equipment efficiency. The decisions in this phase primarily focus on organizational and systemic issues.

Phase V is comprehensive. At this phase, functional areas within manufacturing must be carefully integrated. Essentially, all six TPM activities must be coordinated for effective and efficient equipment operation.

Other issues in implementation

Thilander (1992) conducted a case study of two Swedish firms in an effort to define the benefits of the organizational aspects of TPM. The study shows that well-defined areas of responsibility, one individual who holds the overall responsibility for the maintenance, and direct contact between the operators and maintenance technicians have positive impact on productivity.

There are several case studies of TPM that present the TPM development story or examples of TPM improvement activities in plants (Varughese, 1993; Shimbun, 1995; Steinbacher and Steinbacher, 1993, Chapter 15; Hartmann, 1992; Tsuchiya, 1992; Suzuki, 1992, Chapter 4; and Tajiri and Gotoh, 1992). These studies also include steps for TPM implementation.

Key Concepts: Autonomous and Planned Maintenance; Employee Involvement; Just-in-time manufacturing; Preventive Maintenance; Productive Maintenance; Reactive Maintenance.

Related Articles: Just-in-Time Manufacturing; Lean Manufacturing Implementation; Predictive Maintenance; The Case of Della Steam Plant; Total Quality Management.

REFERENCES

Anderson, M.Q. (1981). "Monotone Optimal Preventative Maintenance Polices For Stochastically Failing Equipment." *Naval Research Logistics Quarterly*, 28, 347–358.

Bain, L. J. and M. Engelhardt (1991). *Statistical Analysis Of Reliability And Life-Testing Models.* Marcel Dekker, Inc., New York.

Chikte, S.D. and S.D. Deshmukh (1981). "Preventive Maintenance And Replacement Under Additive Damage." *Naval Research Logistics Quarterly*, 28, 33–46.

Dada, M. and R. Marcellus (1994). "Process Control With Learning." *Operations Research*, 42 (2), 323–336.

Fine, C.H. (1988). "A Quality Control Model With Learning Effects." *Operations Research*, 36 (3), 437–444.

Gotoh, F. (1991). *Equipment Planning For Tpm: Maintenance Prevention Design*, Productivity Press, Cambridge, MA.

Hartmann, E.H. (1992). *Successfully Installing Tpm In A Non-Japanese Plant.* Tpm Press, Inc., Allison Park, PA.

Jorgenson, D.W. and J.J. Mccall (1963). "Optimal Scheduling Of Replacement And Inspection." *Operations Research*, 11, 723–747.

Lee, H.L. and M.J. Rosenblatt (1989). "A Production And Maintenance Planning Model With Restoration Cost Dependent On Detection Delay." *IIE Transactions*, 21 (4), 368–375.

Marcellus, R.L. and M. Dada (1991). "Interactive Process Quality Improvement." *Management Science*, 37 (11), 1365–1376.

Mccall, J.J. (1965). "Maintenance Policies For Stochastically Failing Equipment: A Survey." *Management Science*, 11 (5), 493–524.

McKone, K., R. Schroeder, and K. Cua (1999). "Total Productive Maintenance: A Contextual View." *Journal Of Operations Management*, 17 (2), 123–144.

McKone, K., R. Schroeder, and K. Cua (1998a). "The Impact Of Total Productive Maintenance Practices On Manufacturing Performance." *Carlson School Working Paper.*

McKone, K. and E. Weiss (1997a). "An Autonomous Maintenance Approach To Cycle Time Reduction: Guidelines For Improvement." Darden School Working Paper Series, Dswp-97-30, University Of Virginia, Charlottesville, VA.

McKone, K. and E. Weiss (1997b). "Analysis Of Investments In Autonomous Maintenance Activities," Carlson School Operations And Management Science Department Working Paper 97–7, University Of Minnesota, Minneapolis, MN.

McKone, K. and E. Weiss (1999). "Managerial Guidelines For The Use Of Predictive Maintenance." Darden School Working Paper Series, University Of Virginia, Charlottesville, VA.

Nachi-Fujikoshi Corporation (1990). *Training For TPM: A Manufacturing Success Story*, Productivity Press, Cambridge, MA.

Nakajima, S. (1988). *Introduction To TPM*, Productivity Press, Cambridge, MA.

Özekici, S. and S.R. Pliska (1991). "Optimal Scheduling Of Inspections: A Delayed Markov Model With False Positive And Negatives." *Operations Research*, 39 (2), 261–273.

Paté-Cornell, M.E., H.L. Lee, and G. Tagaras (1987). "Warning Of Malfunction: The Decision To Inspect And Maintain Production Processes On Schedule Or On Demand." *Management Science*, 33 (10), 1277–1290.

Pierskalla, W.P. and J.A. Voelker (1976). "A Survey Of Maintenance Models: The Control And Surveillance Of Deteriorating Systems," *Naval Research Logistics Quarterly*, 23, 353–388.

Rahim, M.A. (1994). "Joint Determination Of Production Quantity, Inspection Schedule, And Control Chart Design." *IIE Transactions*, 26 (6), 2–11.

Schonberger, R.J. (1986). *World Class Manufacturing: The Lessons Of Simplicity Applied*, Free Press, N.Y.

Shimbun, N.K. (Ed.) (1995). *Tpm Case Studies*, Productivity Press, Portland, Oregon.

Steinbacher, H.R. and N.L. Steinbacher (1993). *TPM For America: What It Is And Why You Need It*, Productivity Press, Cambridge, MA.

Stamatis, D.H. (1995). *Failure Mode And Effect Analysis*, ASQC Quality Press, Milwaukee, WI.

Suzuki, T. (1992). *New Directions For Tpm*, Productivity Press, Cambridge, MA.

Tajiri, M. and F. Gotoh (1992). *TPM Implementation: A Japanese Approach*, McGraw-Hill, Inc., New York.

Tapiero, C.S. (1986). "Continuous Quality Production And Machine Maintenance." *Naval Research Logistics Quarterly*, 33, 489–499.

Thilander, M. (1992). "Some Observations Of Operation And Maintenance In Two Swedish Firms." *Integrated Manufacturing Systems*, 3 (2).

Thompson, G. (1968). "Optimal Maintenance Policy And Sale Date Of A Machine." *Management Science*, 14 (9), 543–550.

Tsuchiya, S. (1992). *Quality Maintenance: Zero Defects Through Equipment Management*, Productivity Press, Cambridge, MA.

Valdez-Flores, C. and R.M. Feldman (1989). "A Survey Of Preventative Maintenance Models For Stochastically Deteriorating Single-Unit Systems." *Naval Research Logistics Quarterly*, 36, 419–446.

Varughese, K.K. (1993). "Total Productive Maintenance." University Of Calgary, A Thesis For The Degree Of Master Of Mechanical Engineering.

Waldman, K.H. (1983). "Optimal Replacement Under Additive Damage In Randomly Varying Environments." *Naval Research Logistics Quarterly*, 30, 377–386.

18. TRANSITION TO CELL MANUFACTURING: THE CASE OF DURIRON COMPANY INC., COOKEVILLE VALVE DIVISION (1988–1993)

JOHN M. BURNHAM
DAVID LAMBERT
CHARLES W. SMITH, Jr.
JOHN A. WELCH
DALE A. WILSON

Tennessee Technological University, Cookeville, TN, USA

Yes, that's true. When you read the original Arthur Andersen "game plan" you note that we should have become all-cellular by the end of 1992. But there are many factors that explain why our *real* progress has been good, but our cell implementation is still not complete.

ROB ADAMS, OPERATIONS MANAGER
DURIRON VALVE DIVISION
NOVEMBER 1992

OVERVIEW

Beginning in 1988, the Duriron Company (DURCO) committed itself to becoming a world-class valve and pump supplier to the worldwide process industries. Chemical, petroleum, and other processors of corrosive or exotic fluids had long been using DURCO products of high-silicon iron and other alloy materials, but global competition created a need to stay competitive. Since the WCM (world-class manufacturer) decision, the company has taken many steps to become more customer responsive. DURCO, as a mature organization, required significant changes. A series of substantial consulting assignments beginning in 1989 led to the strong recommendation to DURCO that the divisions move toward cell manufacturing from the traditional

This is adapted from the case developed by John M. Burnham, David Lambert, Charles W. Smith Jr., John A. Welch, and Dale A. Wilson, of Tennessee Technological University, Cookeville, TN. Copyright 1994 by the National Consortium for Technology in Business, c/o the Thomas Walter Center for Technology Management, Auburn University, Auburn, AL.

batch-job shop environment. This case follows the progress at the Cookeville Valve division (CVD), located in Cookeville, Tennessee.

Since corporate's directive to strive to WCM, CVD took steps to implement the philosophies of *total quality management, people involvement, and just-in-time manufacturing*. The approach has been to focus on manufacturing cells as a means to incorporate these philosophies into the company. The process has been difficult, and there is still a long way to go. See the article on Lean Manufacturing Implementation.

BACKGROUND INFORMATION

The Duriron Company was founded in 1912 in Dayton, Ohio, by John R. Pitman, a former DuPont acid plant manager, William E. Hall, a lawyer and financier, and Pierce D. Schenck, an electrical engineer. The group of men sought to utilize the capabilities of a new corrosion resistant metal they had invented.

Demand for the company's products skyrocketed with the beginning of World War I. Its corrosion resistant metal was essential for munitions production. As a result it was identified as an "essential industry" and almost nationalized. After the war, the company produced its first pipes and fittings, high-silicon iron pumps, and valves to handle chemical solutions.

The main product lines after World War II were pumps, valves, and heat exchanges sold under the DURCO brand name. The next major development was a Teflon-lined valve, first produced in 1965, to handle severely corrosive materials. The Teflon lining was a much cheaper alternative to the expensive alloys previously required to handle such materials.

In 1978 DURCO began a change in strategy with the shift from a centralized to a decentralized structure. They began to sense a change in how their customers evaluated suppliers. The trend was for a customer to purchase valves from a primary group of suppliers. In the selection process, these companies were looking for a commitment to quality and partnership with these suppliers. Therefore, it was critical for DURCO to show buyers what it was doing for continual improvement into the next century. Also, the company felt that it was behind some competitors in keeping pace with technological changes. In response, top management chose the strategy of becoming a world-class manufacturer (WCM) and directed the divisions to take steps toward teaching this goal.

Corporate mission is to serve the needs of the worldwide process industries for fluid handling products, systems, and services. Current product lines include valves and automatic equipment, pumps, filtration devices, and chemical waste systems. DURCO's stated goals for continued growth are to expand into international markets, implement the latest manufacturing technologies, develop new products, and continue acquiring and divesting businesses.

The company is organized by product lines and individual international subsidiaries. These decentralized divisions operate with relatively little detailed direction from corporate and are responsible for their own marketing and sales functions, and profit. The company has shown steady growth over the years and has been identified as a solid investment by many analysts.

VALVE DIVISION

Due to divisionalization in the late 1970s, the Valve Division (CVD) was moved from Dayton, Ohio, to Cookeville, Tennessee. The 160,000-square-foot facility houses all valve activities from preliminary design to sales/marketing and has some 300 direct employees. From 1978 to 1988, production was departmentalized and was characterized by high inventories, long lead times, large lot sizes, and high product costs. Despite these traditional manufacturing characteristics, the division was profitable.

CVD produces a wide variety of valves and accessories to fit almost any customer need. A complete line of butterfly, ball, and plug valves are made in a variety of sizes and alloys.

Even before the corporate WCM decision was made, CVD had noted the market changes, that its customers were beginning to redirect purchasing procedures toward those suppliers with a commitment to customer service. In addition, the division's president had recently toured some Japanese production facilities and decided to reevaluate CVD's production processes.

INDUSTRY CUSTOMERS AND COMPETITORS

DURCO serves the chemical process industry. Its customers include Dow, DuPont, Eastman chemical, oil refineries, and food and beverage manufacturers. These companies are now expanding internationally, so CVD has made efforts to meet their changing needs by designing with metric measurements and the European and Asian standards for valves. CVD has augmented its design and production processes to meet these needs.

Another factor in the move toward WCM has been the increased competition, both domestic and foreign. From the 1960s to late 1970s, DURCO was the leader in the valve market. DURCO's designs were superior, and the company met the customers' basic needs. Service, including lead time, breadth of product line, and quality levels, became the main differentiator. The competitive pressures made Duriron realize the need for comprehensive manufacturing and product design strategies.

Industry demands will continue to change in the future. A large market already exists in high-temperature (over 1000° F) and high-pressure steam applications. The development of synthetic fuels and biochemical industries will create new challenges. Perhaps the biggest shifts will come in the form of increased environmental regulation. New regulations could create stricter requirements on chemical process equipment.

CONSULTANT'S SCHEDULE

The first step in 1990 was to develop an action plan to improve operations at the plant. To provide an outside perspective, the corporation had earlier hired Andersen Consulting to develop plantwide project plans to determine what changes to make and when to make them. When Andersen carried out an assignment at the Valve division, the objective was: (1) to design the cells around a product line, including layout and machine requirement; (2) to create a scheduling system to provide the shortest possible lead times, to level the daily cell production, and to handle emergency orders; and (3) to produce timetables for implementation of the cells.

The consultant's plan called for eight cells of varying sizes. There were to be two T-Line plug valve cells, one for 0.5″ to 2″ valves and another for 3″ to 8″ valves. One cell would make all BTV butterfly valves, and another would produce BX butterfly valves. The remaining four cells would manufacture various sizes of the G4 Plug valve. See Figure 1 for a catalog of typical products.

In addition to the standard plans, the Andersen material called for pay for performance for the workers. The pay-for-performance concept was to be used instead of the seniority process currently used. Also, the Andersen material contained statistical process control (SPC) information, business goals, and management guidelines for cell implementation.

The Andersen implementation plan called for a complete change to focused factories (cells) by October1, 1992. Expected benefits were reduced lead times from six to eight weeks to one to two weeks, floor space reduction of 37 percent, increased inventory turnover from 2 to 12 times per year, improved quality through smaller lot sizes and operator knowledge, and increased capacity and flexibility as a result of the dedicated equipment. In addition, each cell was expected to pay for itself in approximately one year.

IMPLEMENTING MANUFACTURING CELLS

The cells operate as true focused factories. All inventory is maintained within the cell. Cell team members are responsible for each facet of production, including scheduling and performance evaluation. All raw material and work-in-process inventories are maintained within the boundaries of each cell. The processes use a great deal of numerically controlled equipment and are designed for flexibility.

CVD's goals for the cellular manufacturing project were to eliminate work-in-process, reduce finished goods inventory, reduce set-up times, achieve 100 percent on-time delivery, and reduce required floor space and product costs while maintaining required quality levels. To accomplish the goals, manufacturing cells were to be implemented throughout the plant. Steering and implementation committees were formed to get the ball rolling. The steering committee had to choose an initial product on which to test the cellular concepts and to serve as a demonstration project to show success.

The committee chose the 0.5- to 1.0-inch G4 plug valves for the first cell, the reason being that this investment casting contained less variation than other valves. This valve required the fewest number of setup changes, and it was a large-volume valve. The line was not very profitable. In fact, some other manufacturers took a loss on this line to get access to customers. Therefore, improvements coming from the cell could help the company as well as show the potential improvements from cellular manufacturing.

Each cell was to be a totally separate operation. The cell team would schedule the work, do its own purchasing, manufacture the products entirely, perform SPC and

The BTV-2000

The Atomac Lined Ball Valve

The T-Line Plug Lined Plug Valve

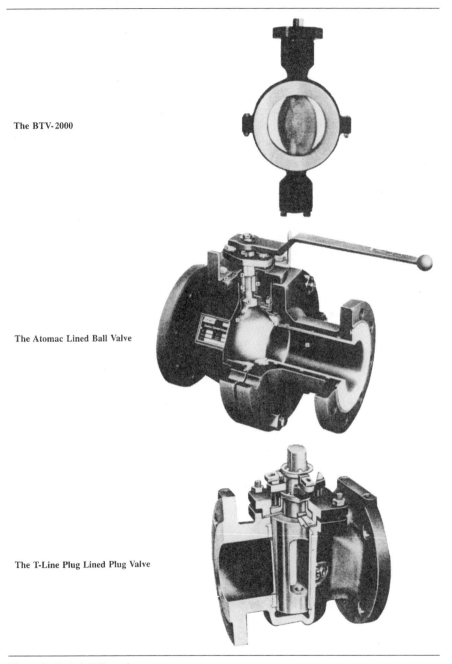

Figure 1. Typical CVD products.

quality inspections, and track its own profits, lead times, scrap rates, inventory turn, and backlog. In addition, the cell would radically reduce the distance that a valve had to travel inside the plant from over one mile (from receiving to shipping) to only 80 feet. Substantially all of the cell equipment was to be added to the plant, since parallel, ongoing "traditional" manufacturing required the same equipment, they could not be dedicated to serve one cell alone. New or used numerically controlled (NC) multifunction machinery made up most of the cell outfit.

EMPOWERMENT

The move to WCM was more than simply implementing manufacturing cells—the cells were only a tool in the process of achieving the company's goals. It meant a change in entire corporate philosophy. The most difficult change was to give hourly employees more control over their working environment. On the production side, workers were given the power to change floor layouts, schedule work, perform inspection and SPC, and improve the process.

Employees previously concerned with running only one machine in the traditional manufacturing environment now had to learn all operations in the cell and work in a team environment. New NC machines required the workers to learn new processes. The change was just as difficult for management to accept as it was for the workers, and engineers and managers had new responsibilities as well. When these facts were combined with the knowledge that the old methods were profitable, there was the potential for organizational resistance to such changes.

Employee empowerment also made the cell team participate in hiring, skill level certification, and promotion. This required a lot of teamwork and shared responsibilities. Coincidentally, foremen under the old methods were not guaranteed positions as cell team leaders. In essence, CVD found resistance at all levels of the organization. Workers were uneasy about the new conditions, and management did not want its powers passed to the hourly employees. Also, managers were no longer controllers. They became facilitators, operations "coordinators" for employees in the performance of their work.

COSTING ISSUES

Cellular manufacturing has also implications for the accounting and finance function. It is much easier to capture real product costs in cells. At the start of cellularization, 50 percent of costs were being allocated as overhead. Currently, this allocated figure is at 28 percent and may reach 10 percent as CVD becomes fully cellular. Less allocation means that more real product costs are being captured.

The department attempted to trace all costs to specific cells. Quality control, manufacturing, management, and shipping and receiving are still allocated. The eight cells in the original plan plus two cells for new products totals 10 planned cells. CVD allocated fixed costs by assigning one-tenth of the total fixed costs to each of the four active cells in 1992, and the remaining 60 percent to traditional manufacturing areas.

Each cell had its own profits and loss (P&L) statement. CVD's controller did include important nonfinancial measures such as inventory turns and throughput on the P&Ls. Corporate also used sales margins to evaluate product performance. Margins fluctuate, however, depending on any discounts given by the marketing department. It appeared that intangible costs, such as the cost of poor quality or failing to meet delivery schedules were not well understood. External reports are less meaningful to the cell team members. The controller utilized two sets of reports to satisfy both the needs of the cell and corporate requirements.

IMPLEMENTATION CHALLENGES AND DELAYS

In the transition from traditional to cellular manufacturing, progress has been hindered by a slowing economy. Valve orders in some segments were down considerably. In order to avoid layoffs, inventories were built in some areas. These increased inventories have prevented CVD from seeing the results originally projected.

Resistance to change had hampered progress. However, one staffer states, "Traditional rules are more of a barrier than the people themselves." Pressures to remain profitable have slowed the cellular implementation. Transition, if not done right, could cause profit to suffer. As a public company, Duriron must keep an eye on the interests of its stockholders.

CVD management stresses that the original Andersen Consulting plan was idealistic. In addition, the plan did not account for any new product additions or revisions. New products have slowed the process considerably. CVD has designed and introduced a second generation of the BTV-lined butterfly valves, with apparent success. The BX valve has been redesigned for cost competitiveness and stands ready for production. Finally, CVD is in the early stages of designing a market replacement for its aging G4 series.

The worldwide recession had great impact on capital expenditures, especially in process industries, and their suppliers such as Duriron were affected. Other Duriron divisions were heavily impacted, and corporate profits depended in large part on CVD. Therefore, cell implementation had to progress cautiously.

Bottom-line considerations led to staff reductions, and 15 percent fewer personnel were available to respond to shop floor needs. Two small voluntary layoffs of the direct workforce also took place during the moves from traditional to cell manufacturing.

HUMAN RESOURCES

WCM has had a major effect on the human resources function within the plant. A natural resistance to change played a significant role in slowing CVD's progress.

From a human relations standpoint, the cells are completely different from traditional manufacturing. The responsibility of the hourly worker increased as cell members are responsible for every facet of production. The workers must interact with each other and suppliers, as well as perform SPC (Statistical Process Control) analysis, preventive maintenance, and performance evaluation. In addition, hourly workers were involved in the design of the BTV and its cell from the beginning.

Hourly workers must bid to move to a new cell. The first cell (G4 valve) had 40 bids. The second cell announcement drew almost twice that many. But successive cells showed considerable foot-dragging as the tales of extra effort and unfamiliar tasks and equipment became more prevalent, along with tales of tighter staffing than in the traditional work settings.

Individual performance evaluations are done by the cell members' teammates. Advancement within a cell is achieved by progressing through a series of five skill blocks. A person who has achieved the final skill rating can operate every machine in the cell, use SPC tools, perform preventative maintenance, and work effectively as a team player. Staff persons sit in on skill block reviews to assure that evaluations are consistent.

Training

As the workers gained more responsibility, advanced training was necessary. Team members were able to ask for specific training in areas where they feel it is needed. In 1993, CVD trained cell members to act effectively as members of a team. A lack of resources slowed the training pace. SPC training was given plantwide. Workers in the traditional areas failed to use this training.

Union

A major human resource issue that has developed along with the WCM plan is the presence of a union interest at CVD. An election concerning representation by the Steel Workers Union was held in 1989. At that time, the complaint of union advocates was that they did not have a voice in the company.

The vote was essentially a tie. The decision was taken to the Sixth circuit Court of Appeals where the recognition election for union representation was found to be valid. In November 1992 CVD notified the union that it would not appeal and was ready to begin negotiations.

CVD management felt that the cellular manufacturing program gives employees a significant voice in determining their own future. It was generally felt that if there were to be another vote, the company would be favored.

Despite the uncertainties that arose from the disputed 1989 election results, it was decided to get on with the job of transforming the plant. Some differences did arise during team meetings. Managers tried not to let the union issue stand in the way of change. However, they do realize the issue has slowed progress toward complete cellularization.

Staff

Demands on staff personnel during the conversion to WCM have been enormous. The division has grown 60–70 percent in sales but has 15 percent fewer salaried people.

Management was forced to wear two hats in managing a plant that operates partly under cellular manufacturing techniques and partly by traditional methods. Members of the staff were also involved as cell sponsors. The sponsors were assigned to a particular cell team. They served as a resource when needed by the team. As the cell progressed past the implementation phase, the team required less and less of the staff resource.

Reaction to WCM

Initially, there was considerable disbelief in the cell concept. The methods went against traditional manufacturing practices that had been followed for years. Some of this sentiment remained within the traditional areas of the plant. There was a lack of knowledge within the traditional areas of the plant concerning cellular manufacturing.

One staff member commented that the training should be more technical. Another staff member suggested that the staff receive cross-training. There were a lot of workers in the traditional areas of the plant who are opposed to or uncertain of the change. Many workers have been with CVD since its move to Cookeville 15 years ago. Some workers see cellular manufacturing as a way for the company to get more work for the same pay.

There was concern that eventually people would be forced into cells. Human resources managers hoped that people in earlier cells could be moved into the new cells so that the holdouts could be placed in existing cells.

G4 CELL

Initially, a steering committee was formed to coordinate the efforts. The steering committee chose the $1/2''$ to $1''$ G4 valve plug valve as its first cellular product. The initial cell was to include only $1/2''$ to $3/4''$ G4 valves with $1''$ G4 valves to be added to the cell later. The small G4 valve was chosen because it was produced from a high-quality casting process with little casting variation, it required the fewest setup changes, and it was a high-volume, low-profit product. The $1/2''$ to $3''$ G4 line accounted for 65 percent of total unit volume and 33 percent of total sales dollars. The steering committee felt the small G4 valve had the highest chance of being successfully produced in a cellular environment.

The G4 project was approved in May 1989. The cell was to serve as a means to evaluate the effectiveness of the cellular concepts. As stated in the proposal, the project had the following objectives:

1. To completely manufacture the $1/2''$ to $3/4''$ G4 valve, including purchasing, scheduling, inventory control, matching, painting, assembly and shipment of the valve direct to the customer.
2. To manufacture valves based on actual customer orders, rather than forecasted inventory made for finished goods stock.
3. To create a manufacturing environment conducive to meeting the customer's needs in a global marketplace.
4. To improve overall customer service by shortening lead time and improving customer delivery time.
5. To reduce inventory levels.
6. To implement a just-in-time (JIT) manufacturing operation.

The $1/2''$ to $3/4''$ G4 cell became fully operational in April 1990. In 1991 the $1''$ G4 was added to the cell. The cell reduced its lead time from 10 weeks to

1–2 weeks. The cell is responsible for scheduling, complete manufacturing, inspection, and performance evaluation.

Performance evaluation of each cell was maintained on a bulletin board within the cell. Included on the storyboard are on-time delivery, schedule adherence, scrap rate, inventory levels, and problem notes.

Suppliers

As part of its cellular manufacturing efforts, CVD attempted to develop close relationships with its suppliers. This was done in an effort to reduce inventory levels and improve quality. Vendors were asked to hold one month of inventory. The quality of suppliers is evaluated quarterly.

Order quantities on castings were fixed for two months with changes allowed on the third month. All inventory was maintained within the cell boundary—a true focused plant.

Layout

The layout of each cell was designed to facilitate smooth and short material flow. A drawing of the original CVD layout (prior to cellular manufacturing) is in Figure 2. Figure 3 shows the extensive flow of materials for G4 valves in the traditional layout. The material handling distance was almost a mile for G4 but was reduced to about 80 feet in the 1/2″–1″ G4 cell in Figure 4. Figure 5 shows the layout and material flow inside the 1/2″–1″ G4 cell.

CELL MANUFACTURING HELPS CONCURRENT ENGINEERING OF BTV VALVES

It became obvious that CVD would have to redesign its lined butterfly valve as the division began to buy into the cellular manufacturing concepts. It should be noted that the BTV was not included in the Andersen Consulting plan.

Prior to the development of the BTV, the engineering department was somewhat removed from the rest of the organization. New design ideas under the previous method usually started with the engineering department. Some development times were in excess of five years. This traditional approach did not allow other departments to have much input into design development, nor was there much support by upper management until well into the design process. It was not unusual to have an unmanufacturable design reach the manufacturing engineering department that could not be "built-per-print." In turn, this led to some cumbersome design changes.

BTV was to be the pilot program for concurrent engineering. Concurrent engineering allowed for input from all parts of the organization at the earliest stages of development. Development times are reduced and designs come to the manufacturing floor in a form that can be produced with few design changes. The cell was designed along with the product. This tends to ensure that the best available manufacturing methods are used where possible.

The first step in the development process was team selection. The team comprised of handpicked members from sales, marketing, engineering, quality engineering,

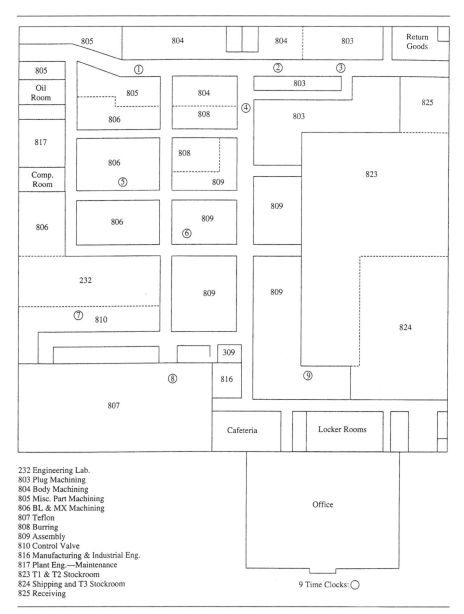

Figure 2. Traditional layout.

232 Engineering Lab.
803 Plug Machining
804 Body Machining
805 Misc. Part Machining
806 BL & MX Machining
807 Teflon
808 Burring
809 Assembly
810 Control Valve
816 Manufacturing & Industrial Eng.
817 Plant Eng.—Maintenance
823 T1 & T2 Stockroom
824 Shipping and T3 Stockroom
825 Receiving

9 Time Clocks: ○

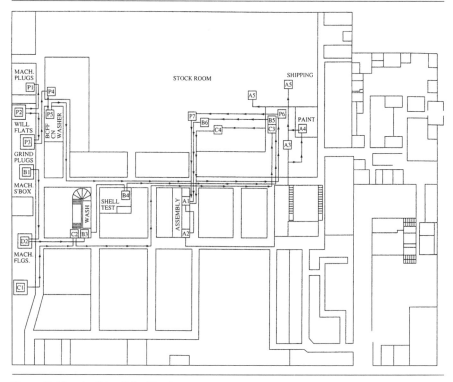

Figure 3. Flow (traditional) for G4 valves.

manufacturing engineering, as well as accounting, supervisors, and hourly personnel. The team was urged by corporate to bring the product online quickly, to meet
marketing and capital plans.

A primary goal of the design team was to design a product that facilitated manufacturability. As an example of this approach, consecutive valve sizes use identical shaft
diameters, bearings, and seals. This reduces setup time, tooling costs, and bearing and
seal inventories.

Only one piece of equipment was transferred from the traditional manufacturing
area to the BTV cell; the remainder was bought new and used. The focus was placed on
more flexible machinery. Also, much of the new equipment was considerably smaller
than those in the traditional areas, to suit just the 3″–12″ valves.

BTV-2000: THE NEW BUTTERFLY VALVE

During most of 1990, one G4 cell was operating with the BTV cell in the prototyping
stage. In late 1990 the additional cell (for the redesigned 3″ to 12″ Butterfly valve)
was created by a concurrent engineering project team made up of representative from
sales, marketing, engineering, quality, manufacturing engineering, foremen, hourly
workers, and accounting. With the goal of making the next generation (BTV-2000)

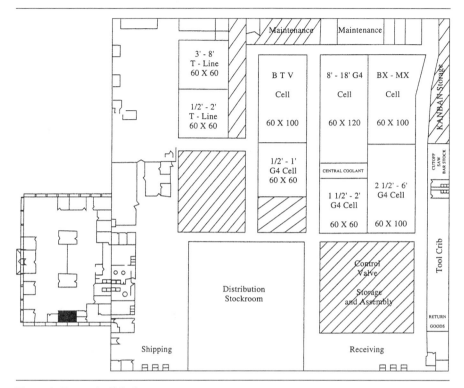

Figure 4. Proposed cellular layout.

valve, the team worked together to create the design and worked on marketing, tooling, cell layout, and inspection techniques at the same time. After production began, many team members stayed with the product for cell support.

As before, CVD used the latest manufacturing technologies, so numerically controlled machines were chosen wherever possible. To make things more challenging, the development and capital budgets were tight, and the cells were pushed toward production to generate revenues as soon as possible.

ISO 9000 AT THE PLANT

ISO was also seen by CVD as a way of gauging success and guiding continuous improvement. The division was ISO certified in December 1992.

Documentation systems for engineering required by ISO 9000 would make it much easier to review the design history of a valve.

PROGRESS

A progress report for each of the four cells was completed for a corporate Audit/Finance Committee meeting in July 1992. A summary of the progress report is in the Appendix.

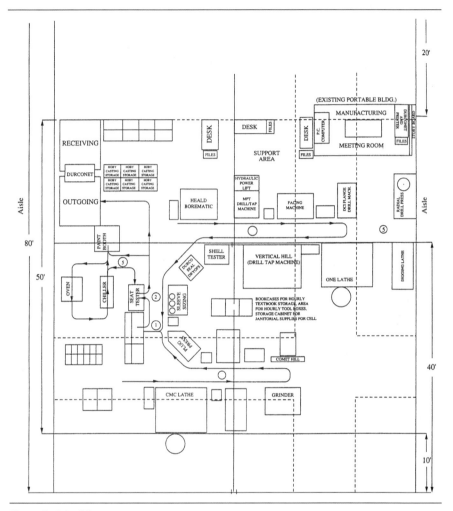

Figure 5. G4 cell layout.

Despite obstacles, CVD made considerable progress toward implementing cellular manufacturing. Five cells were fully operational while two more were near implementation by early 1993.

APPENDIX: STATUS REPORT SUMMARIES, JULY 20, 1992

Status reports for the various cells were prepared for the corporate Audit/Finance committee Meeting on July 20, 1992. The following are summaries of these reports.

G4 valve (half-inch to three-quarter-inch)

The project was approved in May of 1989 to convert the manufacture of small G4 valves from traditional to cellular techniques.

The project has spent 5.8 percent more than the goal of $295,000. On-time delivery is averaging 95 percent. Lead time have been reduced from 10 weeks to 1–2 weeks. The projected ROI has not been achieved due to unreached inventory goals.

The cell became fully operational in April 1990. As the first cell, it served as a test of the cellular manufacturing concepts. In 1991 the 1″ G4 was added to this cell. Currently, the cell contains 13 hourly employees.

BTV-2000 cell

This project sought to develop a totally new product to replace the existing BL valve, which had lost its market leadership. Capital expenditures for the project have been exactly the goal amount of $3.2 million. Sales results have been significantly higher than expected. In addition, cost of sales has been slightly lower than expected and on-time delivery is 96 percent.

T-Line (half-inch to two-inch)

This project was to change the manufacture of the small T-Line valves to a cellular operation.

At $447,000, actual capital expenditures for the project are 12 percent lower than the goal. Inventory reduction goals have not been reached. The goal was to reduce production costs by 6 percent, while actual reductions have been 7 percent. The cell is averaging 97 percent on-time delivery. Lead time has been reduced from 10 weeks to 1–2 weeks. Finally, ROI is 34.2 percent, which is 2.3 percent higher than the goal.

The T-Line cell became fully operational in November 1991, and currently has a total of 15 employees.

One-inch G4 valve (added to existing G4 cell)

This project sought to add the 1″ G4 to the existing small G4 cell. Capital expenditures have been 48 percent less than expected. Inventory reduction goals have been reached. Cost reduction audit results are inconclusive. ROI, on-time delivery, and current status are the same as the 1/2″–3/4″ G4 valves.

Atomac cell

This project sought to transfer the manufacturing and technology of lined ball valves from Atomac in Germany. All 1″–3″ lined ball valves to be sold in the United States were to be produced at CVD.

Capital expenditures have been slightly less than expected at $1.6 million. Sales are well below goal. Cost of sales is 6 percent lower than goal. ROI figures are well below goal due to a low level of incoming business. On-time delivery is currently at 100 percent. Currently the cell uses 17 hourly and salaried employees.

Related Articles: Just-in-Time Manufacturing; Lean Manufacturing Implementation.

19. PREDICTIVE MAINTENANCE: THE CASE OF DELLA STEAM PLANT[1]

P.K. RAJU

Department of Mechanical Engineering
Auburn University, AL 36849

CHETAN S. SANKAR

Department of Management
Auburn University, AL 36849

POWER GENERATION AT THE DELLA STEAM PLANT

Figure 1. The Della steam plant.

[1] See glossary of terms for description of technical terms. This case study is based on work from field study and was prepared for class discussion rather than to illustrate either an effective or ineffective handling of maintenance strategies in this company. Names of individuals and companies have been changed. This case study is based upon work partially supported by the Thomas Walter Center for Technology Management and the Division of Undergraduate Education, National Science Foundation under Grant No. 9752353. Any opinions, findings, and conclusions or recommendations expressed in this material are those of the author(s) and do not necessarily reflect the views of the National Science Foundation. Abridged by Paul M. Swamidass, Auburn University. For the unabridged version, contact Taveneer Publishing Company, Anderson, SC 29625, phone 864-287-2559.

Figure 2. Turbine generator blades.

Della Steam Plant is one of the power plants owned by a major company in the southern part of the USA. Located between a free-flowing river and nearby coal mines, Della steam plant started generating electric power in 1917. The plant has five units in operation. It is an efficiently operated steam power plant with a capacity of about 1,000 Megawatts.

The steam power plant burned coal that was available from nearby mines and from the barges that pulled up next to the plant. Trucks and rail were also used to bring coal from other mines. Water from the nearby river was fed into the boiler and was converted to steam by burning coal. The high-pressure steam was fed into a turbine, struck the blades of the turbine and turned them. The turning of the blades rotated the generator as both of them were mounted on a shaft using sets of journal bearings. The rotation of the generator produced electric power.

A typical turbine generator unit in this plant was as heavy as 120,000 pounds and was over 100 feet long. A partial view of an unit is shown as Chart. This unit was so tall that it had to be installed on a four-story bay. Eight sets of bearings were used to support the turbine-generator unit. When the turbine-generator was running, a film of oil separated the running shaft and the stationery bearing housing. The clearance between the two was measured in thousandths of an inch, called mils. Since the shaft was rotating and the unit was very heavy, any contact between the shaft and bearing housing could lead to catastrophic failure of the shaft or the bearing.

IMPORTANCE OF THE TURBINE GENERATOR UNITS IN POWER GENERATION

In power generation, the power produced was consumed immediately. There was no ability to store power and supply it later. If the power plant lost a unit, power must be supplied from another source to meet the demand. Many of the units at the Della Steam Plant were base load units that ran all the time because they were cheap to run. When base load units failed, the company obtained power from peaking units

Figure 3. Turbine generator.

that were much more expensive to run. The electricity was supplied to customers but it cost the company several hundred thousand dollars more per day because the peaking units were less efficient. Therefore, the strategy to make more profits was to run cheaper units continuously without any outages.

COMPETITIVE PRESSURES FACING THIS INDUSTRY

This electric utility company was undergoing a period of great change driven by the global economy, increased competition, and fast-changing technology. The top management of the company expected that this industry would be deregulated. They did not expect to be guaranteed a certain rate of return on investments like they had received in the past. There was pressure on this company to reduce operations and management costs. The President of this company stated:

No one really knows what the industry will look like and who its players will be in the next century. To be competitive in the future, we've got to reduce the real price of the company's electricity at least 10 percent. All of us would be asked to do this without reducing or sacrificing customer service or investor value. Therefore, freezing the operations and management budget and reducing the capital budget is just one more step in the effort to identify what the company's targets have to be for success beyond year 2000.

The company's strategic leadership council had announced that operating and maintenance (O&M) budget forecasted for the year 2000 must be reduced by $50 million. To do this, the company was freezing its O&M budget at the 1995 level through 2000. In addition, projected capital expenditures were expected to be reduced by $250 million between 1995 and 2000. The deregulation of the utility industry had forced this company to find innovative tools and processes to improve productivity and extend the useful life of existing facilities.

PREVENTIVE AND PREDICTIVE MAINTENANCE
OF UNITS AT DELLA STEAM PLANT

Figure 4. Maintenance practices at Della steam plant.

The company had traditionally serviced its turbine-generator units based on fixed time intervals, called preventive maintenance. A major preventive maintenance outage was scheduled every five years, whereas, minor maintenance outages were scheduled once a year. During a major outage, the unit was torn apart, parts were checked, faulty parts were replaced, and the unit was put back together. This outage normally took about two months depending on the nature of the work and was scheduled during Spring or Fall seasons. During a minor outage, any problems identified during the past would be fixed and filters would be changed. This would take between a few days to a week. Corrective maintenance was performed when a unit failed unexpectedly and had to be fixed.

This company created a centralized predictive maintenance group to further improve the maintenance operations during 1990. Predictive maintenance was the scheduling and maintenance of equipment based upon the operating conditions of the unit. Just as routine dental checkups were usually less painful and less costly than a root canal, the same reasoning suggested that it was more desirable to monitor the health of plant equipment, and act accordingly (predictive maintenance), than it was to repair or replace a component that had catastrophically failed (corrective maintenance), or to perform unnecessary maintenance on healthy equipment (preventive maintenance). Experience, in several electric utilities and in other industries, had shown that predictive maintenance measures were usually more cost-effective than time-based preventive maintenance or corrective maintenance.

Figure 5. Oil analysis.   **Figure 6.** Infrared thermography. **Figure 7.** Motor current analysis.

But predictive maintenance was a new industry practice and the benefits were still being questioned by the engineers in the plants. It required implementation of new measurement technologies and highly skilled personnel to collect and interpret data from the measuring systems. Some of the most extensively used technologies in predictive maintenance were vibration monitoring, oil analysis, infrared thermography, and motor current analysis. Vibration analysis was an important technique used in predictive maintenance since there was a direct correlation between vibration signatures and specific machinery defects. Many machinery problems such as misalignment, imbalance, bearing wear, gear problems, cavitation, soft foot, structural defects, electrical defects, and resonances could be detected using vibration signature analysis.

Predictive maintenance also needed skilled personnel who were able to interpret the results from the Fast Fourier Transform (FFT) analysis produced by proximity probes attached to the unit. This analysis traced a vibration level to its precise cause and related it to frequency measured in hertz. A problem on the machinery produced signals that combined into a composite waveform; it was this composite that was detected and displayed by instruments such as a shaft rider probe. An FFT spectrum analyzer divided the composite signal produced by a proximity probe into its components and presented the operator with the precise stage of damage. With this data, the engineer could predict the remaining service life of a damaged component and plan shutdowns to the best advantage.

Six years back, the plant had a vibration analysis program, but with no direction or equipment to perform predictive maintenance studies. There were a group of engineers who were taking the vibration readings but no one was responsible for the program. They wrote a report, filed it, and forgot about it. Steve was then assigned responsibility for the program and he convinced the management to buy proximity probes and computer software to conduct predictive maintenance studies. He was active in applying predictive maintenance techniques in the plant.

VIBRATION OF TURBINE GENERATOR UNIT #9 AT 7:56 A.M.

The turbine-generator unit #9 in the Della plant was a base load unit that was expected to be operational as long as possible since it was cheap to produce power using this unit. During the months of March–April 1995, this unit was taken out of service for major preventive maintenance. The maintenance personnel inspected all the tubes inside the boiler, replaced turbine blades, and lubricated bearings. Lucy Stone, the engineer representing the manufacturer of the unit (RLS), was in charge of the tearing down and putting together of this unit. At the end of the maintenance operations, Steve Potts, the plant engineer, balanced the shaft of the unit.

In order to measure the vibration levels of the turbine-generator unit while it was in operation, Lucy's company had installed single shaft-rider probes on the shaft of the unit at each bearing. Readings from these probes were transmitted to the control room. These probes were common on older generator units and measured the absolute displacement measurement of the shaft. A screen in the control room was dedicated to showing the measurements from these probes on a continuous basis. The software

Figure 8. Eddy-current proximity probes.

Figure 9. Installation of shaft-rider probes.   **Figure 10.** Monitoring devices in the control room.

also stored the results for the past maintenance periods and she could recall the charts at any time.

Steve, in his enthusiasm to adopt predictive maintenance practices, had placed two eddy-current proximity probes (one horizontal and one vertical) at each bearing on the turbine generator during this maintenance period. These probes were small in size, cost less, and measured the relative shaft displacement. The vibration data from these probes were recorded on an instrument similar to a digital tape recorder. The recorder stored data when the turbine-generator unit was either started or stopped. The information from unit #9 was fed into a laptop computer. A specialized software package compiled the information and produced plots of time, machine speed, vibration amplitude, and vibration frequency. Steve expected these plots to be used in evaluating the operating condition of the unit. This evaluation could determine whether the unit would operate reliably until the next scheduled preventive maintenance outage or whether maintenance was required earlier.

After all the maintenance operations were completed, the unit was started on April 21st at 3:00 a.m. The unit was brought up to 30 megawatts, about 10% of the load.

Chart 1. Readings from the shaft rider probe compared with readings from startups in the past.

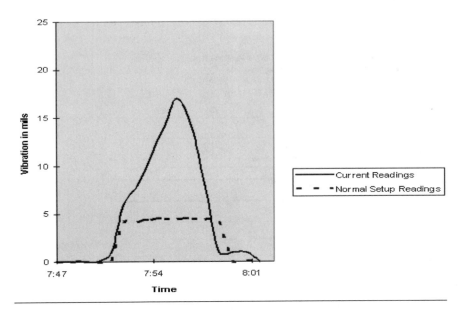

Shaft Rider Probe Readings

Steve and Lucy were pleased that their maintenance work was done on schedule and the unit became operational. Steve left at 6 a.m. for home.

About 7 a.m., the RLS engineer, Lucy Stone, checked the overspeed trip mechanism attached to the turbine generator. This was a mechanical spring-loaded device that was located on the shaft of the turbine itself. Its purpose was to stop the unit when the speed exceeded 3,900 rpm for overspeed protection. The normal operating speed of the turbine generator was 3,600 rpm. As the unit was brought above its overspeed limit, at 7:56 a.m., it started vibrating and the building started to shake. The employees around the unit thought it was beginning to come apart and they started moving away. Fortunately, the unit tripped and coasted down to a stop in the next few minutes. Sam Towers, the plant manager, was immediately informed of the problem and he rushed to the spot. He did not see any visible damage to the plant or the unit and wondered what he should do next.

SHUT DOWN UNIT FOR A WEEK AND TROUBLE-SHOOT PROBLEM?

Lucy Stone, the RLS engineer, checked the vibration levels from the shaft-rider probes her company had originally installed on the turbine generator unit. She had supervised the tear down and the rebuilding of the unit herself. Her ten years of experience had made her double-check everything. She checked the screen that showed the overall vibration level of this unit at 7:50 a.m. and compared it with the last time this unit was started and tested for overspeed without any accompanying major vibration (Chart 1). The comparison showed that the vibration level was 17 mil during the current problem

Chart 2. Readings from vertical proximity probe time versus displacement (bottom), time versus phase angle (top).

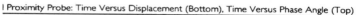

I Proximity Probe: Time Versus Displacement (Bottom), Time Versus Phase Angle (Top)

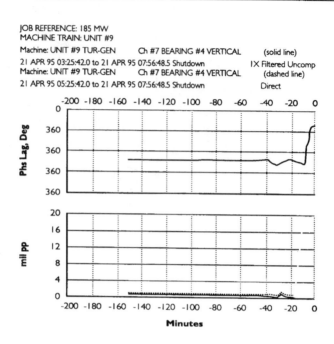

JOB REFERENCE: 185 MW
MACHINE TRAIN: UNIT #9
Machine: UNIT #9 TUR-GEN   Ch #7 BEARING #4 VERTICAL   (solid line)
21 APR 95 03:25:42.0 to 21 APR 95 07:56:48.5 Shutdown   IX Filtered Uncomp
Machine: UNIT #9 TUR-GEN   Ch #7 BEARING #4 VERTICAL   (dashed line)
21 APR 95 05:25:42.0 to 21 APR 95 07:56:48.5 Shutdown   Direct

compared to 4 mil vibration level when there were no problems. The current vibration level was very close to the 22 mil clearance between the shaft and the bearing.

She reasoned that if the unit was restarted, the vibration might become severe, and the shaft might hit the bearing and make the whole unit come apart. Already some of the parts within the unit might have been broken due to the heavy vibration. From her past experience, she knew that similar units had failed at least 30 times due to a fault in retainer rings. She suspected a similar problem in this unit and did not want the unit to be restarted until the retainer ring and other parts were rechecked. Any mistake in her recommendation might alter the credibility of RLS Inc. with the power plant management. Therefore, she decided to recommend to the plant manager that the generator be torn down and all the parts checked.

Bob Make, the day shift assistant engineer, used the data collected by the proximity probes installed by Steve to double-check Lucy's interpretation. He printed the chart that showed the vibration levels between 5:25 and 7:56 a.m. (Chart 2). This chart

Table 1 Financial implications of recommendation 1, stop unit and fix problems

(1) Cost of operating this unit per day	:	$100,000
(2) Cost of operating a peak unit per day	:	$200,000
(3) Excess cost of operating a peak unit instead of unit #9 per day = (Row 2 − Row 1)	:	$100,000
(4) Number of days unit #9 will be out of operation	:	7 days
(5) Excess cost of operating a peak unit for 7 days = row 3 * row 4	:	$700,000
(6) Labor and material cost to fix unit #9	:	$200,000
(7) Total cost if this recommendation is followed = row 5 + row 6	:	$900,000

depicted the time in minutes versus the vibration amplitude in mils in the bottom chart for the vertical probe. The top chart showed the time versus the phase lag in degrees. Bob noticed that the vibration amplitude was very near 16 mil at − 3 minutes (i.e., 7:53 a.m.). Also, looking at the legend of the chart, he was unsure whether the dashed line represented the 1X filtered reading or a direct overall reading. Given the difficulty in interpreting this chart and the high vibration amplitude registered at shutdown, Bob agreed with Lucy's decision that the turbine should be torn down and the problem should be identified.

Based on Lucy's and Bob's recommendation, Sam Towers decided that the unit should be taken apart, all parts rechecked, and the balancing redone. He was aware that it might take more than a week to perform this maintenance activity and it would cost the company $900,000 since peaking units had to supply the power (Table 1).

RESTART THE TURBINE THE SAME DAY?

Steve Potts, the plant engineer, came to know of Lucy's recommendation at 1:00 p.m. when he called Bob, the day-shift engineer, to find about the status of the unit. Bob told him that the generator was being taken apart due to RLS's recommendation. At 2:00 p.m., Steve arrived in the plant and started checking the vibration data.

Steve, based on his ten years experience in the field, felt that Bob's attempt to identify the problem from Chart 2 was not correct since it did not provide sufficient details. He printed details of what happened at 7:50 a.m. using frequency and speed plots (Charts 3 and 4) for the vertical probes on bearing #4 of the generator. The bottom part of chart 3 represented a plot showing the vibration amplitude in mils against the running speed of the time in rpm (rotations per minute). The top part of the chart represented a plot showing the variation of the phase lag in degrees with the speed of the turbine in rpm. The dashed line showed the overall vibration of the unit measuring the average amplitudes of vibration at different running speeds whereas the solid line indicated the filtered reading at the running speed. The chart showed that the overall vibration level was 17 to 18 mil and the filtered reading at running speed was less than 3 mil. The charts from the vertical and horizontal probes on other bearings showed similar differences between the overall and filtered running speed vibration levels.

Chart 4 showed the frequency in kilocycles per minute (horizontal axis), versus the machine speed in rpm (left vertical axis), versus amplitude of vibration in mils (right vertical axis) for bearing #4 as measured by the vertical probe. This was a three-dimensional waterfall type chart that indicated the amplitude of vibration (mils)

Chart 3. Readings from vertical proximity probe speed versus displacement (bottom), speed versus phase angle (top).

Vertical Proximity Probe: Speed Versus Displacement (Bottom), Speed Versus Phase Angle (Top)

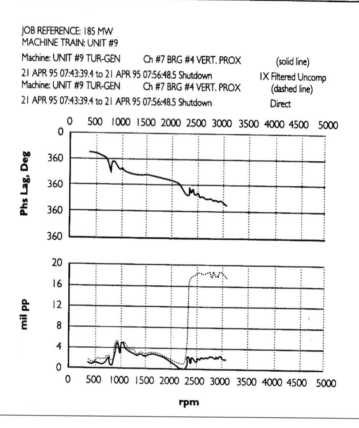

JOB REFERENCE: 185 MW
MACHINE TRAIN: UNIT #9
Machine: UNIT #9 TUR-GEN   Ch #7 BRG #4 VERT. PROX   (solid line)
21 APR 95 07:43:39.4 to 21 APR 95 07:56:48.5 Shutdown   1X Filtered Uncomp
Machine: UNIT #9 TUR-GEN   Ch #7 BRG #4 VERT. PROX   (dashed line)
21 APR 95 07:43:39.4 to 21 APR 95 07:56:48.5 Shutdown   Direct

versus frequency (kcpm) at different running speeds of the turbine (rpm). The 1X, 2X, and 3X lines in the chart showed the amplitude of vibration of the turbine versus frequency at one times, two times, and three times the running speed of the turbine. After studying Charts 3 and 4, Steve said:

When I started looking at this plot, I realized that oil whip took place and the vibration level at the running speed was very low. If the shaft or the bearing tore apart, then the vibration level should have been very high at the 1X running speed. But, at this running speed, it can be seen from chart 3, that the vibration level went up to a maximum of only 3 mils at 2600 rpm. The high vibration level that is seen in Chart 4 is toward the leftmost and occurs at a frequency of 1 kilocycle per minute which, I think, is due to oil whip.

When a turbine is started, the oil in the bearing starts whipping producing high levels of vibration. The whipping normally comes down once the oil is heated up over a period of time.

Chart 4. Readings from vertical proximity probe frequency versus displacement versus machine speed.

Vertical Proximity Probe: Frequency Versus Displacement Versus Machine Speed

JOB REFERENCE: 185 MW
MACHINE TRAIN: UNIT #9

Machine: UNIT #9 TUR-GEN   Ch #7 BRG #4 VERT. PROX
21 APR 95 07:43:34.0 to 21 APR 95 07:55:53.2 Shutdown   UNCOMP

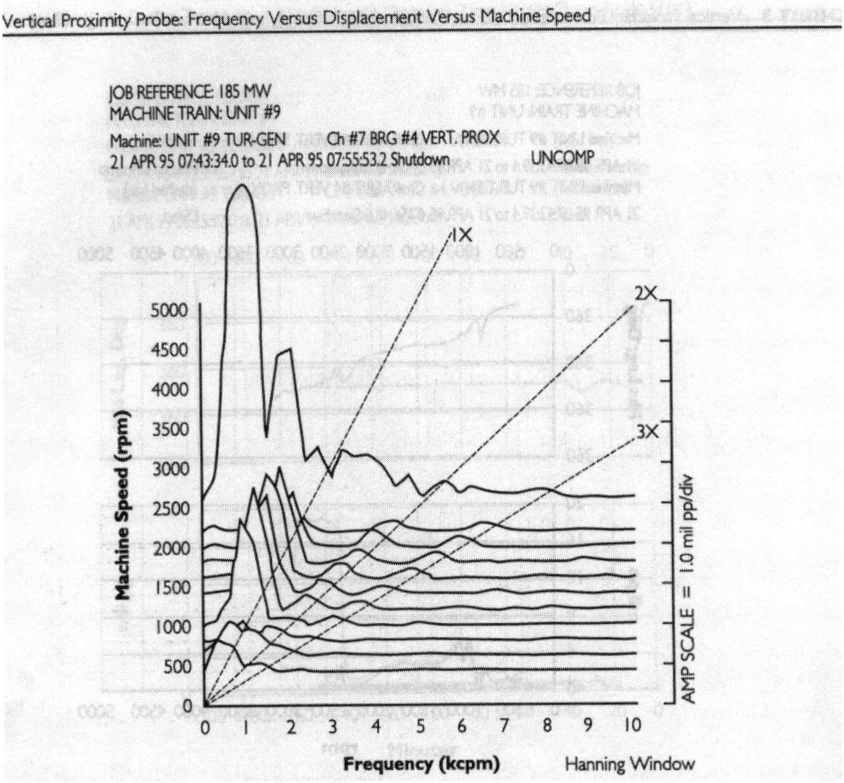

The turbine ran for only five hours, which is not sufficient time for the oil to heat up. At least 24 hours of run time are needed so that the oil and the other parts could heat up. When the turbine is aligned in a cold condition, the bearings are set at different heights so that thermal expansion causes them to align themselves in a straight line. The bearings were not up to proper temperature at the time the RLS engineer wanted to over speed it. Given the measurements shown in Charts 3 and 4 and the lack of sufficient time to heat the oil, the high vibration level in this case is due to oil whip.

Based on his reasoning, Steve recommended to Sam that the unit be restarted immediately. He did not expect the oil whip problem to re-occur if proper warm-up procedure was followed.

SHOULD THE UNIT BE SHUT DOWN FOR A WEEK OR RESTARTED IMMEDIATELY?

Sam Towers called a meeting of the plant engineers and maintenance personnel. Lucy Stone, the RLS engineer, made a strong case for tearing down the turbine generator

Table 2 Financial implications of recommendation 2: Restart the unit immediately

Alternative (a): Unit worked well; No failure		
(1) Excess cost of operating a peak unit per day instead of unit #9	:	$100,000
(2) Number of days for which the unit failed	:	0 days
(3) Total cost if this recommendation is followed row 1 ∗ row 2	:	$0.00
Alternative (b): Unit failed, new unit ordered and installed		
(1) Cost of operating this unit per day	:	$100,000
(2) Cost of operating a peak unit per day	:	$200,000
(3) Excess cost of operating a peak unit instead of unit #9 per day = row 2 − row 1	:	$100,000
(4) Number of days unit #9 will be out of operation	:	180 days
(5) Excess cost of operating a peak unit for 180 days = row 3 ∗ row 4	:	$18 million
(6) Cost to buy a new unit	:	$1 million
(7) Labor and material cost to install unit #9	:	$0.5 million
(8) Total cost if this recommendation is followed = row 5 + row 6 + row 7	:	$19.5 million

unit, trouble-shooting the problem, replacing any defective parts, and then restarting the unit. Steve Potts, the maintenance engineer at the plant, made an equally strong case for restarting the unit immediately. Lucy Stone responded by stressing the safety issues:

Steve, if we restart the unit and the parts start flying out, then we might be damaging not only the unit, but also other parts of the plant. The retainer rings as well as other parts might have been damaged already and it is important to replace them. A new unit costs approximately $1 million and we do not have any in stock. It will take us at least six months to a year to replace this unit, if it breaks.

Sam, if the unit breaks, the revenue loss for your company will be extremely high since this unit generates power worth $200,000 per day. Your company might be losing as much as $19.5 million if the unit failed (Table 2). Given that it would cost $900,000 to follow my recommendation, you have to trust Steve's recommendation is correct with a probability of more than 95%. I suggest that you do not take such a high risk. It is important to tear down the unit and have it thoroughly checked once again.

Sam Towers was concerned about possible failure of the turbine generator unit if it was restarted immediately. He knew that Steve and Lucy had equal years of experience in maintenance and both were concerned about the safety and cost effectiveness of their recommendations. He asked Steve to clarify his recommendation. Steve was confident about his recommendation and stated:

I am fully confident about the readings from the proximity probes and the interpretation of the data, even though they have been installed for the first time as part of the predictive maintenance trial. The shaft-rider probe installed by RLS Inc., gives only the overall vibration level of the unit, whereas, the proximity probes provide specific frequency information. I have used this information to pinpoint that the problem is due to oil whip.

As far as the probability Lucy mentioned we have just gone through preventive maintenance and there were no indications of any damage to the unit. Even if something unforeseen were to happen during restart, the unit has an over-speed trip mechanism that will automatically shut the unit off thereby preventing further damage to the unit. Therefore, we do not have to worry about safety of people. We should not waste money and time fixing what is not broken.

Sam was not thoroughly familiar with the measurement technologies that generated the charts used by Lucy and Steve and requested them to generate more charts so that he can better judge the consequences of deciding to go with either recommendation.

Sam Towers summarized the recommendations made by Lucy and Steve and stated the issues he had to consider before making the final decision:

Lucy, if I agree with your recommendation, it is a less risky situation for our companies, although it would cost us $900,000 to have the unit shut down for a week. In the past, Steve and you had agreed on the recommendations. This is the first time you both have conflicting recommendations. Are you suggesting we spend $900,000 even if there was no possible damage? What is the utility of deploying predictive maintenance equipment in our units if we are going to play it safe each time a similar problem occurs in the future?

Steve, if I agree with your recommendation, it would provide a boost to our plant's maintenance record. We would show to top management that the investments on the predictive maintenance practices are worthwhile. We would not have wasted any costs associated with this shut-down. If your recommendation works, we will be highlighted in our company publications as innovators. My major concern is that, if your recommendation does not work, and the unit breaks during restart, it would cost us as much as $19.5 million for a new unit to be put back in operation. This estimate does not include the costs of any litigation if employees are injured during the restart. We will be risking our jobs if we go with your recommendation. How confident are you about the problem being oil whip given that this is the first time you had installed the measurement devices on the turbine-generator unit?

EPILOGUE

Sam Towers, eventually, decided in favor of restarting the unit immediately as recommended by Steve. (Readers: Do you agree with him?)

IX. PRODUCT DESIGN
AND DEVELOPMENT REDEFINED

20. PRODUCT DESIGN FOR GLOBAL MARKETS

K. RAVI KUMAR
University of Southern California, Los Angeles CA, USA

GEORGE C. HADJINICOLA
University of Cyprus, Nicosia, Cyprus

ABSTRACT

The last two decades saw a phenomenal increase in the globalization of business. More manufacturers now have a market or manufacturing presence outside their native country. Implications of this product strategy are many. One of the most important competitive consideration for manufacturing firms in this economy is the design of their product for different markets.

In this article the authors discuss the efforts of the following manufacturers in designing their products for global markets: **Bausch & Lomb; Boeing; Coca-Cola; Colgate; Fuji-Xerox; Heineken; Hewlett-Packard; Honda; Ikea; Lego; Levi; Marriott; Mars; McDonalds; Motorola; Pizza Hut; Polaroid; Subway; Tyco.**

INTRODUCTION

Contemporary business firms are facing increased globalization of markets. Companies that have traditionally dominated their domestic markets, recognize that they cannot survive unless they establish their presence in all major markets. Various factors, termed globalization drivers, are forcing companies to compete in the global market place. Market factors include the consumers with similar income levels, spending habits, educational background, and life-style. Cost factors, stemming from the rising cost of developing a new product, force firms to enter major markets so as to achieve production and distribution economies of scale. Environmental factors contributing to globalization include the removal of nationalistic barriers and protectionism, the establishment of the General Agreement of Tariffs and Trade (GATT) and World Trade Organization (WTO) that promote free trade, the formation of economic communities such as the European Union and the North American Free Trade Agreement. Competitive factors make the need for a global market an imperative since firms need to disseminate their new product ideas worldwide before competitors mimic their products in other markets.

Two of the strategic decisions that International Enterprises (IEs) operating in the global market have to make are the type of product policy to adopt, and the management of the design process for the global marketplace. Product policy encompasses a number of issues such as physical product design, packaging, labeling, brand, warranty, and after sales service. Research on product policies in a global environment originated in the early sixties, when the issue was raised as to whether IEs should provide customers around the world the same standardized product or a country-tailored product.

PRODUCT POLICIES IN A GLOBAL ENVIRONMENT

Levitt (1983) noted that advances in technology and more specifically, improvements in communication, transport, and travel, are making consumers around the world to be aware of, and demand the same set of products. While facing this phenomenon termed "globalization of markets," Levitt states that "companies must learn to operate as if the world were one large market—ignoring superficial regional and national differences," and as a result, IEs should "sell the same things in the same way everywhere." Offering a standard product design worldwide embraces the belief that homogeneous markets exist across countries. This belief has been challenged by the traditional paradigm that consumers across countries have different needs and therefore, demand different sets of products (Boddewyn et al., 1986; Douglas and Wind, 1987). The two viewpoints of offering a standardized product versus offering a custom-tailored product seem to hold their own. Both sides can report examples of companies who have succeeded or failed when they adopted either viewpoint. Such examples can be found in Levitt (1983) and Kashani (1989). Walters (1986) provides an extensive discussion on the product customization versus standardization debate.

In general, the issue of product policies across countries has become a debate of polar opposites, namely complete uniformity versus full localization of the product policy. Some researchers, though, have pointed out that extreme points of view should not be the focus of the debate. The effort should be directed in finding ways to adapt the global marketing concept to suit the company's needs and provide it with a sustainable competitive advantage (Takeuchi and Porter, 1986; Quelch and Hoff, 1986). Product uniformity is simply one strategy in an array of strategic choices that an IE may pursue. The crucial point is to find the product policy that suits the needs of the organization in the environment in which it operates.

Which types of product policy is appropriate? Takeuchi and Porter (1986) identified three basic types of international product policies: *the universal, the country-tailored,* and *the modified.* Two of the above three product policies, the universal and the country-tailored product policies, represent the polar positions in the product standardization versus customization debate. The third policy, modified product, deals with products that are relatively similar in overseas markets, but minor adaptations are made to these products to meet basic market needs or customer idiosyncrasies. We provide below a more detailed discussion of the three international product policies.

Universal product policy

A universal product, also referred to as standardized product or uniform product, is physically identical in all overseas markets in which it is sold, with the exception of such elements as labeling and the language used in the manuals. Studies by Sorenson and Wiechmann (1975), and Hill and Still (1984) provide evidence that product policy is an area where IEs have the highest propensity for standardization. Products such as basic materials, components, high technology products, industrial products, steel, chemicals, plastics, ceramic castings used in memory chips, aircraft turbine engines, and cameras are often universal products. Jain (1989) states that industrial and high technology products are more suitable for standardization than consumer products.

The primary benefits derived from product standardization are cost savings. There are five main reasons for to these cost savings. First, offering a standardized product across countries implies higher production volumes that enable the IE to reduce the unit cost due to economies of scale. Fixed costs such as wages, insurance costs on the plant and equipment, property taxes, and interest costs are distributed over a larger number of products causing the unit cost to decrease. Such economies of scale are referred to as short-term economies of scale.

Second, cost savings can be derived from the design of a single product instead of a number of products. Third, the production of a standardized product enables the IE to increase the volume of the purchased raw materials and components. This assists the IE to increase its bargaining power over its suppliers, and eventually enjoy better purchasing contracts. Fourth, the production of a standardized product may allow the IE to centralize the production of the product to a few specialized facilities. The construction of fewer but bigger facilities results in the cost savings. Lastly, promoting a standardized product across countries results in economies of scale in marketing. Advertising the standardized product may require the same type of campaign in each country so as to maintain a common global image of the product. The commonality in the advertising campaigns results in cost savings. In general, savings from standardization allow the company to position itself as a low-cost producer.

The higher production volume of the standardized product, as discussed above, leads to low production costs that eventually ensure lower prices for customers. Nevertheless, this benefit may come at the risk that the standardized product may not meet the needs of customers in a specific country. The universal product approach can result in overstandardization of a product, and may prove to be catastrophic for some companies. Kashani (1989) describes the negative experiences of overstandardization in the Danish toy-maker Lego. Lego was marketing around the world its educational toys in the same fashion. In the United States, Lego faced fierce competition after Tyco, its major competitor, started marketing its toys in plastic buckets that could store the toys. Lego's universal approach to packaging was the elegant transparent boxes, which was not appealing to American parents, who preferred the functional packaging approach of toys in buckets. Lego's eroding market position alarmed the U.S. management but headquarters refused the idea of introducing toys in buckets, claiming that this could harm Lego's reputation for high quality. Nevertheless, massive losses forced Lego to design its own bucket for storing toys. The result reversed its eroding market.

Walters and Toyne (1989) describe three methods that an IE can follow to develop standardized products. The first method is *product extension* under which the product developed for a single market, usually the home market of the IE, is sold to foreign countries. Companies that adopt this method have an ethnocentric orientation such as companies that are at the first stages of their internationalization and have started exporting. Textile and fitted carpet manufacturers are examples of such companies. Fuji-Xerox designed and targeted its high performance plain paper copiers exclusively for the Japanese market. Subsequent expansion to foreign markets was simply limited to exporting the copiers designed for the Japanese market. The probability of success increases when the foreign markets are "similar" to the home market of the enterprise.

The second method is referred to as *premium prototype*. The universal product developed for this method meets the needs of the most demanding customers whether the group is overseas or in the home market. The product is also designed to withstand the most severe conditions of use. Such a high quality product can be appealing to customers in foreign markets, but it may often include superfluous design features that are not necessary for the average foreign customer. The high quality design and, perhaps, high production cost due to the enhanced nature of the product may result in high prices and discourage same potential buyers.

The third method, known as *global common denominators*, strives to identify a *global segment* of international customers with homogeneous preferences and conditions of product use. After identifying such a segment, a product is developed to meet the common needs of customers in that segment. The new hotel chain Courtyard by Marriott introduced by Marriott is an example of the effort of a company to satisfy a global segment. Marriott tries to identify the attributes that travelers value most in a hotel in designing the hotel.

Country-tailored product policy

A country-tailored product is designed and produced to meet the idiosyncratic needs of the country (or group of countries) for which it is intended for. Czinkota and Ronkainen (1993) present various factors that encourage the design of products that are exclusively targeted for certain countries. Such factors include: (1) different conditions of use depicted by different climatic conditions; (2) government regulations that have to be met in each country, (3) different consumer behavior patterns; and (4) an attempt to meet local competition when competing firms customize their products to the local market. The primary benefit obtained from designing products specifically for certain countries or regions is that a customized product more fully meets the needs of local consumers. This leads to higher demand, higher market share, and eventually higher profitability.

The case of Fuji-Xerox and its high performance plain paper copiers is an example of a product designed exclusively for the Japanese market. Bausch & Lomb developed 25 new products in 1986 with only one custom-tailored for foreign markets. Since then, Baush & Lomb has changed its strategy. In 1991 more than half of its products were developed for foreign markets. In Europe, Bausch & Lomb's Ray-Ban glasses are made flashier, more avantgarde, and are costlier, in order to accommodate the high fashioned European consumer. Ray-Ban glasses in Asia have been redesigned to better suit the Asian face with its flatter bridge and higher cheekbones (Fortune Magazine, 1992: 76).

Consider the case of Mars producing its traditional chocolate bar in Europe and the United States. The two chocolate bars, even though they have the same brand name, have different tastes, texture, and packaging in order to meet the different tastes of consumers. Colgate toothpaste offers a more spicy toothpaste, which has been exclusively designed for the Middle East market. In fact, many consumer products, especially food and household products, are country-tailored products because of their strong culture dependency.

Designing a customized product that meets the consumer preferences in a specific region has a drawback, the cost of the making product may be pushed higher. The higher cost is caused primarily by two reasons. First, the design of localized products significantly affects the new product identification and development costs. Costs increase because market research has to be conducted in each country or region and in addition, more funds must be allocated to design more than one product. As a result, the customized product designed for a specific country has to absorb relevant developmental costs. If the product was to be marketed in several countries, this cost would have been distributed over more units demanded by the various countries. Second, since each customized product is sold in one country, the smaller production volumes may prevent the company from exploiting economies of scale.

Another drawback of the country-tailored product policy is that the logistic operations for distributing the customized products may be inefficient. Take for example the experience of Hewlett-Packard and its deskjet printers. Initially, Hewlett-Packard was producing and packaging printers for the United States and European countries. Customization involved the installation of the appropriate power supply, software, and manuals in the specific language. Inaccurate sales predictions left the distribution centers with inventories of printers meant for some countries while suffering shortages in some others. The distribution centers were forced to rework the packages and printers to customize products for the countries where demand was not satisfied.

Modified product policy

A modified product is substantially similar in each country in which it is sold but minor alterations/adaptations are made on the basic product to conform to local needs and regulations. Modified products are considered products whose adaptations represent a "modest" percentage of the total cost (Takeuchi and Porter, 1986). The modified product policy enables an IE to reap the benefits of standardization and centralized production, and at the same time, respond to local needs. Modified products include industrial products such as CT scanners, broadcast video cameras, mainframe computers, copiers, precision testing equipment, and construction equipment. In consumer goods, modified products include cars, motorcycles, calculators, and microwave ovens. One form of modification of products is product adaptations. Product adaptations are classified as *involuntary* or *voluntary*.

Involuntary adaptations

Involuntary adaptations are made to meet legal requirements of the foreign government on packaging, labeling, metric system, and also local climatic and economic conditions. Examples include the use of different paper trays for copiers in Japan, the United States, and Europe since all three regions use papers of different sizes. Similarly, car manufacturers must modify the speedometer on cars sold in the United States to display in miles per hour where as car speedometers in Europe provide the reading in kilometers per hour. Manufacturers of electronic appliances must install the proper power supply module in order to comply with the country's voltage requirements.

Voluntary adaptations

Voluntary adaptations are those made as part of the company's marketing strategy to meet local needs and consumer idiosyncrasies such as color, the product's size, and its accessories. For example, the color of the package, especially for consumer products, is very important since different colors are perceived differently across countries. Packages in the Middle East tend to include the green color that is preferred by consumers in this region.

McDonalds, besides the standard products that it serves, tries to adapt to local needs by selling wine in France, vodka in Russia, and beer in Germany. Besides the world famous sandwiches that it serves, Subway prepares sandwiches that contain cheeses and ham made in the country of operation. Similarly, Pizza Hut prepares pizzas not only with the toppings that have traditionally seen in the USA, but also uses toppings that include cheese and processed meat found in the local country. As such, these two companies modify their overall products to meet local tastes. Heineken, the Dutch beer manufacturer, offers its beers in bottles of various sizes depending on the country of operation. Finally, an example of product modification involving car accessories is the cup-holder, which is almost a standard feature in cars in the United States but not so commonly found in cars destined for the European countries.

An international enterprise many modify products for different markets using the *modular approach*, and *core-product approach*. Either approach permits the firm to reap the benefits of standardization, and at the same time, respond to local needs and differences in the conditions of product use across countries (Walters and Toyne, 1989).

Modular approach: The notion of modular design was pioneered by Starr (1964: 137), who suggested that products should be made of interchangeable modules to supply "consumers with apparent variety even though the production output is based on the concepts of mass production." The modular approach entails the design and production of a range of standard parts that can be assembled in a variety of combinations yielding a series of products. These products can be assembled from the standard components to meet, as closely as possible, the preferences of local consumers in a particular foreign market. Ikea, the Swedish furniture manufacturer, is an IE that has adopted the modular approach. Specifically, Ikea is selling furniture that can be assembled from a standard collection of pieces.

Cost savings can be achieved when the components are mass-produced in a single facility, and the number of variant product combinations is limited. Nevertheless, under the modular approach, one must consider the tradeoffs between the number of product variants that enhance the attractiveness of the product to local markets (due to customization) and the number of components produced that promote economies of scale. Apparently, such a solution resides in the specifics of each firm.

Core-product approach: The core-product approach is based on the same principles as the modular approach. Under this approach, a uniform "central/core" product is designed that can accept a number of standard attachments, parts or components. The combination of attachments to the core product allows it to meet performance criteria and local consumer preferences in a country. The core product is of higher value relative to the total value of the finished product, whereas the components in the modular approach are of lower value relative to the total value of the finished product. As such, the core product represents a significant proportion of the total value of the finished product.

The production of a standard core product enables a firm to reap the benefits of economies of scale and thus enjoy lower production costs. On the other hand, the flexibility allowed by the attachment of components to the core product helps a firm localize its product and increase its attractiveness. Yip (1989) argues that "in a global strategy, the ideal is a standardized core product that requires minimal local adaptation." An additional advantage of the core product approach is that it facilitates delayed product differentiation, which enables marketers to react to market demand.

Across countries, a manufacturer of agricultural machinery was confronted with differences in climatic conditions, topography, and customer needs. To overcome these differences, the IE produced a robust and standard product that could operate in a variety of conditions. This core-product was also able to accept a number of attachments that gave the basic product different performance characteristics to meet local requirements and conditions of use (Walters and Toyne, 1989). In a similar fashion, car manufacturers employ the core-product approach by using a basic product, e.g. the chassis of the car, and then attach to it features in order to customize it for specific countries and even consumer needs. Such attachments include bigger engines, air conditioning units, electric windows, and other accessories. The core product is also referred to as platform product in the automobile industry, or as a generic product in the electronics industry (Lee et al., 1993). McGrath and Hoole (1992) state that each new product development should revolve around the design of a core product.

Honda recognizes the need for localized products since customers in different countries use them in different ways. To suit the needs of their customers and conform to the socio-economic conditions of the diverse marketplace, Honda uses its common basic technology to develop different types of motorcycles for different regions. In fact, the core product concept reduced the cost of Honda's products, which "cheaply built its new car lines from the platforms of existing cars" (Fortune Magazine, 1996: 32). Yip (1989) describes the product modification made by Boeing on its 737 in order to revive leveling sales. Boeing recognized that its current design (early 1970) was not appealing to developing countries because the shortness and softness of their runways, combined with the lower technical skills of their pilots made the planes bounce upon landing. Bouncing resulted in brake ineffectiveness. Boeing modified its 737 design by adding more thrust to the engines, redesigning the wings and landing gear, and installing lower pressure tires.

Lee et al. (1993) describe Hewlett-Packard (HP)'s use of the concept of generic/core product. Specifically, HP uses the principle of "postponement in product differentiation" on its deskJet printers. Postponing the point of product differentiation means delaying, as much as possible, the stage at which customization occurs before the final product is completed. Under this strategy, a generic/core product is manufactured at Vancouver, Canada, and Singapore that is shipped to the distribution centers in the U.S., Far East, and Europe. The distribution centers perform the product customization by inserting into the packaging boxes the language-specific manuals, power supply modules, and software. HP has even adopted the concept of packaging postponement under which final packaging is performed at the distribution center and not at the manufacturing facilities. The above concept requires the design of a generic/core product that can accommodate and facilitate the postponement of the customization process. Lee et al. (1993) refer to this product design as *design for localization* or *design for customization*.

FACTORS AFFECTING THE CHOICE OF A PRODUCT POLICY IN A GLOBAL ENVIRONMENT

The need to compete globally is becoming an imperative for firms aspiring to be key players in their industry and for those interested in entering new markets, and exploiting economies of scale. Competing globally requires IEs to choose one of the product policies previously described. This choice for an international product policy is influenced by various factors that can be broadly categorized into market-related, production-related, and international economic and political factors. These factors must be analyzed before reaching a decision on the type of international product policy to adopt. We provide below a partial list of such factors and their repercussions on the product policy choice.

Market-related factors

- When the "similarity" of customer preferences across countries is significant and creates a homogeneous global segment, IEs tend to produce standardized products. Levi's jeans, Coca-Cola, and Colgate toothpaste are often cited as universal products (Levitt, 1993). Nevertheless, there are still substantial differences in consumer preferences across countries (Boddewyn et al., 1986). The following market-related factors may prevent product standardization:

- The nature of the product is important; industrial products are more suited for standardization than consumer products (Boddewyn et al., 1986; Jain, 1989; Cavusgil and Zou, 1993).
- Where the product is marketed; developed or a less-developed country (Hill and Still, 1984).
- The size of the market in various countries and their potential for growth.
- The level of experience of the IE in the global marketplace. That is, whether the IE is simply an exporting company, or has a multinational or global orientation (Cavusgil and Zou, 1993).
- The presence and capabilities of international and local competitors (Jain, 1989).

International economic and political factors

- Economic recession in the countries where the IE promotes its products (Boddewyn et al., 1986).
- The exchange rate and the fluctuations of the currency of the various countries where the IE operates (Hadjinicola and Kumar, 1997).
- Quotas and tariffs set by local governments.
- The presence of nationalistic feelings among consumers.
- National government regulations that products must conform to.
- The political stability and similarity of the legal environments of the countries the IE operates.

Manufacturing factors

The international product policy adopted by an IE is closely related to its manufacturing *configuration* and *coordination* (Porter, 1990). Configuration refers to the degree of concentration/dispersion of activities in the value-added chain. On the other hand, coordination of the activities around the globe deals with information sharing among the activities, the allocation of responsibility, and alignment of effort. The three product policies described above require the support of a distinct manufacturing strategy, and accordingly, the right configuration and coordination. The universal product policy is facilitated by centralized production, whereas the country-tailored product policy is often promoted through localized production. A modified product policy calls for the adoption of manufacturing rationalization, implying factory specialization and closely integrated flow of materials among facilities in various countries. Some manufacturing factors that affect the choice of a product policy are presented below.

- The cost of setting up facilities in the various countries. When the cost of setting up product facilities is high, the IE will tend to centralize production in a small number of countries. This may lead to the design and production of standardized products.
- The level of centralization/decentralization of the production function. Centralization tends to result in less country-tailored products (Jain, 1989).

- The expected production volume and the savings derived from economies of scale. If the expected benefits from high production volumes are significant, the IE may tend to design standardized products (Hadjinicola and Kumar, 1997).
- The coordination and the cost of the new product development process. High development costs for new products may force the IE to design a standardized product.
- High transportation costs may lead the IE to set up production facilities in each country, which eventually may lead to the design of customized products.
- Maintaining a few global suppliers to ensure consistency in the components may lead the IE to design standardized products. On the other hand, maintaining multi-regional suppliers with different capabilities improves relations and may lead the IE to design country-tailored or modified products.

THE PRODUCT DESIGN PROCESS IN A GLOBAL ENVIRONMENT

Developing new products in a global environment often requires the various collaborating teams to be located in different parts of the world. Reasons for dispersed product design activities include: 1) the need to tap into the skilled human resources in some countries; 2) to take advantage of financial incentives offered by the host countries; 3) to exploit cost benefits from local labor; 4) to be close to academic and business centers; 5) to exploit reduced costs in equipment acquisition and building installations; and 6) to take advantage of the technology and infrastructure of the foreign country.

Having research and development facilities dispersed around the world allows the IE to roll out the product in less time since the design process go on for 24 hours a day in the various time zones. Of course, the choice of locating a new design center also depends on the competitive advantage acquired by locating a design center in that region. An inhibiting factor for designing a product in various locations around the world is the need for transferring technology abroad. Companies are often reluctant to transfer their research and development technology abroad to prevent its leakage to competitors.

The management of the design process in a global context is truly a challenging one since it involves the coordination of people in different locations, organizations, and cultures. The design process relies heavily on advanced communication technology since face-to-face interactions between the dispersed team members cannot be frequent. Differences in language, leadership/management styles, risk aversion, propensity to meet deadlines, organization and planning are only a few of the cultural issues that need to be resolved in order to ensure the success of the new product design process in a dispersed environment. Rosenthal (1992) provides an extensive discussion on the impact of cultural differences on the effectiveness of new product design teams.

HOW MOTOROLA DID IT

The case of Motorola and the development of the Keynote pocket pager highlight some of the issues raised above concerning the design process in a global environment (Rosenthal, 1992). Motorola decided to split the design activity of the Keynote pager between Singapore and Boynton Beach in Florida. Boynton Beach was responsible for the conceptual design and factory introduction, while Singapore was assigned such activities as the detailed design, the sourcing of parts in Asia, and pilot runs. Three basic reasons led Motorola to create a design center to Singapore. First, it was meant to off-load work from the design center in Florida. Second, the design center was to be located close to the manufacturing facility in Singapore. As a result, management believed that the close contact with production people would assist the design process. Third, from the strategic point of view, Motorola wanted to establish a design center in Asia in preparing for a global market, where products had to be designed in the vicinity of the customer.

As it turned out, the biggest challenge in the design of the Keynote pager was the management of the geographically separated teams. The 12-hour time difference meant that during normal working hours there was no communication between the members of the two teams. Issues that arose had to be resolved using the company's electronic mail system. This meant a delay in resolving even the minor problems that appeared in the design process. Language was also a barrier for the two teams since Singapore has four official languages. Another problem that arose was the difference in priorities and perceptions of the two design teams in Florida and Singapore. Working in a manufacturing environment, the priorities of the people in Singapore were meeting delivery times and controlling the budget. There was a natural aversion towards new products since they perturb the system. People in Singapore were trying to balance daily operations with the new product design. As a result, the design team in Florida perceived that management in Singapore was not fully committed to the design process. Furthermore, designers in Florida were apprehensive of the idea of delegating work to offshore sites, for fear that their work would disappear.

From Motorola's experience, two key factors ensure the success of the design process. First is the selection of design teams with prior experience, and their success in designing new products. If possible, teams should have some experience on design collaborations with teams located in other countries. Second is the choice of the manager with necessary technical and personality traits to lead the design process, act as a coordinator and cohesive agent among the dispersed teams.

Other ways to increase the probability of success of the design process in a global context is to allow designers from different locations to visit other research facilities so as to assimilate the culture and establish personal relationships with the rest of the team members.

De Meyer (1991) describes six broad approaches that IEs are adopting to solve the communication problem among dispersed laboratories. Such approaches include socialization efforts through temporary assignments and travelling, the establishment of rules and procedures to enhance formal communication, the presence of people serving as a liaison among the laboratories, the centralized control of the design activity, and the extensive use of electronic communication.

LESSONS

The success of an international product policy does not only lie in the thorough analysis of the factors discussed above but also in the way products are designed. Consider the experience of Polaroid, when the company was planning to introduce a camera that had a voice chip to simulate the photographer saying "cheese" or "smile" in order to get the target to smile. In this case, in which language should the company program its chip given that the camera was going to be distributed around the world? Initially, Polaroid introduced customized cameras with chips programmed with various languages. The number of different cameras led to inventory problems, similar to those experienced by HP described before, consequently, Polaroid offered a standardized product with three phrases/sounds. In other words, Polaroid alleviated its operational problems at the expense of complete customization by offering a standardized product. An intermediate solution could have been to think of the camera as the core-product and allow customers themselves to record the phrase/joke in their own language. This idea has also been used in dolls that allow customers to record the sounds that they would like the dolls to produce. As such, the above example illustrates the importance of the design process that can allow the IE to reap the benefits of customization and at the same time exploit operational efficiencies.

Key Concepts: Core-Product Approach; Design for Customization; Design for Localization; Global Common Denominators; Global Marketing Concept; Involuntary Product Adaptations; Premium Prototype; Product Extension; Universal Product Policy; Voluntary Product Adaptations.

Related Articles: Concurrent Engineering; Implementing Mass Customization; Mass Customization; Product Development through Concurrent Engineering.

REFERENCES

Boddewyn, J.J., R. Soehl, and J. Picard (1986). "Standardization in international marketing: Is Ted Levitt in fact right?" *Business Horizons*, November–December, 69–75.

Czinkota, M.R. and I.A. Ronkainen (1993). *International Marketing*. Dryden Press, Fort Worth.

De Meyer, A. (1991). "Tech tack: How managers are stimulating global R&D communication." *Sloan Management Review*, Spring, 49–58.

Douglas, S.P. and Y. Wind (1987). "The myth of globalization." *Columbia Journal of World Business*, Winter, 19–29.

Hadjinicola, G.C. and K.R. Kumar (1997). "Factors affecting international product design." *Journal of the Operational Research Society*, 48, 1131–1143.

Hill, J.S. and R.R. Still (1984). "Adapting products to LDC tastes." *Harvard Business Review*, March–April, 92–101.

Jain, S.J. (1989). "Standardization of international marketing strategy: Some research hypotheses." *Journal of Marketing*, 53, 70–79.

Kashani, K. (1989). "Beware of the pitfalls of global marketing." *Harvard Business Review*, September–October, 91–98.

Lee, H.L., C. Billington, and B. Carter (1993). "Hewlett-Packard gains control of inventory and service through design for localization." *Interfaces*, 23, 1–11.

Levitt T. (1983). "The globalization of markets." *Harvard Business Review*, May–June, 92–102.

McGrath, M.E. and R.W. Hoole (1992). "Manufacturing's new economies of scale." *Harvard Business Review*, May–June, 94–102.

Porter, M.E. (1990). *The Competitive Advantage of Nations*. The Free Press, New York.

Quelch, J.A. and E.J. Hoff (1986). "Customizing global marketing." *Harvard Business Review*, May–June, 59–68.

Rosenthal, S.R. (1992). *Effective Product Design and Development*. Business One Irwin, Homewood.

Sorenson, R.Z. and U.E. Wiechmann (1975). "How *multinationals* view marketing standardization." *Harvard Business Review*, May–June, 38–54.

Starr, M.K. (1965). "Modular production—a new concept." *Harvard Business Review*, November–December, 137–145.

Takeuchi, H. and M.E. Porter (1986). "Three roles on international marketing global strategy." In M.E. Porter (Ed.) *Competition in Global Industries*. Harvard Business School Press, Boston, 111–146.

Walters, P.G. (1986). "International marketing policy, a discussion of the standardization construct and its relevance for corporate policy." *Journal of International Business Studies*, 17, 55–69.

Walters, P.G. and B. Toyne (1989). "Product modification and standardization in international markets: Strategic options and facilitating policies." *Columbia Journal of World Business*, Winter, 37–44.

Yip, G.S. (1989). "Global strategy . . . in a world of nations?" *Sloan Management Review*, 31, 29–41.

21. CONCURRENT ENGINEERING

MORGAN SWINK

Michigan State University, Michigan, USA

ABSTRACT

Recently, global competition has led to shorter product life cycles and in-creased technological sophistication. Products are becoming more complex due to rapid technological developments and increasing consumer demands for lower costs, greater variety, and greater performance. At the same time the proliferation of new technologies is rendering products obsolete at an increasingly rapid pace. These market and technology trends place great de-mands on the new product development (NPD) process.

Several studies have shown that delivering a product to market on-time but over budget is more profitable than delivering the same product on budget but six months late. In fact, a recent study estimated that a delayed new product entry reduces a firm's market value by $120 million on average (Hendricks and Singhal, 1997). On the other hand, getting new products to market quickly provides numerous competitive advantages. In many in-dustries, product life cycles are currently only two years long or even less. First entrants in these markets typically capture large portions of market share and get a head start down the manufacturing learning curve. These advantages enable lowered manufacturing costs through economies of scale and improved manufacturing efficiency. Further, frequent new product in-troductions can help a firm dominate its markets. This was demonstrated in the motorcycle market in the 1980s. In an eighteen month period, Honda introduced eighty-one new models in Japan, while Yamaha could introduce only thirty-four. As a result Honda gained strategic advantage and market share at Yamaha's expense.

In the new millennium, product development lead time is becoming an even more important basis for competition. The auto industry provides a good example. Pressure from Japanese firms has forced U.S. auto manufacturers to seek ways to reduce product development times. By offering new models more often, auto makers can more quickly incorporate newly developed technologies into product designs. In addition, they are able to more quickly respond to changes in market tastes or economic conditions. Because these abilities hold the promise of enormous competitive advantage, reducing product development time has become a key priority in many industries. Chrysler corporation (before the merger with Daimler) announced its goal of developing a new car in as little as twelve months! Current development times in the auto industry vary from three to five years. Concurrent engineering practices of the following companies are mentioned in this article: **Boeing; Chrysler** (before the merger); **Cummings Engine; Ford; General Motors; Hewlett-Packard; Honda; Texas Instruments; Thomson Consumer Electronics; Yamaka.**

WHAT IS CONCURRENT ENGINEERING?

Concurrent engineering (CE) is defined as the simultaneous design and development of all the processes and information needed to manufacture a product, to sell it, to distribute it, and to service it. Other terms sometimes used in place of CE include "simultaneous engineering," "design-for-manufacturability," and "integrated product development." Concurrent engineering represents an important evolution in new product development (NPD) practices. Two aspects that distinguish CE from conventional approaches to product development are cross-functional integration and concurrency. Conventional NPD programs execute concept exploration, product design, testing, and process design activities serially. Each of these development activities is typically under the control of one functional organization at a time (e.g., marketing, engineering, manufacturing). As one function completes its design and development tasks, it then hands over control and responsibility to the next organizational function.

In the CE approach, integrated, multi-functional teams work together, simultaneously attacking all aspects of new product development. Control and responsibility are shared among functions and development activities overlap. For example, manufacturing process designers do not wait until product specifications are completed before developing tooling and processing equipment designs. Instead, they work closely with product designers to concurrently develop product and process concepts. In doing so, manufacturing personnel influence the product design in ways that make the product less costly or more producible. Other functions are similarly integrated into the design process so that their concerns may also be addressed.

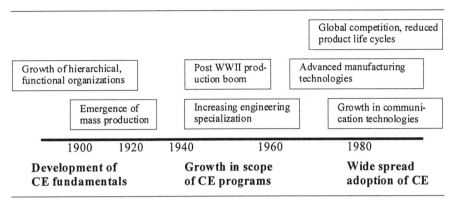

Figure 1. Timeline of major events in the development and adoption of CE.

BACKGROUND

Figure 1 illustrates the development and growth of CE in the twentieth century. The undergirding principles of concurrent engineering have been around for a long time, and were discussed as early as the beginning of the twentieth century (see Smith, 1997 for an interesting history of CE). However, CE has grown to become a dominant product development management approach only in the past two decades. Smith (1997) suggests several reasons for this growth. First, the need for CE increased because engineering training became intensely specialized, emphasizing engineering science over engineering practice. Second, the advent of useful information and communication technologies has enabled and lowered the cost of implementing CE. Third, changes in the competitive environment have increased the importance of reducing product development lead time and improving product quality.

While the fundamentals of CE are not new, there has been a marked growth in the scope of CE. According to Nevins and Whitney (1989), early CE approaches focused only on identifying part fabrication issues early in the product development process. Over the years, this focus was expanded to include assembly issues and groups of parts in design decisions, until finally all production and product support processes were addressed. For example, the U.S. Department of Defense emphasizes "cradle-to-grave" considerations in CE programs with the primary objective of coordinating decisions between different engineering functions (MIL-HDBK-59A-CALS, 1995). Market-oriented advocates of CE also stress the need for integrating the "voice of the customer" and marketing strategies into design decisions, emphasizing information exchange between marketing and R&D personnel. Others suggest an even broader view of CE which addresses environmental and societal cost issues (Alting, 1993).

Performance goals associated with CE have grown as well. Early CE developments were aimed at improving quality or minimizing product acquisition costs, while more recent programs have emphasized reductions in product development time. The result of this growth in CE concepts is a more holistic, strategic view of CE.

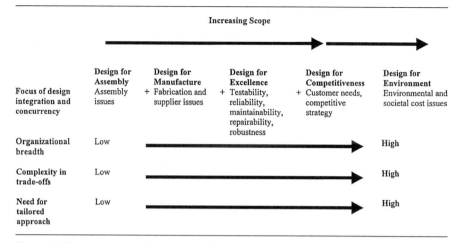

Figure 2. The varying scope of concurrent engineering programs. Adapted from a figure in Clark (1992).

IMPLEMENTATION

There are significant differences in the ways CE is conceived and implemented in different projects, companies, and industries. Some programs address only narrow product producibility issues (e.g., ease of assembly). More comprehensive CE programs address the impacts of product design decisions on competitive issues and product life-cycle considerations. Figure 2 illustrates the types of issues that add to the scope of a CE program. As the scope of CE increases, additional stakeholders (e.g., designers, manufacturers, suppliers, customers) must be involved early in design and development processes. As the group grows larger, the complexity of design decisions and related trade-offs increases, creating a greater need for a customized management to meet the specific needs of the NPD program. For example, managers can implement standard protocols across a wide variety of product development programs in order to insure that consistent product assembly issues are addressed early in product design. However, the protocols and approaches that managers use to address broad, strategic issues will differ, depending on the NPD program importance, the complexity and newness of the product, and the intended uses and markets for the product.

One of the difficulties of implementing CE is deciding what activities should be done concurrently and establishing where the most important points of integration are. Program priorities should drive these decisions. Customer desires and competitive threats influence the relative priorities placed on design quality, product costs, and product introduction speed. For example, lowering product cost is often a high priority in incremental product redesign projects. When new technologies or other product differentiation dimensions are absent, product cost becomes a primary basis for competition. Close interactions among designers, suppliers, and manufacturing personnel are important in this situation.

Comparable arguments can be made regarding other NPD outcomes in various circumstances. All functional groups are important although no single group contains

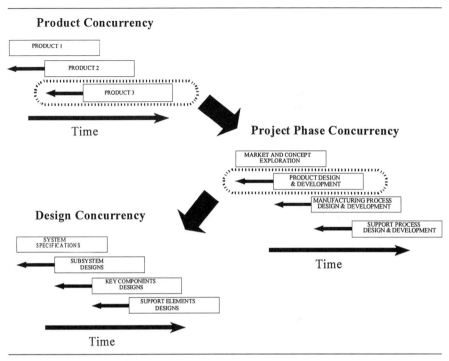

Figure 3. Different types of concurrency.

all the knowledge needed to complete NPD. However, the degree of influence that each group exerts in NPD should reflect program priorities and project characteristics. Leadership roles should be played by functional representatives who are most able to reduce uncertainty in the areas of greatest need.

Concurrent execution of design and development activities stimulates needed cross-functional interactions.

Figure 3 illustrates relationships among three different types of concurrency. *Product concurrency* is the overlap of separate but related new products requiring coordination between NPD programs. Product concurrency exists in the concurrent development of first generation and next generation products, or in the development of product variants. For a given product development effort, *project phase concurrency* involves simultaneously developing market concepts, product designs, manufacturing processes, and product support structures. Within any project phase, *design concurrency* involves the overlap of design disciplines (e.g., system, software, electrical, and mechanical engineering) so that system level and component level designs are produced concurrently. Greater the concurrency, greater the reduction in development lead time. However, greater concurrency also increases the number of functional relationships that must be managed simultaneously and increases the severity of design risks undertaken.

While the scope and design of CE programs varies across firms and across NPD projects, two management initiatives remain central. First, project managers must organize personnel, policies, and procedures in ways that improve cross-functional integration and communication. Second, CE requires design analysis and decision making methods that foster design excellence. CE implementation requires cross-functional integration, and systematic design analysis, which are discussed below.

(1) Improve Cross-Functional Integration: In order to concurrently address the many decisions involved in NPD, persons from different organizational functions must be willing to collaborate, share information, and resolve conflicts quickly and effectively. This is accomplished through the following actions:

- Promote Teamwork.
- Set and analyze goals.
- Direct and control integration.

Promote teamwork: The use of cross-functional teams is a fundamental method for fostering communication in CE projects. Susman and Dean identify a number of integrative mechanisms that can be considered prerequisites to the success of CE teams (see Susman, 1992, Chapter 12).

- Status parity: All members of the team must have equal perceived value and relevance. This also means having equal voting power in making decisions.
- Project focus: A project structure should be elevated over functional structures. The project manager should have real power and the project members should not be too tightly controlled by or rewarded only by functional department heads.
- Number of levels: The firm should seek to minimize the number of organizational levels that exist between project-level personnel in design and manufacturing and the first manager who has authority over both.
- Frequency of rotation: Frequently rotating people among positions in design and manufacturing will improve CE.

While team arrangements vary, three organizational levels of teams frequently appear in NPD programs: a program management team, a technical team, and design-build teams. Figure 4 illustrates relationships among the teams.

Program management teams typically include the program manager, marketing manager, finance manager, operations manager, aftermarket manager, and design managers. This group provides management oversight and planning, approves large resource allocations, approves and controls the project budget, and manages the project schedule.

The technical team reports to the program team and provides technical oversight, approves key design decisions, and maintains consistency between design elements. Engineering managers from design and functional support areas are typically members of the technical team along with representatives from marketing, service, manufacturing, quality assurance, test engineering, drafting/documentation, and key customers and suppliers.

Membership of design-build teams replicates the technical project team with responsibility for components at the lower levels of the product structure. Each team

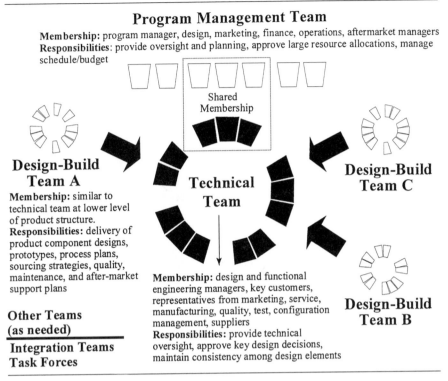

Program Management Team
Membership: program manager, design, marketing, finance, operations, aftermarket managers
Responsibilities: provide oversight and planning, approve large resource allocations, manage schedule/budget

Shared Membership

Design-Build Team A
Membership: similar to technical team at lower level of product structure.
Responsibilities: delivery of product component designs, prototypes, process plans, sourcing strategies, quality, maintenance, and after-market support plans

Technical Team

Design-Build Team C

Design-Build Team B

Other Teams (as needed)

Integration Teams Task Forces

Membership: design and functional engineering managers, key customers, representatives from marketing, service, manufacturing, quality, test, configuration management, suppliers
Responsibilities: provide technical oversight, approve key design decisions, maintain consistency among design elements

Figure 4. Concurrent engineering teams.

is oriented around a particular product component, with responsibility for delivering designs, prototype hardware, process plans, sourcing strategies, quality engineering, maintenance plans, and after-market support plans. These teams are frequently co-led by design and manufacturing engineers who maintain high degrees of design authority and budgetry control. Suppliers and counterpart engineers from partnering companies also get involved at this level.

Managers often mention that it is difficult to get team members to work together as an actual team beyond a mere "information exchange group." The key to fostering teamwork is to vest the team with responsibility for its deliverables and the authority to obtain needed resources. Without these, most people will focus on their individual technical responsibilities and will view team membership as irrelevant to their "real jobs."

Performance measurement is another critical issue. Different functions typically emphasize different performance measures in NPD. Sometimes these differing priorities and their associated measures create conflicts. For example, product design engineers are frequently rewarded for reducing product material costs. On the other hand, operations personnel are usually rewarded for reducing labor costs. Choices made in product design can force trade-offs between these two priorities. Total product unit cost would be a better measure of performance, as this would encourage both functions to work together to produce an overall optimal design. Project managers should carefully consider how measures used in the development program might motivate collaboration or conflict within project teams.

Most firms use recognition as the primary incentive for motivating individuals on teams. For example, some firms hold team celebrations, competitions, or formally recognize teams at the corporate level. Others document product design improvements or cost-saving suggestions generated by teams and reward them with gifts or trophies. Team participation may be also assessed as part of the normal annual performance review process, addressing each employee's ability to participate in teams or to lead teams.

Another way companies encourage teamwork is by increasing each team member's sensitivity to the needs of the other team members. Larger projects establish liaison positions for personnel who act as go-betweens for major functional groups. Other approaches give team members a broader base of experiences by enlarging their job roles, or by rotating personnel through different job assignments and functional departments. In many cases, co-location is considered an important prerequisite to team success. Placing the work sites of personnel physically close to one another encourages team building and communication through daily contact and socialization.

Set and analyze goals: Clearly defined goals reduce nonconstructive conflict. High-level program goals also motivate cross-functional integration by providing rallying points for team members from different functional areas. For example, a product development goal of cutting product weight and cost by 30 percent motivates interactions among all affected functions (e.g., product design, manufacturing, purchasing, etc.). Clear, pre-specified goals help resolve conflicts between functional concerns because they focus the team's emphasis on common goal achievement.

Furthermore, great benefits can be gained when project goals and product performance specifications are clearly linked to specific customer needs. Project team members are more likely to pull together to meet project timing and budget targets when they believe that customer satisfaction is on the line. In setting product specification goals, quality function deployment (QFD) is widely regarded as a useful tool for decomposing customer needs into engineering design parameters. Essentially, QFD requires a cross-functional team to generate and evaluate customer attributes and the requisite engineering specifications (Hauser and Clausing, 1988).

The priority given to various customer needs and project goals can also influence both the degree of interaction between certain persons and the nature of their interactions. For example, setting aggressive product cost goals is likely to stimulate intense manufacturing-design interactions, since both parties are needed to generate and evaluate cost reduction ideas. Development project managers should carefully identify the greatest needs for functional integration and design program goals accordingly.

Direct and control integration: Project managers must also design procedures that direct and control functional interactions. Some firms formally document the major phases and general deliverables for any product development program in the form of a program contract. A "blank" program contract contains templates for establishing the "who, what, where, when, and how" of specific, key tasks on the project. Intense preprogram training may be connected with the development of the program contract including discussions that identify the most important project management aspects and product design requirements. The effectiveness of program contracts and training processes lies in the ability to adapt broad organizational rules and procedures to the specific needs of the program at hand.

Corporate initiatives such as quality and productivity improvement programs are often useful for directing integration. A number of NPD projects have effectively integrated key components of corporate initiatives into their program management plans. For example, one program manager leveraged a company-developed quality improvement procedure by establishing it as the basis for team interaction and problem solving on his project. The potential danger of relying heavily on corporate-wide initiatives is that project specific goals may be overshadowed by the initiative.

(2) Improve Design Analysis: An excellent product design maximizes customer satisfaction while effectively utilizing the company's production, sales, and service capabilities. The conventional way of insuring that good design practices are followed is to give functional engineering representatives approval authority for all designs. A designer creates a part design, for example, that is then serially reviewed and revised by a number of functional experts. While this approach has merit, it is also inefficient because rework is institutionalized and very little learning is transferred to the original designer.

A more productive approach is for designers to incorporate good design practices in the initial process of generating the design. In order to minimize design rework, progressive firms are seeking innovative ways to train designers in good design practices. Training can include general design axioms, guidelines, or specific design rules pertaining to commonly used manufacturing processes. Design rules and best-practices can also be embedded into computer-aided-design (CAD) systems. These systems are discussed in more detail in the Technology Perspective section.

Another way to reinforce design rules and best practices is to create an internal consulting group, staffed by skilled and experienced engineers, and separated organizationally from other functional units. Representatives from the group can be assigned to NPD projects to assure that state-of-the-art design practices are followed and to serve as liaisons to functional experts. Descriptions of other tools and methodologies for facilitating manufacturing process design analyses are found in Nevins and Whitney (1989), Susman (1992), Kusiak (1992, 1993), Dorf and Kusiak (1994), and Ulrich and Eppinger (1995).

A summary of CE best practices is provided in Table 1. To maximize the effectiveness of CE, managers need to customize NPD processes to meet specific NPD program challenges, priorities, and product characteristics. Managers should consider company culture, the level of experience program personnel have with CE, the complexity of the product, the level of innovation attempted in the development effort, and the technical risk involved in the project. Product complexity is reflected by the numbers of people, functional specialties, and outside suppliers involved in NPD and by the degree of interdependency. Project teams and integration systems should be designed to ameliorate the negative effects of this complexity while dealing with the constraints

Table 1 Checklist of best practices in concurrent engineering

- The design of the CE program is guided by a clear set of goals with attached priorities. These are communicated to team members in the form of a program contract.
- The design of the CE program takes into account the effects of the firm's organizational culture and prior experience with CE.
- Overlapped activities in NPD are supported by incentives, performance measures, technologies, and organizational arrangements that motivate and enable rich communications among the appropriate team members.
- Cross-functional teams are given clear objectives along with budget control, authority, and resources needed to meet those objectives.
- Barriers to CE are removed. These barriers can be created by functional silos, the sequential scheduling of activities, lack of access to information, and lack of co-location of personnel.
- Advanced technologies are used to support design, analysis, and communication. Managers plan for the additional time team personnel need to implement and become proficient in the use of new technologies.
- Project personnel are trained in the CE process. Each team member has a clear understanding of his/her role in achieving NPD program goals and priorities.
- Members of design-build teams have equal status and awareness of the requirements of other team members.
- The NPD project manager has greater power than that of functional heads.
- For any two team members from different functions, only two to three organizational levels must be bridged before reaching a common manager.
- A high level champion helps to create a sense of urgency, uniqueness, or importance for the project.
- Methods have been established for capturing and conveying technical discoveries and learning gleaned from the various design and development activities.

and risks of innovation. For example, the design of radically new product components sometimes call for limited concurrency early in NPD due to an increased need for product testing and validation. At the same time, breakthrough technological developments increase the need for interactions among designers and team members inside and outside the firm who have the expertise needed to reduce technical uncertainties. (see Swink et al., 1996a).

ENABLING CONCURRENT ENGINEERING

Technology can be a strong enabler of concurrent engineering. Many new computer applications improve the richness of communication among NPD team members, and improve the quality of product designs. These technologies can be broadly classified as follows:

- Physical facilities: The building layout and technologies facilitate co-location of personnel, team meetings, information display, and an "open" atmosphere.
- Communications Technologies: These technologies improve the speed, richness, and distance over which communications can occur.
- Computer-Aided-Design Systems and Databases: These systems provide digital design capabilities and a centralized source of design data that can be accessed electronically.

- Computer-Aided-Engineering and Analysis Systems: These systems provide simulation capabilities and easy access to engineering information that is useful to designers who are making choices among design alternatives.
- Group Technology and Coding Systems: These systems provide a scheme for classifying existing component designs and sources of components (e.g., suppliers), thereby reducing the need for creating new designs or finding new sources.

The physical layout of facilities used to house NPD team members can promote greater communication and teamwork. The Chrysler Corporation's new research and technical center has been highly touted for its ability to co-locate engineering and pilot production operations. Group discussions are enhanced by an open building concept that is sprinkled with numerous high-technology meeting rooms. Each floor of the circular building houses a single product development team.

Electronic mail and communications networks are powerful tools for rapidly communicating information and for providing to wide audiences easy access to product and project data. Information can also be stored on centralized computer-aided-design (CAD) databases. Data captured in these systems can be accessed by persons located around the world for use in product design, process planning, and computer-aided manufacturing. Boeing made extensive use of these systems in the design of its 777 aircraft, the first plane to be completely "digitally" designed.

Computer-aided-engineering (CAE) tools are frequently linked to CAD systems in ways that reinforce good design practices. These sophisticated systems create and analyze three-dimensional models of parts and assemblies, reducing the need to build expensive and time consuming physical prototypes. For example, CAD/CAE systems can automatically analyze assembly designs to identify areas of potential interference between parts. Further, many CAD systems embed process information and design rules directly into the design software so that they may be linked to certain design features. For example, when a designer draws a hole, he can then select a pull-down window of information providing a list of processes that could create the hole, typical dimensional tolerances, defect rates associated with each process, and any other design rules related to the feature. Some companies have developed "expert systems" that aid the evaluation of design choices. In addition, numerous off-the-shelf CAE systems address stress and thermal analyses, mechanical assembly, printed circuit board design, and integrated circuit (IC) design.

In large organizations, designers often waste time and resources by unknowingly recreating existing designs. CAD systems can be linked with databases that contain information on preferred components, existing designs from other products, and suppliers of purchased items. Group technology-based classification and coding systems enable designers to easily search design databases for existing designs which meet their current needs. Similarly, databases which prioritize certain components and vendors can speed up a designer's search for suitable parts. Coding systems also allow manufacturing planners to identify "families" of parts that have similar design or processing characteristics. These approaches reduce design time and reap enormous manufacturing benefits because fewer unique parts must be fabricated and inventoried, less special

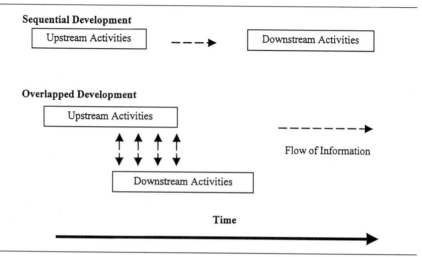

Figure 5. Sequential and overlapped product development activities.

tooling is needed, production scheduling is simplified, and less disruption is experienced. For more information on enabling technologies in CE, see Nevins and Whitney (1989), Kusiak (1992, 1993), Dorf and Kusiak (1994), and Ulrich and Eppinger (1995).

BENEFITS

CE helps product development managers meet growing competitive challenges by producing higher quality products in less time. CE improves product quality in at least two dimensions. Design quality, the match between customers needs and product attributes, is improved by emphasizing customer needs in the concurrent execution of market research and product design activities. Product conformance quality is similarly improved. Manufacturing defects are reduced by considering manufacturing process capabilities while product design and process design activities are concurrently executed. Concurrent engineering efforts in various industries have improved process capability performance for critical product dimensions by as much as 100 percent, frequently producing C_{pk}[1] levels of 1.5 to 2.0 (Handfield, 1995).

Development lead time is reduced by the overlap of NPD activities coupled with intense communication between upstream and downstream development activities (see Figure 5). Also, when CE is successful, fewer time-consuming manufacturing problems are encountered in the initial stages of production.

Concurrent engineering processes are sometimes viewed as expensive in the short term, requiring resources and levels of management and worker commitment that may not be readily available. However, when appropriately used, CE approaches can

---

[1] C_{pk} is a commonly used measure of process capability that reflects the ratio of product specification limits (e.g., tolerances) to process variation limits. Values greater than 1.0 are desirable.

produce higher quality, more producible products, in less time. Further, CE produces organizational benefits that far exceed the profits associated with any single product. For example, some companies have documented savings in overall product development costs of 20 percent, reductions in development time of 50 percent, and reductions in engineering design changes of 45–50 percent (Swink et al., 1996a). More importantly, successive CE development programs build upon one another. Improved design processes lead to improved products and services. CE experiences produce engineers and managers who are more aware of the company's goals and of the needs of other functional areas. Combining this type of organizational learning with effective CE processes can improve competitive advantage.

CAVEATS

Numerous cases of concurrent engineering success have been documented in a variety of industry and product development contexts. Even so, researchers continue to debate the appropriateness of CE in different situations. On-going research is investigating the dynamics of functional integration and concurrency by addressing such issues as to who should be integrated and when and how this is best achieved.

Some researchers and users of CE argue that it is especially important when there is stiff competition, when new manufacturing processes are being used, and when reducing development lead time is very important. Indeed, managers may have little choice but to implement CE when presented with a lucrative market opportunity or confronted by a particularly aggressive competitor. At the same time, there are indications that an over-emphasis on CE can hurt NPD performance outcomes in some circumstances. Heavy concurrency may increase development cost and lead time when NPD activities are risky with a high potential for failure. Because process design activities are begun earlier and with less complete information, there is an increased probability that some designs will need frequent rework.

Recent empirical research suggests that heavy manufacturing influences early in NPD can be detrimental to product innovation, and may unnecessarily increase lead time in NPD projects when they involve little new technology (Swink, 1997). For example, manufacturing personnel may sometimes be locked-in to the firm's current processing capabilities or remain unknowledgeable about new technologies or options outside the firm. In this case, their influence in NPD may work against the adoption of innovative new product features. In addition, R&D personnel sometimes complain that the inclusion of downstream process personnel early in NPD creates confusion and slows down decision making, especially at very early stages of NPD when design concepts are not well-defined (Gerwin, 1993). These findings are tentative and need to be confirmed. However, it is important to remember that the primary goals of NPD should govern the training and participation of various functional groups and the usage of cross-functional integration methods.

Good communication among marketers, product designers, process designers and manufacturing personnel is always crucial. However, communications across certain functional groups should be prioritized according to the objectives established for the project (Swink et al., 1996a). For example, the use of an important new manufacturing technology would necessitate rich communications among the designers using the technology, the designers and installers of the technology, and the users of the technology (i.e., manufacturing personnel). Further, each CE team member should realize that the strategic importance of certain NPD program outcomes (e.g., project timing, innovative features) might sometimes outweigh the importance of his or her own function's design guidelines.

EXAMPLES OF USERS

Many well-known companies in different industries have adopted CE approaches including Thomson Consumer Electronics, Cummins Engine, Chrysler, Ford, Motorola, General Motors, Texas Instruments, Hewlett Packard, and Intel. Descriptions of these and other cases are found in Clark and Fujimoto (1991), Mabert et al. (1992), and Swink et al. (1996a).

Haddad (1996) provides an interesting account of a major CE implementation in an automotive company. The firm had experienced poor profitability and market share losses for a number of years. Executives at the firm realized that their Japanese competitors were able to bring new products to market in half the time. As a result, in 1990 the company launched a dramatic reorganization of engineering, purchasing, manufacturing, and planning activities into cross-functional, product-focused teams. The new arrangement required that personnel from these various functions, including marketing and first tier suppliers, be dedicated to a new product platform team. The overall platform team was then divided into groups known as "product action teams." Each group was headed by an executive engineer and was responsible for a portion of the platform (e.g., the interior of the automobile). These groups were further divided into departments responsible for components in the car (e.g., seats or controls). Groups averaged around 16 people in number. Groups were given significant authority and encouraged to make decisions by consensus.

The organizational change was supported by several changes in corporate standards, human resource policies, and technology. A new policy dictated that no more than five levels of organizational hierarchy should exist from vice presidents down through executive engineers, managers, and supervisors. Human resources managers developed training sessions and evaluation systems aimed at motivating team behavior. For example, the old, performance appraisal system was replaced by a new system that emphasized a broader set of criteria, which included the acquisition of skills in the areas of teamwork, communication, planning, process improvement, problem solving, and so on. Technological improvements included more extensive use of CAD, CAE, and electronic networking technologies. In addition, a new facility was built to co-locate vehicle engineering, design, pilot production, and training activities.

These changes by the automobile firm produced handsome returns. The first two new platform automobiles introduced using the new processes were brought to market in about three years. Prior models had required 5 years or more to develop. The company earned healthy profits in 1992 while its competitors showed losses for the same period. In 1993 and 1994 the company posted record earnings, and their stock increased sixfold in the period from 1991 to 1994. The changes brought high praise from market analysts, and workers inside the firm experienced many qualitative benefits of the change. Workers felt that the development projects had greater focus and less confusion than in the past. In addition, there appeared to be a higher level of worker commitment and buy-in.

While the change to CE produced great benefits for the company, it was not without its problems and challenges. Union relationships were strained due to decisions by the firm to increase the amount of outsourced design work. Some workers expressed feelings of greater stress and a lack of recognition for their contributions. The change to CE was hampered by shortage of design engineers who had meaningful manufacturing experience. Also, functional experts were spread across too many platform teams. Finally, newly hired engineers had difficulty knowing who the experts in other platform teams were, since they had no opportunity to work with them on prior projects.

This case study points to the challenges and benefits of implementing CE in a very large organization. However, similar benefits have been recorded in smaller firms who did not make the kinds of sweeping organizational and technological changes described in this study. Regardless of the form CE takes, the key benefits are derived from the empowerment and teamwork of program personnel.

LESSONS

The involvement of corporate-level management is important to the success of CE, especially for initial attempts at using the approach. To increase the probability of success, managers must:

- Elevate the project.
- Elucidate goals.
- Eliminate barriers to integration.
- Elaborate the CE processes (Swink et al., 1996b).

(1) Elevate the Project: For most firms, CE is an unconventional approach to product development, requiring personnel to behave in non-traditional ways that violate prevalent habits and management philosophies. Consequently, during early attempts at CE, it is vital to get people to view the project as "special." Creating a crisis environment is one way to break people out of conventional practices. A sense of urgency can encourage team members to overcome barriers formed by the corporate culture and standard operating procedures. Furthermore, when team members perceive that the project is unique, they often become more interested, take greater ownership, and are more willing to take risks.

Top management commitment is also important. A high level champion for the project provides the energy and enthusiasm needed to get people to change. Committed top level managers are more willing to fight for necessary project resources. This is especially important since CE development programs can be initially expensive to implement. Initial programs may require resources dedicated to the development of CE concepts for the firm, extensive training for team members, and the purchase and implementation of communication systems. In addition, program managers should expect to spend more money early in the NPD effort since downstream personnel are asked to participate in up-front decision-making and analysis.

(2) Elucidate Goals: In CE projects it is important to define goals that stimulate interactions between the right groups of people, and to the right extent. Shared, globally-identifiable, customer-initiated goals produce a common language and sense of purpose for the project. In addition, team-based performance measures and feedback that are related to super-ordinate goals allow team members to share the success or failure of their efforts.

(3) Eliminate Barriers to Integration: A fundamental barrier to integration in most large organizations is bureaucracy. Standard operating procedures often create functional barriers which must be broken down. In these situations, it may be useful to separate the project from the larger organization by dedicating resources and by realigning reporting relationships and reward structures, at least temporarily.

The forming of teams can cause conflicts and upset well-established domains of power. Team members may be put in situations where they must choose between benefiting the project and benefiting their own home functions. Functional leaders may try to exert control over portions of the project that they consider to be within their domains. Project leaders should try to anticipate and manage these situations.

(4) Elaborate Concurrent Engineering Processes: Team members require more than technical skills and information access to be effective players in a CE environment. They also need a clear understanding of the CE process itself. A well-defined process must be established and clearly articulated to all team members inside and outside the firm.

Defining the process means developing a shared vision of the important dimensions of CE. Managers need to explore how CE approaches can be tailored to meet specific project needs. Furthermore, managers need to capture and leverage learning from CE experiences by peppering NPD projects with people who are experienced in the process. Sharing project documentation and post mortem project analyses across divisions within the firm can be helpful in disseminating learning.

Key Concepts: Cross Functional Team; Cross-Functional Integration; Design Concurrency; Design-for-Manufacturability (DFM); Functional Organization; Integrated Product Development; Process Capability; Product Concurrency; Product Life Cycle; Product Producibility; Project Phase Concurrency; Quality Function Deployment (QFD); Simultaneous Engineering.

Related Articles: Integrated Product Development: The Case of Westinghouse Electronic Systems (1991–1993); Product Development and Concurrent Engineering; Teams: Design and Implementation.

REFERENCES

Alting, L. (1993). "Life-Cycle Design of Products: A New Opportunity for Manufacturing Enterprises." *Concurrent Engineering*, A. Kusiak (ed.). Wiley, New York.
Clark, K.B. and T. Fujimoto (1991). *Product Development Performance*. Harvard Business School Press, Boston.
Clark, K.B. (1992). "Design for Manufacturability at Midwest Industries: Teaching Note." Note no. 5-693-007, Harvard Business School Press, Boston.

Dorf, R.C. and A. Kusiak (1994). *Handbook of Design, Manufacturing, and Automation*. Wiley, New York.

Gerwin, D. (1993). "Integrating Manufacturing into the Strategic Phases of New Product Development." *California Management Review*, Summer, 123–136.

Haddad, C.J. (1996). "Operationalizing the Concept of Concurrent Engineering: A Case Study from the U.S. Auto Industry." *IEEE Transactions on Engineering Management*, 43 (2), 124–132.

Handfield, R.R (1995). *Re-engineering for Time-Based Competition*, Quorum Books, Westport, CT.

Hauser, J.R. and D. Clausing (1988). "The House of Quality." *Harvard Business Review*, May–June, 62–73.

Hendricks, K.B. and V.R. Singhal (1997). "Delays in New Product Introductions and the Market Value of the Firm: The Consequences of Being Late to the Market." *Management Science*, 43 (4), 422–436.

Kusiak, A. (1992). *Intelligent Design and Manufacturing*. Wiley, New York.

Kusiak, A. (1993). *Concurrent Engineering: Automation, Tools, and Techniques*. Wiley, New York.

Mabert, V.A., J.F. Muth, and R.W. Schmenner (1992). "Collapsing New Product Development Times: Six Case Studies." *Journal of Product Innovation Management*, 9, 200–212.

MIL-HDBK-59A-CALS (1995). *Military Handbook on Computer-Aided Acquisition and Logistics Support*.

Nevins, J.L. and D.E. Whitney (1989). *Concurrent Design of Products and Processes*. McGraw-Hill, New York.

Susman, G.I. (1992). *Integrating Design and Manufacturing for Competitive Advantage*. Oxford University Press, New York.

Smith, R.P. (1997). "The Historical Roots of Concurrent Engineering Fundamentals." *IEEE Transactions on Engineering Management*, 44 (1), 67–78.

Swink, M.L., J.C. Sandvig, and V.A. Mabert (1996a). "Customizing Concurrent Engineering Processes: Five Case Studies." *Journal of Product and Innovation Management*, 13 (3), 229–244.

Swink, M.L., J.C. Sandvig, and V.A. Mabert (1996b). "Adding Zip to Product Development: Concurrent Engineering Methods and Tools" *Business Horizons*, 39 (2), 41–49.

Swink, M.L (1997). "An Exploratory Study of Management Approaches and Their Relationships to Performance in Incremental, Radical and Breakthrough Product Development Projects." Working paper, Indiana University.

Ulrich, K.T. and S.D. Eppinger (1995). *Product Design and Development*. McGraw-Hill, New York.

22. PRODUCT DEVELOPMENT AND CONCURRENT ENGINEERING

CHRISTOPH H. LOCH

INSEAD, Fontainebleau, France

CHRISTIAN TERWIESCH

University of Pennsylvania, Philadelphia, PA, USA

ABSTRACT

Concurrent Engineering (CE) represents a key trend in product development over the last decade. It has changed academic and industrial approaches of looking at the product development process. It can be defined as an integrated "new product development process to allow participants, who make upstream decisions to consider downstream and external requirements." Central characteristics of a concurrent development process are activity overlapping, information transfer in small batches, and the use of cross-functional teams (Gerwin and Susman, 1996). Some related practices of **Toyota** are described here.

INTRODUCTION

Many elements of CE have been recognized since the first half of this century (Smith, 1997). They include the formation of cross-functional teams, an important role for manufacturing process design in product development, and a focus on fast time-to-market. The term CE, however, was not adopted until the 1980s. Coming from manufacturing engineering, for example, Nevins and Whitney (1989) were among the earliest to demand concurrent design of the product and the manufacturing system. Others emphasize the cross-functional aspect by observing a "...need to transfer decision making authority from managers to teams" (Gerwin and Susman, 1996).

Concurrent engineering became popular in operations management literature with the work by Imai et al. (1985) and Takeuchi and Nonaka (1986). These studies contrasted two practices in product development: on the one hand, most Western organizations followed a more sequential process, similar to running a relay race with one specialist passing the baton to the next. This approach has also been referred to as "batch processing" (Blackburn, 1991; Wheelwright and Clark, 1992). In contrast, some high performing development organizations, most of them in Japan, followed a "rugby team" approach with a strong emphasis on cross-functional integration (a cross-functional team running in a staggered fashion). These early examples were motivated mainly from the camera and automobile industries (Takeuchi and Nonaka, 1986).

In a landmark study on the automobile industry, Clark and Fujimoto (1991) extended this pioneering work, and provided a detailed empirical analysis of product development processes, their impact on development performance and their differing use across regions. This study operationalized many important variables of CE, including measures for task overlapping and cross-functional communication. Clark and Fujimoto observed that companies that succeeded in short time-to-market were found to combine activity overlap with intensive information transfer, a practice they referred to as "integrated problem solving." These ideas were refined in further studies, e.g., Wheelwright and Clark (1992).

The managerial framework of CE pose four key managerial challenges (see Figure 1).

- Defining development tasks based on the product architecture.
- Timing of activities with emphasis on task overlapping.
- Defining coordination and integration mechanisms among development tasks, which ensure adequate information exchange.
- Establishing support processes and an appropriate organizational context.

The framework with its four managerial challenges is summarized in Figure 1.

TASK DEFINITION IN COMPLEX DESIGN PROBLEMS

When faced with product development on a large scale, often hundreds of people work simultaneously on separate portions of the overall development effort, organized in many small teams (Eppinger et al., 1994). The division of the complex overall design task into smaller, separately manageable portions depends on couplings among these portions, which are largely determined by the product architecture.

The *product architecture* identifies the basic building blocks, their relationships to the functionality of the product ("what they do"), and the interfaces among them. A modular architecture is one with standardized interfaces and a one-to-one correspondence between building blocks and functions: a module roughly corresponds to a function. The other extreme is an integrated architecture, with complex interfaces and a one-to-many correspondence between blocks and functions: functions are spread out over several building blocks, or several building blocks share one function (Ulrich, 1995).

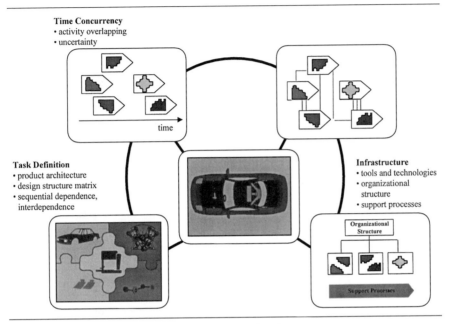

Figure 1. Operations management issues in concurrent engineering.

In terms of the final product, a modular architecture offers the possibility of component standardization, easily achievable product variety, and add-ons. An integrated architecture, on the other hand, may increase the compactness of the product, or allow better adaptation to specific customer needs, higher performance, or longer life through the tight integration of its building blocks. Moreover, the architecture heavily influences the product development process, since modularity reduces interactions among product components, and thus complexity (Fitzsimmons et al., 1991). In a modular architecture, product components can be developed in a decomposed and parallel manner (Ulrich, 1995).

The goal of task definition is to minimize interactions among activities, or at least make them explicit so as to clearly understand the information transfer necessary. Two types of couplings exist, defined by information flows among activities. First, *sequential dependence* refers to a one-way information flow from an upstream to a downstream activity. In this case, one activity is the information supplier and the other the information receiver. Second, task interdependence refers to several tasks requiring information input from one another.

The process of decomposing the overall design problem can be supported by operations management methods. The most important support tool is the design structure matrix (DSM), which groups activities so as to minimize the interdependencies among the groups (Steward, 1981; Eppinger et al., 1994; Smith and Eppinger, 1997). A brief example of the application of this tool is below.

	A	**B**	**C**	**D**	
Task A	A				A-B: sequential
Task B	x	B			B-C: independent
Task C			C	x	C-D: coupled
Task D	x		x	D	A-D: sequential

Figure 2. Design structure matrix (Ulrich and Eppinger, 1995: 262).

Based on the product architecture and the resulting component interfaces, the DSM captures interdependencies among the development tasks (corresponding to components), in the sense that tasks need input (physical or informational) from other tasks in order to be completed.

As is shown in Figure 2, information-receiving tasks are listed along the columns, and information-supplying tasks along the rows. Crosses (X) mark information dependencies. Task B is *sequentially dependent* of task A, as the information flow goes only one way. Tasks B and C are *independent*. Tasks C and D require mutual information input and are *coupled* (interdependent). The matrix suggests a plan for the order of the tasks: A, then B in parallel with C and D, the latter two being performed in a closely coordinated way.

TIME CONCURRENCE AND DESIGN ACTIVITY OVERLAPPING

Task overlapping originated in the automotive (notably, Toyota and Honda), camera (e.g., Canon), and aerospace industries (Takeuchi and Nonaka, 1986). It has also been adopted in electronics (Krishnan et al., 1997) and the software industry, and recently even in making movies (*New York Times*, May 5, 1997: D1).

Once development tasks have been defined in the overall design problem, their execution over time must be planned. While it is natural to execute interdependent tasks in parallel, it is easiest to perform sequentially dependent tasks in their logical sequence. This results, however, in a lengthy sequential process similar to a relay race. To reduce the time needed, CE attempts to, at least partially, overlap the sequentially dependent activities.

In the context of classical project planning terms, overlapping proposes, in effect, to shorten the critical path of a project by "softening" precedence relationships, and by conducting sequential activities in parallel. This offers a fundamental time advantage, but also has drawbacks. In a fully sequential process, downstream starts with finalized information from upstream, whereas in an overlapping process, it has to rely on preliminary information. This approach can be risky if the outcome of the upstream activity is too uncertain to be accurately predicted. Under these conditions, overlapped activities create uncertainty for the downstream activity. The uncertainty would not exist in a sequential process. Thus, a trade-off exists between time gained from parallel execution and rework caused by uncertainty in the project. An optimal balance between parallelity and rework has been derived via analytical models and confirmed by several empirical studies showing that concurrence benefits do decrease with increasing project uncertainty.

Studies have shown that overlap in product development is not equally applicable in all situations. This is supported by several empirical studies. In their study of the world computer industries, Eisenhardt and Tabrizi (1995) identify substantial differences across different market segments. For the stable and mature segments of mainframes and microcomputers, the authors find that overlapping development activities significantly reduces time-to-market. However, in rapidly changing markets such as printers and personal computers, overlapping is no longer found to be a significant accelerator. Eisenhardt and Tabrizi argue that compressing the development process through activity overlaps only yields a time reduction if the market environment is stable and predictable.

Terwiesch and Loch (1996) find that project uncertainty in general (caused by the market or the technology) reduces the benefits from overlapping. Their study is based on data from 140 completed development projects across several global electronics industries. Although overlap helps to reduce project completion time overall, it is less helpful in projects with late uncertainty reduction than it is in projects with early uncertainty reduction. Intensive testing and frequent design iterations emerge as effective measures to reduce completion times in projects with late uncertainty reduction.

The effects of *uncertainty* in CE are compounded by product *complexity*. Complexity of a product refers to the number of elements in the system and the level of interactions among these elements (Fitzsimmons et al., 1991). The higher the complexity, the more interactions among subsystems exist in the design structure matrix, and the more components are affected by uncertainty manifested in ECs. This makes intensive coordination among the parallel tasks even more important, which is discussed next.

INFORMATION EXCHANGE

When the subtasks of the overall product design problem are executed, information exchange is required to keep the tasks coordinated. In simple cases, the information exchange may occur prior to the start of the downstream activity. Such *ex-ante* coordination can be achieved by, for example, design rules or preferred parts lists, which are often used in the context of design-for-manufacture. These methods help to anticipate problems in the manufacturing process as early as possible, but they are not sufficient when the design problem is complex (Wheelwright and Clark, 1992). In this case, ongoing coordination is required during task execution. This is consistent with Clark and Fujimoto's (1991) finding that successful task overlapping is supported by intensive information exchange.

Loch and Terwiesch (1998) show that the more significant the number of changes from upstream and the dependency of the downstream task, the more intensive communication is required. High communication capabilities are thus an important enabler of task overlapping. If shared problem solving activities are performed before the actual design tasks begin, uncertainty levels during development may be significantly reduced in routine development projects.

These analytical results are consistent with empirical evidence: Morelli et al. (1995) examined communication patterns during the development of electrical computer board connectors, and they found that most communication took place among activities that were heavily interdependent. In a more general context, the relationship among dependencies, uncertainty, and communication has been extensively studied in the organizational sciences (e.g., Thomson, 1967).

Coordination format

In addition to the frequency of communication, the format of information exchanged between concurrent activities is important. Ward et al. (1995) report that Toyota in its product development process pursues a seemingly excessive number of parallel prototypes and "communicates ambiguously." They conclude that Toyota uses "set-based" concurrence, where whole sets of possible design solutions are explored in parallel, and over time, the set of solutions considered is narrowed down (as a rule, not broadened). For example, key body dimensions are not fixed until late in the development process. However, when a fixed solution is reached, it no longer changes unless absolutely necessary. Ward et al. contrast the set-based approach to "iterative" or "point-to-point" design, where design teams fix solutions for their respective components and then coordinate among teams by iteration.

One cannot, however, conclude that a set-based approach is always superior. Terwiesch and Loch (1997) examined climate control system development of an automobile manufacturer and encountered both approaches, set-based and iterative, the choice based on the nature of the component developed. The trade-offs involved in the choice are summarized in Figure 3, where the two communication approaches are classified according to the nature of the information exchanged. Information precision refers to whether it is communicated as a range or a precise value, while stability refers to whether the information tends to remain stable afterwards, or whether it is subject to change.

If the cost of making the wrong choice, and if the number of candidate solutions that must be pursued is relatively small, then the set-based approach offers advantages. On the other hand, if the development of a component is very flexible, i.e., engineering changes are very easy to incorporate, then the iterative approach is preferable.

Thus, the choice between set-based and iterative concurrence depends on the nature of the component developed, and often a combination is used. The choice is influenced by the technological and process capabilities of the organization. This is examined further in the next Section.

SUPPORTING INFRASTRUCTURE FOR IMPLEMENTATION

The three aspects of CE (task definition, time and information concurrence) must be viewed within the context of a supporting infrastructure (Fitzsimmons et al., 1991).

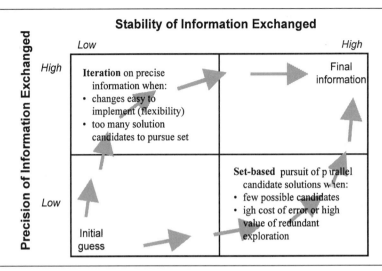

Figure 3. Set-based and iterative approach to communication.

Three infrastructure components are discussed below: technologies, project organization, and support processes.

Technology and tools

New problem solving technologies can play an important role not only in diminishing the cost of iteration, but also in anticipating and eliminating project uncertainty in the first place. Examples of such new technologies are CAD, simulation tools, and rapid prototyping. These technologies have the potential of fundamentally changing the trade-offs involved in time and information concurrence. Thomke (1996) studied integrated circuit design and distinguished between flexible design technologies, which allow incorporation of design changes quickly and cheaply, and inflexible design technologies. He found that projects using flexible design technologies are more efficient than projects using inflexible technologies. The difference stems from the lower cost of direct iterations (design changes), but also from the reduced need for costly resource investments to reduce the risk of design changes.

Thomke's (1996) study offers evidence of how new technologies shift the information exchange trade-off away from set-based toward iterative concurrence (see Figure 3). In particular, powerful CAD systems allow the frequent information sharing and coordination necessary for activity overlapping in complex projects (a widely quoted example is the Boeing 777 jet, which was developed with heavy CAD usage).

Communication technologies also have the potential of influencing CE practices (e.g., Allen, 1986). E-mail, portable phones, and shared databases all have the effect of making communication more time-efficient and independent of personal meetings between busy parties. However, even without considering implementation problems of incompatible data formats and protocols, communication technologies do not make communication instantaneous and painless.

Project organization

Often, the structure of the project organization mirrors the logical structure of the design problem: the architecture determines communication patterns and physical proximity.

Commonly, four main types of project organizations are distinguishable (e.g., Wheelwright and Clark, 1992, Chapter 8). In a *functional structure*, people are grouped by discipline, and projects are coordinated via *a priori* agreed-upon specifications complemented by occasional meetings. In *a lightweight structure*, each function has a representation in a coordinating committee, led by a project coordinator. A *heavy weight structure* makes the coordinator a real project manager, with control over budgets and direct supervision of the functional representatives. Finally, the *autonomous structure* pulls the whole project outside of the regular organizational structure, and the project manager becomes akin to a line manager, who controls all resources and evaluates all project employees.

The strength of the autonomous structure is focus and speed. However, it often suffers from integration problems with the rest of the organization, which may result in rejected output and disrupted career paths, as well as lost commonality (of parts and designs) across different products. The functional structure, on the other hand, is weak in its ability to coordinate complex projects with tight time-to-market goals because communication and coordination is difficult across functional boundaries. However, it supports deep expertise in specialty disciplines, which is important in research projects with high technical uncertainty.

The project organization also influences behavioral aspects of the parties involved. Clark and Fujimoto (1991: 212f.) point out the importance of *attitudes* in CE: an attitude of information sharing and cooperation is preferred. Hauptman and Hirji (1996) confirm this observation in a multi-industry study of development projects using CE. They find that CE overcome the differences resulting from functional specialization: the parties involved must understand and trust one another sufficiently to be willing to share information. Once two-way communication happens, it tends to improve trust and employee satisfaction. For example, misunderstandings may arise from two engineering groups speaking different technical languages and having different perceptions of the same problem (e.g., Terwiesch and Loch, 1998).

In addition, Hauptman and Hirji find that CE requires a tolerance for releasing and using preliminary information, information that is imprecise and/or unreliable. Workers tend to minimize the risk of having to re-do tasks by suppressing information, procrastinating, etc. Active management is necessary to overcome such risk-avoiding behavior in implementing CE.

Support processes

CE requires a number of effective support processes, interfaces with suppliers, engineering change orders, and interfaces among multiple development projects.

Clark and Fujimoto (1991) observed that close *integration with suppliers* in product development is necessary for applying CE. This was further studied by Liker et al. (1995), who identified important success patterns.

1. Simplify the coordination problem by outsourcing self-contained chunks.
2. Communicate clearly and often (e.g., by having resident engineers at the supplier's site.
3. Use a simple, stable development process that is well understood by both sides.
4. Suppliers participating in CE to develop their capability of solving technical problems without external help ("full service capability").

A general principle is that the investments and open information sharing required for involving suppliers in CE only work in long-term relationships with mutual commitment that prevents abuse and cheating. Such mutual commitment can be supported by investments in common infrastructure and systems (such as CAD or order management) and by establishing a mutually agreed-upon and understood development process. The process should describe who is responsible for which deliverables, what information must be exchanged, and what milestones must be met.

Engineering change orders

Engineering change orders are an almost unavoidable consequence of CE, as many design tasks started on the basis of preliminary information that is subject to change. ECs arise also during component integration, and during fine-tuning of the design to improve market success. EC's often consume significant resources and therefore deserve careful management. Five principles can help make the EC management process effective (Terwiesch and Loch, 1998):

- Give clear responsibilities and avoid unnecessary handoffs in the process.
- Manage capacity in order to reduce congestion. Technical personnel often have multiple responsibilities, which can result in congestion and delays.
- Set-ups occur every time an engineer switches from one activity to another. To reduce set-ups, engineers (just as in manufacturing) batch their work, which can lead to substantial delays for jobs in the queue.
- A complex product with a highly integrated architecture makes EC management more difficult, as ECs will have a "snowball" effect on other components. In this situation, it is important to educate all personnel involved about the key interactions among components, and give strong incentives for fast problem communication and resolution.

Multiple development projects

In many companies, engineering work is often performed in functional organizations of specialists, who serve several projects at the same time. On the one hand, projects compete for the same resources, leading to congestion and interference with the intensive communication and coordination within the project (Adler et al., 1995). This requires a clear prioritization of projects, not only when projects are initiated, but also during execution, when resource conflicts arise. Clear rules are necessary (e.g., degree of lateness) to settle conflicts speedily.

On the other hand, the pursuit of multiple projects in parallel builds expertise in people. Nobeoka and Cusumano (1995) find that overlapping development projects for consecutive product generations, rather than sequential development projects, facilitate improved sharing of specifications and the transfer of knowledge.

Key Concepts: Autonomous Teams; Convergence Rate of a Project; Downstream Sensitivity; Flexible Design Technologies; Heavy Weight Project Manager; Light Weight Project Manager; Modular Architecture; Product Architecture; Time to Market; Work Transformation Matrix.

Related Articles: Concurrent Engineering; Integrated Product Development: The Case of Westinghouse Electronic Systems; Mass Customization; Product Design for Global Markets.

REFERENCES

Adler, P.S., A. Mandelbaum, V. Nguyen, and E. Schwerer (1995). "From Project to Process Management: An Empirically Based Framework for Analyzing Product Development Time." *Management Science*, 41, 458–484.

Allen, Th.J. (1986). "Organizational Structure, Information Technology, and R&D Productivity." *IEEE Transactions on Engineering Management*, EM 33, 212–217.

Blackburn, J.D. (1991). *Time Based Competition*, Homewood: Business One Irwin.

Clark, K.B. and T. Fujimoto (1991). *Product Development Performance: Strategy, Organization and Management in the World Auto Industry*. Harvard Business School Press, Cambridge.

Eisenhardt, K.M. and B.N. Tabrizi (1995). "Accelerating Adaptive Processes: Product Innovation in the Global Computer Industry." *Administrative Science Quarterly*, 40, 84–110.

Eppinger, S.D., D.E. Whitney, R.P. Smith, and D.A. Gebala (1994). "A Model-Based Method for Organizing Tasks in Product Development." *Research in Engineering Design*, 6, 1–13.

Fitzsimmons, J.A., P. Kouvelis, and D.N. Mallik. "Design Strategy and Its Interface with Manufacturing and Marketing: a Conceptual Framework." *Journal of Operations Management*, 10, 398–415.

Gerwin, D. and G. Susman (1996). "Special Issue on Concurrent Engineering." *IEEE Transactions on Engineering Management*, 43, 118–123.

Gulati, R.K. and S.D. Eppinger (1996). "The Coupling of Product Architecture and Organizational Structure Decisions." MIT Working Paper 3906.

Ha, A.Y. and E.L. Porteus (1995). "Optimal Timing of Reviews in Concurrent Design for Manufacturability." *Management Science*, 41, 1431–1447.

Hauptman, O. and K.K. Hirji (1996). "The Influence of Process Concurrency on Project Outcomes in Product Development: an Empirical Study with Cross-Functional Teams." *IEEE Transactions in Engineering Management*, 43, 153–164.

Imai, K., I. Nonaka, and H. Takeuchi (1985). "Managing the New Product Development Process: How the Japanese Companies Learn and Unlearn." In: Clark, K.B., R.H. Hayes, and C. Lorenz (eds.). *The Uneasy Alliance*. Harvard Business School Press, Boston.

Jaikumar, R. and R. Bohn (1992). "A Dynamic Approach to Operations Management: An Alternative to Static Optimization." *International Journal of Production Economics*.

Krishnan, V., S.D. Eppinger, and D.E. Whitney (1997). "A Model-Based Framework to Overlap Product Development Activities." *Management Science*, 43, 437–451.

Liker, J.K., R.R. Kamath, S.N. Wasti, and M. Nagamachi (1995). "Integrating Suppliers into Fast-Cycle Product Development." In J.K. Liker, J.J.E. Ettlie, and J.C. Campbell (eds.). *Engineered in Japan: Japanese Technology Management Practices*. Oxford University Press, Oxford.

Loch, C.H. and C. Terwiesch (1998). "Communication and Uncertainty in Concurrent Engineering." *Management Science* 44, August.

Morelli, M.D., S.D. Eppinger, and R.K. Gulati (1995). "Predicting Technical Communication in Product Development Organizations." *IEEE Transactions on Engineering Management*, 42, 215–222.

Nevins, J.L. and D.E. Whitney (1989). *Concurrent Design of Products and Processes*. McGraw Hill, New York.

Nobeoka, K. and M.A. Cusumano (1995). "Multiproject Strategy, Design Transfer, and Project Performance." *IEEE Transactions on Engineering Management*, 42, 397–409.

Pisano, G. (1997). *The Development Factory*. Harvard Business School Press, Boston.

Smith, R. (1997). "The Historical Roots of Concurrent Engineering Fundamentals." *IEEE Transactions on Engineering Management*, 44, 67–78.

Smith, R.P. and S.D. Eppinger (1997). "Identifying Controlling Features of Engineering Design Iteration." *Management Science*, 43, 276–293.

Steward, D.V. (1981). *Systems Analysis and Management: Structure, Strategy and Design*. Petrocelli Books, New York.

Takeuchi, H. and I. Nonaka (1986). "The New Product Development Game." *Harvard Business Review*, 64, 137–146.

Terwiesch, C., C.H. Loch, and M. Niederkofler (1996). "Managing Uncertainty in Concurrent Engineering." Proceedings of the 3rd EIASM Conference on Product Development, 693–706.

Terwiesch, C. and C.H. Loch (1997). "Management of Overlapping Development Activities: a Framework for Exchanging Preliminary Information." Proceedings of the 4th EIASM Conference on Product Development, 797–812.

Terwiesch, C. and C.H. Loch (1998). "Managing the Process of Engineering Change Orders: The Case of the Climate Control System in Automobile Development." Forthcoming in the *Journal of Product Innovation Management*.

Thomke, S. (1996). "The Role of Flexibility in the Design of New Products: An Empirical Study." Harvard Business School Working Paper 96-066.

Ulrich, K. (1995). "The Role of Product Architecture in the Manufacturing Firm." *Research Policy*, 24, 419–440.

Ulrich, K. and S.D. Eppinger (1995). *Product Design and Development*. McGraw Hill, New York.

Ward, A., J.K. Liker, J.J. Cristiano, and D.K. Sobek II (1995). "The Second Toyota Paradox: How Delaying Decisions Can Make Better Cars Faster." *Sloan Management Review*, Spring, 43–61.

Wheelwright, S.C. and K.B. Clark (1992). *Revolutionizing Product Development*. The Free Press, New York.

23. MASS CUSTOMIZATION

REBECCA DURAY

University of Colorado at Colorado Springs, CO, USA

ABSTRACT

Mass customization has become popular in the nineties. Companies are striving to provide customized products and services to their customers at low costs. No industry is immune from the desire to produce low cost customized products. Mass customizers can be found in both consumer and industrial markets. As the concept of mass customization gains popularity, more companies may venture into the uncharted territory of mass customization without a clear idea about the nature of mass customization and without the appropriate manufacturing system to support this new marketing concept. Looking at companies that have successfully implemented mass customization provides little guidance. The diversity of the products and production system for mass customization does not lead to easily generalizable concepts.

Although mass customization is usually thought of in terms of marketing or competitive strategy, mass customization should be viewed as a competitive capability that resides in marketing, manufacturing, and engineering functions. The integration of these functional areas through product design is the key to breaking the paradox of mass customization. A mass customizer must identify a market for low-cost customization, determine the customizable features required of that market, design a product that can provide customization with mass production, and manufacture the product in a cost effective manner. This view of mass customization fits the resource-based view of business strategy (Wernerfelt, 1984; Barney, 1991) and suggests that the resources needed to provide a mass customization strategy reside in the integration of marketing, manufacturing, and engineering strategies.

Several types of mass customization are explained in this article. Further, mass customization in the following companies are discussed here: **Bally Engineered Structure; Dell; General Motors; Levi-Strauss; Ross Controls and many more in Table 1.**

The practice of mass customization does not fit the mindset associated with conventional manufacturing methods. Historically, companies chose to produce either customized, crafted products or mass produced, standardized products. This traditional practice means that customized products are made using low volume production processes that cope well with a high variety of products. Similarly, mass production process is chosen for making standardized products in a high volume, low cost environment. Mass customization, as defined by Davis (1987) provides a one-of-a-kind product manufactured on a large scale. Customers are able to purchase a customized product for the cost of a mass produced item. Thus, mass customization represents an apparent paradox by combining customization and mass production, offering unique products in a mass produced, low cost, high volume production environment.

WHAT IS MASS CUSTOMIZATION?

The essence of mass customization, as distilled from literature (Pine, 1993; Pine, Victor and Boyton, 1993; Kotha, 1995) is providing personalized products at reasonable prices or mass production prices. Although this description invokes the essence of mass customization, it does not help us identify companies that are mass customizers. Literature on mass customization labels a broad range of companies as mass customizers, but the diversity of these companies can be confusing to the observer. The variety of the attributes possessed by the example companies requires a definition that establishes rigorous conceptual boundaries of mass customization and presents a means to distinguish among the vast array of mass customizers.

The boundaries of mass customization can be more clearly established by separating two issues. The first issue is the nature of customization. When should a product be considered customized, and not just another version of product proliferation? The second issue is the "mass" in mass customization. How can unique products be developed in a mass production fashion? How can high-volume, low-cost customization be implemented? Pine (1993) suggests that modularity is the key to achieving mass customization. Modularity is critical to gaining scale or "mass" in mass customization. The issues of customization and modularity are discussed below for a more comprehensive development of the idea of mass customization.

Customization issues

Customized products are uniquely produced for each customer, which requires that the customer be involved in the design process. Mintzberg's (1988) defined customization as taking three forms: *pure, tailored* and *standardized*. The role of the various parts of the value chain and the degree of uniqueness of the product differ in the three forms of customization.

A *pure* customization strategy furnishes a product developed from scratch for each unique customer. This type of customization affects the entire chain, from design, fabrication, assembly and delivery, and provides a highly unique product. Construction projects provide many good examples of pure customization: an architecturally designed house, an office building, or a highway overpass.

Tailored customization alters a basic design to meet the specific needs of the customer. The customer affects the value chain at the fabrication level, where standard products are changed. Eyeglasses are a good example of tailored customization. Customer select from standard frames, but each lens is fabricated to the unique optical prescription of the customer.

In *standardized* customization, a final product is assembled from standard components. Here, the customer penetrates the assembly and delivery processes through the selection of the desired features from a list of standard options. A familiar example of standardized customization can be seen when one orders a pizza. By selecting items on a pizza from a prescribed list, the customer enters the assembly stage of production which produces a standardized custom pizza.

Mass production issues

Mass customization requires that unique products be provided in a cost effective manner, in essence, it requires volume production. To address the issues of mass production, mass customizers must adopt some form of modularity in order to provide both economies of scope and economies of scale simultaneously. This is true for all types of manufacturing systems. Even in group technology, family groupings are determined by component commonality which incorporates some form of modularity providing both economies of scope and scale. The production of standardized modules is the key to high volume "mass" customization. For mass customization, Ulrich and Tung (1991) recognized several types of modularity shown in Figure 1.

Component sharing modularity uses common components in the design of a product. Products are uniquely designed around a base unit of common components. *Component swapping* is the ability to switch options on a standard product; modules are selected from a list of options to be added to a base product. *Sectional modularity* is similar to component swapping, but focuses on arranging standard modules in a unique pattern. *Mix modularity* also is similar to component swapping, but is distinguished by the fact that, when combined, the modules lose their unique identity. *Bus modularity* is the ability to add one or more modules to an existing base. *Cut-to-fit modularity* alters the dimensions of a module before combining it with other modules. That is, products are

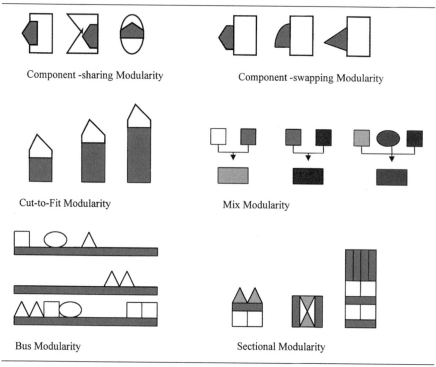

Figure 1. Modularity types.

customized by varying dimensions such as length, width, or height. Various types of modularity can be used separately or in combination to provide a unique end product.

TYPES OF MASS CUSTOMIZATION

When known mass customizers are examined, it is difficult to determine commonalties in manufacturing systems because each company uses a manufacturing systems appropriate to their product, process and industry. For example, Dell Computers manufactures a mass customized product with an assemble-to-order process. Each base computer is fitted with the appropriate features selected by the customer from a common list of options. In contrast, Levi Strauss provides personalize jeans for women through a process that cut a new pattern for each pair of customized jeans. Both manufacturers should be considered mass customizers, yet their manufacturing systems are very different. Mass customization includes a vast array of products made with appropriate manufacturing systems to deal with the mass customization paradox.

The previous section argues that the model of mass customization that emerges from the literature uses two critical definers: customization and modularity. Therefore, mass customization can be defined as:

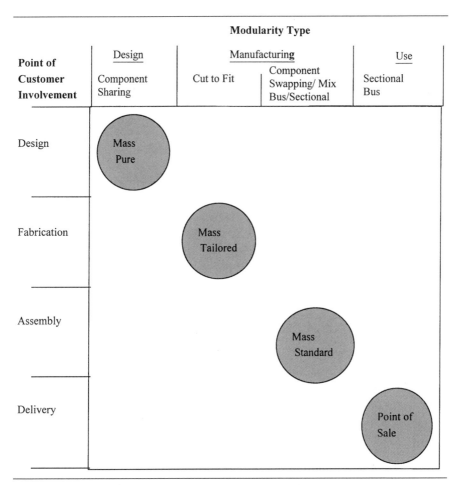

Figure 2. Dimensions of mass customization.

A mass customized product provides the end-user with specific customization achieved through the use of modularity of components.

Practical distinctions can be made between mass customizers based on the point of customer involvement in the process and the type of modularity employed. Figure 2 shows the archetypes of mass customization based on the point of customer involvement and type of modularity. Table 1 shows company examples for each mass customization archetype.

In Figure 1, in the case of *mass pure customization*, modularity takes the form of design modularity with customers involvement from an early stage of design. Core components are easily shared, and custom built features are designed around the common modules. For example, a common engine type may be shared in specialty machines, such as a crane or mining equipment, but the actual application may require special

Table 1 Case examples of mass customization types

Mass pure	Mass tailored	Mass standard	Point of sale
Sign-tic (Kk)	Bally Engineered Systems (PVB, P, PP)	Motorola (PVB, P, Do)	Personics cassettes (P,SP)
Ross Controls valve systems (PPR)	Lenscrafter (P)	TWA Getaway Vacations (P)	Electronic components —Stereo —Computers (D, GK)
	Andersen Windows (PPR)	Lutron lighting fixtures (SP)	Ikea Modular Furniture (D)
	National Bicycle (Ko)	Hallmark/American Greetings (PPR)	Legos (P)
		Peapod Virtual Supermarket (PPR)	
		R.R. Donnelley/ Farm Journal tailored to subscriber (PPR)	

D = Davis (1987); Do = Donlon (1993); GK = Garud and Kumaraswamy (1993); Ko = Kotha (1995); Kk = Kubiak (1993); P = Pine (1993); PP = Pine and Pietrocini (1993); PPR = Pine, Peppers, and Rogers (1995); PVB = Pine, Victor, Boyton (1993); Sp = Spira and Pine (1993)

features. Ross Controls provides mass customized valve systems for General Motors using a common platform that is individually customized for each stamping press. These valve systems, which are mass customized, perform better than the systems they replaced at one-third the price (Pine, Peppers, and Rogers, 1995).

Both *mass tailored* and *mass standard* customizers are modular in production; standard components are assembled to create unique products. However, *mass tailored* customizers use cut-to-fit modularity, where standard modules are altered to fit unique customer specifications. Customers are involved at the fabrication stage of production, where modules must be altered to customer specifications. Products that are produced to customer specified size or length fit this category. Bally Engineered Structures manufactures walk-in coolers, refrigerated rooms, and clean rooms. Product modules are cut-to-fit specific dimensions provided by the customer to make unique rooms manufactured from modular components (Pine, Victor, and Boyton, 1993).

Mass standard customization also occurs at the production stage, and requires modular components that can be joined in production to make unique end items. Component swapping, mix, sectional, and bus modularity could all be applied in this customization configuration. The unique nature of the product is developed from the many possible combinations of the components adapted to suit the unique needs of the customer. Customers are involved in the assembly stage of production as options are chosen from a limited set of module choices. Motorola pagers, are the result of "mass standard" customization. Pagers are designed to a customer's specification from a wide range of options that are introduced at the production phase (Pine, 1993; Donlon, 1993).

Point of sale mass customization uses completed modules or standard products that are combined by the consumer to create a unique product. Customers provide the customization through their use and selection of modular components. Stereo components (receivers, CD players, cassette deck and speakers) are combined by the consumer to provide a customized audio system. Each customer selects and combines the required modules to form unique products. Modular furniture and bookshelves are also point of sale customized products. Bus modularity also may be employed as finished modules can be added to an existing structure. Track lighting (Pine, 1993) is a familiar example of bus modularity. Different lighting fixtures are added to an electrical track making the track units unique. A child's yard play-set with a base frame to which swings, bars, horses, or rings are added can be considered a point of sale mass customized product.

WHAT IS NOT MASS CUSTOMIZATION?

Only those manufacturers that employ modularity and involve the customer in the design process for specific units should be considered mass customizers. It is interesting to note that some companies used as examples of mass customizers do not fit this mass customization typology. Davis (1978) uses Cabbage Patch dolls as an example of mass customization that gained a competitive edge from the unique nature of its products. Each doll is a unique end item that customers select at a retail outlet. Although customers do select a doll, they do not have a doll made to their specifications such as, eye or hair color. Customers do not participate in the design process; therefore, the Cabbage Patch doll is not an example of mass customization.

Similarly, Pine (1993) misclassifies Swatch Watch as an example of mass customization. Swatches are unique products, and the wide selection gives customers many choices. Although this provides great variety, customers do not have the ability to request changes. Therefore, this product is not an example of a mass customized product.

Key Concepts: Bus Modularity; Component Sharing Modularity; Component Swapping Modularity; Cut-to-Fit Modularity; Mass Standard Customizers; Mass Tailored Customizers; Mix Modularity; Pure Customization; Sectional Modularity; Standardized Customization; Tailored Customization.

Related Articles: Mass Customization and Manufacturing; Manufacturing Flexibility; Product Design for Global Markets.

REFERENCES

Barney, J. (1991). "Firm Resources and Sustained Competitive Advantage." *Journal of Management*, 19 (1).

Davis, S.M. (1987). *Future Perfect*. Reading, Massachusetts, Addison-Wesley Publishing Company.

Donlon, J.P. (1993). "The Six Sigma Encore; Quality Standards at Motorola." *Chief Executive*, November, 90.

Duray, R. (1997). *Mass Customization Configurations*, Ph.D. Thesis, The Ohio State University, Columbus, Ohio.

Garud, R. and A. Kumaraswamy (1993). "Technological and Organizational Designs for Realizing Economies of Substitution." *Strategic Management Journal*, 16, 93–109.

Kotha, S. (1995). "Mass Customization: Implementing the Emerging Paradigm for Competitive Advantage." *Strategic Management Journal*.

Kubiak, J. (1993). "A Joint Venture in Mass Customization." *Planning Review*, July, 21 (4), 25.

Mintzberg, H. (1988). "Generic Strategies: Toward a Comprehensive Framework." *Advances In Strategic Management*, 5, 1–67.

Pine II, B.J. (1993). "Mass Customizing Products and Services." *Planning Review*, July, 21 (4), 6.

Pine II, B.J. and T.W. Pietrocini (1993). "Standard Modules Allow Mass Customization at Bally Engineered Structures." *Planning Review*, July, 21 (4), 20.

Pine II, B.J., D. Peppers, and M. Rogers (1995). "Do You Want to Keep Your Customers Forever?" *Harvard Business Review*, 103–114.

Pine II, B.J., B. Victor, and A.C. Boyton (1993). "Making Mass Customization Work." *Harvard Business Review*, 71, 108–119.

Spira, J.S. and B.J. Pine II (1993). "Mass Customization." *Chief Executive*, March, 83, 26.

Ulrich, K. and K. Tung (1991). "Fundamentals of Product Modularity." *Proceedings of the 1991 ASME Winter Annual Meeting Symposium on Issues in Design/Manufacturing Integration*. Atlanta.

Wernerfelt, B. (1984). "A Resource-based View of the Firm." *Strategic Management Journal*, 5, 171–180.

24. MASS CUSTOMIZATION AND MANUFACTURING

PRATAP S.S. CHINNAIAH
SAGAR V. KAMARTHI
Northeastern University, Boston, MA, USA

ABSTRACT

The concept of mass production was introduced in the 1800s and reached its full development in the 1920s. However, craft production, where artisans built products to a customer's specifications continued catering to some select market niches. In the 1970s the competition intensified among producers and at the same time the variety of products proliferated. As a result, lean production which combines the advantages of craft and mass production, while avoiding the high cost of the former and the inflexibility of the latter, was born (Womack et al., 1990). In the 1990s, while some markets are reaching their saturation limits for products and customers are becoming more demanding, we see the rise of mass-customized production (Anderson, 1998; Davis, 1987; ICM Conferences Inc., 1996; Kotha, 1995; Lau, 1995; Preiss, 1995; Roos, 1995).

The concept of mass customization (MC) focuses on satisfying individual customer's unique needs with the help of technologies such as agile manufacturing, flexible manufacturing systems, computer integrated manufacturing, and information and communication systems. In MC, the needs of an individual customer are translated into design, accordingly produced, and delivered to the customer. Such mass-customized production (MCP) systems are also referred to as one-of-a-kind production (OKP) systems. MC draws from other production strategies such as personalized production, focused job shop, agile manufacturing, and virtual enterprise (Elantora, 1992; Wortmann, 1992).

Changing global patterns of customer demands are beginning to give MC the competitive edge over lean production. For many companies, MC is becoming an inevitable means for survival in the market place. An observation of companies engaged in MC indicates that most of them have evolved over an extended period of time into this production scenario rather than by systematically implementing a well planned and well documented strategy. Lutron Controls is a good example of a company that entered into mass customization as a matter of survival. The company found that, by providing products according to each customer's unique specifications, they could survive and grow in the competitive market of lighting controls (ICM Conferences Inc., 1996).

The practices of the following companies are described here: **Andersen Windows; Bechtel; Black and Decker; Coca-Cola; Compaq; Custom Clothing Technology Corporation; GE Fanuc; GE; Gillette; Hewlett-Packard; IBM; Individual, Inc.; Lego; Levi-Strauss; Lutron Electronics Co.; Matsushita Bicycle Company; Mercedes Benz; Motorola; Nissan; Reebok; Space Electronics, Inc.; Telco, India; Toyota.**

VARIOUS OF PRODUCTION SYSTEMS

Table 1 illustrates the differences between craft, mass, lean and MCP systems. A mass production system focuses on producing a single product for a mass market. A lean production system focuses on producing a finite number of variants of a single product or a finite set of standard products designed to meet demands of segmented markets. Lean production can offer a considerably large variety of products but this variety is attained without individual customer involvement. The Japanese method of fast paced introduction of an array of products sometimes has a downside. In the nineties, Toyota announced that it would reduce the varieties of the corolla model from 11 to 6. Nissan planned to reduce its variety of engines by 40% in five years (Lau, 1995). GE Fanuc discovered that customers don't want more choice but want what they need. While lean production focuses on the elimination of waste in processes, MC aims at the elimination of waste in products by eliminating the features unwanted by customers. An MCP system attempts to produce just as many variants of a finite set of products as customers indicate they want.

When a production system produces only one product (e.g. GE's electric bulb, Coke's soda can, Gillette's Sensor razor), we can call it a mass production system. A mass production system is not usually designed to respond to a variety of customer demands. The Telco 1010 truck production line in India can be considered as a typical example of a real-world mass production system. On this production line thousands of trucks are produced with no variation in design. When a production system is capable of producing several variants of a product (e.g. the present range of automobiles available

Table 1 Comparison of craft, mass, lean, and mass-customized production systems

Characteristic	CRAFT PROD.	MASS PROD.	LEAN PROD.	MCP
Anybody	Yes	Yes	Yes	Yes
Anything (Any product, Any design)	Yes	No	No	Yes
Any where (in the world)	No	Yes	Yes	Yes
Any time	No	Yes	Yes	Yes
Any volume	No	No	No	Yes
Main theme	Individualized products at unavoidably high cost.	Products for all at lowest possible cost.	One product for each market segment possibly at mass production cost.	One product for each customer possibly at mass production cost.
Nature of products	Nonstandard goods with a lot of variety.	Standard goods of negligible variety with expected tolerance for defects.	High quality standard goods with considerable variety.	High quality, individualized products and services with unlimited but manageable variety.
Productivity aspect	Human labor and skills.	Plant and equipment utilization at the expense of maintaining extra supplies, space, manpower, inventories, etc.	Elimination of nonvalue adding process and operations (process waste) in the value chain.	Increasing the value of product offered to customer by eliminating nonvalue adding features (product waste).
Technologies	Skill based technologies.	Special purpose fixed automation equipment. Narrowly skilled operators.	Programmable and flexible automation equipment: CNC, FMS, Robots, AS/RS, CAD, CAM, CIM, LAN, WAN.	Highly flexible and integrated automation equipment. Internet, ERnet, EInet, CAD, CAM, Digital & Rapid prototyping, FMS, CIM, Flexible tooling/fixturing, EDI, Configurators, Multimedia workstations, Electronic shopping, Electronic kiosks, Personalized smart cards, Virtual reality, Experience simulators, Portable computers.
Markets	Rich class one-to-one market.	Mass market.	Segmented markets driven by customer surveys.	Mass one-to-one market.

in the market, often produced on the mixed-model lines), then such a system can be called a lean production system. In fact, the variety of products that are requested by innumerable customers may be unlimited but the production system tries to satisfy their demands by providing a finite variety of products. A company that can manufacture shoes of any model, any size, any profile, with the user's choice of material is considered an MCP system. That means, an MCP system can almost (if not completely) satisfy customer needs for a selected group of products.

Mass customization is defined as *a business strategy for profitably providing customers with anything they want, anytime, anywhere, any volume, in any way.* A simple intuitive definition of MC is captured in the 5A's of Nissan's vision for the year 2000: anybody, anything, any volume, any time, and anywhere (ICM Conferences Inc., 1996). However, the ideal mass customization represented by this definition is unachievable by even the most dedicated firms that hope to realize the ideal mass customization. A more realistic definition is that *mass customization is the use of flexible processes and organizational structures to produce varied and often individually customized products and services at the price of standardized, mass-produced alternatives.*

To understand the differences among craft, mass, lean production, and MCP systems, the concept of *customer sacrifice* must be understood. Customer sacrifice is the gap between what each customer truly wants or needs and what the company can supply. Companies have to go beyond aggregate customer-satisfaction and seek individual customer satisfaction (Pine II et al., 1995). For example a woman tries, on average, thirteen pairs of jeans before making a purchase; and even after this considerable amount of effort, approximately 30% of the women do not buy pants at all because they cannot find the pair that fit them. The cost of customers' time and their disappointment or dissatisfaction of having to live with imperfect pants is called, customer sacrifice. Levi-Strauss is capitalizing on such differences between customer needs and product functionalities by mass customizing blue jeans for women using the technology supplied by Custom Clothing Technology Corporation. After a woman customer has her measurements taken in a store, the information is then sent to the factory for prompt production.

In the process of manufacturing products according to customer requirements, Ford's truck plant in Kentucky offers 2.5 million variants of trucks, and Motorola offers 29 million varieties of pagers (Henricks, 1996). When a company advertises, *any design, anybody, any volume, anywhere, anything* or *whatever, whenever,* or *wherever,* it is an indication that the company has the capability to undertake mass customization (Pine Ii et al., 1993). Even though craft producers could meet the challenges of *anybody* and *anything* but they would not claim *any volume, anywhere* and *any time.* Crafts producers made in small volumes, costs were very high. In contrast, MC attempts to give customers what they want at the cost of mass production.

CLASSIFICATION OF MASS CUSTOMIZED PRODUCTION (MCP) SYSTEMS

A careful review and analysis of several products produced under the MC protocol demonstrates that, while some products are designed, produced and delivered as customized products, other products are assembled from modules to achieve customization. Some products are produced with embedded technology, which will allow the customers to customize product features to satisfy their individual requirements. In some cases, a small amount of manufacturing or customization is done at the delivery end to suit customers' unique needs. In this way, several opportunities for customization of products exist. They can be classified into five MC production systems depending on customer requirements and customer involvement in the production system.

Type 1 MCP: Make-to-stock MCP (mass-produced and mass-customized products)
Type 2 MCP: Assemble-to-order MCP (with or without a small amount of manufacturing at the delivery end)
Type 3 MCP: Make-to-order MCP (manufacture to order)
Type 4 MCP: Engineer-to-order MCP (design and manufacture to order)
Type 5 MCP: Develop-to-order MCP (R&D, design, and manufacture to order)

This classification has some similarities with the generic types of process choices such as continuous processing, line, batch, jobbing, and project. Products with built-in embedded technology are produced by Type-1 MCP systems. Production systems, which assemble products from a fixed set of pre-manufactured modules, are classified under Type-2 MCP systems. A Type-3 MCP system encompasses production systems, which manufacture products after receiving a customer order. Products that involve design and manufacturing are produced by Type-4 MCP systems. If products involve both research and development in addition to usual design and manufacturing activities, they are delivered by Type-5 MCP systems. Each one of these five MCP systems is discussed in detail with several examples in the following sections.

Type I: Make-to-Stock MCP. In a make-to-stock MCP system, a few standardized products which have a provision for easy customization are produced to stock in large volumes and with long production runs. Products with embedded technology that enable customization by the customer to suit his or her needs are outcomes of make-to-stock MCP. To a large extent, and for all practical purposes, these customizable products are mass produced. Therefore, the products of this kind are known as mass-produced and mass-customized products. This is the simplest kind of MC production because such a production system derives the benefits of mass production and at the same time meets the unique needs of individual but anonymous customers. The following are some of the examples of mass-produced and mass-customized products (Pine Ii, 1993a):

1. A pair of Reebok shoes with an air pump: These shoes contain a pocket of air which can be pressurized or depressurized by the wearer to get the desired amount of cushion. Reebok shoes with air pump are mass produced but the embedded technology satisfies each customer's unique needs.
2. A Mercedes-Benz model with 13 different environmental controls: These built-in controls allow the driver to adjust the car's environment to suit his or her personal comforts and preferences.

What is common in the above examples is that the customers have the convenience of modifying or customizing the products to suit their needs by virtue of the products' embedded technology. This kind of MC requires a careful study of customer requirements to embed matching technologies at the product design stage.

Type 2: Assemble-to-Order MCP. In this type of production system, assembly activity starts after receiving a customer order. These products are assembled from modules that were already manufactured based on product demand forecasts. Therefore a module manufacturing shop is driven by manufacturing resources planning, and the module assembly activity is driven by customer orders. The assemble-to-order MCP can take two different variations. In one situation, the products are assembled from their modules exploiting the concept of modularity in product design. In another situation, after products are assembled from modules, a small amount of manufacturing is employed at the delivery point. These situations are discussed below.

Products assembled from modules: Modularity in product design helps to minimize costs and maximize customizability. Low product costs are achieved through economies of scale of the product modules rather than the products themselves. There are several examples of products that are constructed from modules.

1. Automobiles built with options like all-wheel drive, anti-lock brakes, air bags or navigation systems.
2. Newscasts on the Internet that are assembled together to suit individuals by sections on sports, travel, auto, real estate, or fashion. For example, Individual, Inc. provides published news stories selected to fit the specific and ever changing interests of each client.

From the above examples it can be seen that in such MCP systems, a customer's order typically initiates the assembly of modules into final products.

Products assembled from modules plus a small amount of manufacturing: In this case, the products are first assembled from the standardized modules at the production site and a small amount of manufacturing is carried out at the delivery end for incorporating a cosmetic, or an ergonomic feature into the products. The following are some examples of this type of MC.

1. Watches: The dial, hands, case, and strap are the appearance parts in a watch. If these visual features are customized, a customer can feel that his or her watch is unique. With CAD/CAM installed at dealer show rooms, such a features can be extended to customers.
2. Tennis Rackets: They are mass produced at central factories but can be subjected to a final customization step before their delivery to customers to give an individualized comfortable grip.

Type 3: Make-to-Order MCP. In this class of MC, manufacturing of the product starts after receiving a customer order. Therefore in this production system, customer orders drive component manufacturing, their assembly into products, and the delivery of products. For example,

1. Chemical equipment suppliers produce equipment such as reactors, centrifuges, dryers, filters, etc. after receiving the customer orders.
2. A general-purpose machine tool builder may start manufacturing machine tools only after receiving customer orders.

The production of specialty chemicals and fertilizers which are custom-blended for each hectare of a farm according to the type of soil, salinity, and nitrogen content, calls for redesign, movement, and relocation of process subsystems for each new order (Wortmann, 1992). IBM currently runs 95% of its business on a make-to-order basis. Hewlett-Packard conducts 80% of its business, and Compaq runs almost 100% of its business in a similar fashion (Sullivan, 1995).

The concept of evolving products also comes under this category. Products such as computers have traditionally been of the evolving type. In this type of product, both current and future requirements of the customers are considered together. Now this concept is being applied to other products as well. For example, Nissan has explored the concept of the 'evolving car' that owners could bring to the dealer for the add-on latest innovations, such as improved exhaust system, or restyling features every a few years. Design of these types of products can be very complex because the future needs and expectations of the customers have to be charted in advance. These future needs and expectations must be within the capabilities of the value chain.

Type 4: Engineer-to-Order MCP. In this type of MCP system, products are designed from scratch, then manufactured and delivered to a customer. A customer order initiates the design and manufacturing. A job shop which can design, produce, and deliver according to customers' specifications and requirements falls into this category. The main building-blocks of this type of job shop are general-purpose machine tools, CAD, CNC, FMS, CIM, and a highly skilled workforce. Examples are:

1. An FMS supplier, after receiving a customer order, designs, manufactures, and then delivers a flexible manufacturing system to the customer.
2. Building construction and refurbishment.

Following two years of reengineering, Bechtel established an organizational structure, work processes, and information systems necessary to implement MC which enabled Bechtel to design and construct engineer-to-order power generation plants at a lower cost. Some percentage of orders of Lutron Electronics company are of the engineer-to-order type. Space electronics, Inc. designs and builds one-of-a-kind satellite electronic systems based on orders.

Type 5: Develop-to-Order MCP. In a develop-to-order MCP system, a customer requests that the company carry out research and development, prototype development, process development, and subsequently manufacture and delivery of the product. Typically, research and development institutions undertake these types of jobs. A customer order may conceptually specify the functions expected of the product. Then, intense research and development is carried out, usually over several years, to develop the product and in some situations the production methods also. Each order is treated as a project. Examples are:

1. Spacecraft design and development.
2. A project related to the development of a new and innovative packaging method. It may require a research and development effort, prototype development, testing, and manufacturing process development.

FACTORS THAT INFLUENCE AN MCP SYSTEM

As discussed in the previous section, MC takes on many different forms. In the following sections, a set of factors, which influence different types of MCP systems are discussed. The important factors that influence MCP systems are customer involvement, product variety, production volume, bill-of-material coefficient, tooling and process changes, quality system changes, MIS modifications, learning rate, customer-order decoupling point, manufacturing at delivery site, personnel, advanced technology, the basis for production planning and control, order promising, and the handling of demand uncertainty. The studies of Terry Hill (Hill, 1994) have identified these factors. An understanding of these factors will be most useful for the companies that are adopting a mass customization philosophy or restructuring themselves to achieve an effective MC.

Customer involvement. In make-to-stock and assemble-to-order MCP systems, thorough research and development efforts are required to finalize a product's design. In some cases there is no customer involvement at the design and manufacturing stages but such an involvement is required at the point of sales or assembly stage when an order is placed. In other cases, customer involvement is required at the design stage or the manufacturing stage. In order to acquire customer specifications and requirements, an MCP system must have facilities which might call for an additional investment in technology, time, infrastructure and human skills. For example, Matsushita Bicycle Company, which manufactures unique bicycles, has ergonomic frames at each store for measuring physical dimensions of customers (Moffat, 1990). Companies also learn and capture the information about customer requirements using special software. Customer involvement is low or distant in Type-1 (make-to-stock) MCP and it is very high in Type-5 (develop-to-order) MCP.

Product variety. In the case of make-to-stock and assemble-to-order MCP systems, the product variety, from the customer's point of view, is low compared to make-to-order, engineer-to-order, and develop-to-order MCP systems. In an ideal MC situation the ratio of the product variety to the number of customers is 1:1. The product variety varies from low in Type-1 (make-to-stock) and Type-2 (assemble-t-order) MCP systems to high in Type-5 (develop-to-order) MCP systems.

Production volume. In make-to-stock and assemble-to-order MCP systems the production volumes and lot sizes can be very high and in other MCP systems the customer to lot size ratio tends toward 1:1. Thus, the lot sizes are truly small in make-to-order, engineer-to-order and develop-to-order MCP systems. At Lutron's mass customization plant over 95% of its products have annual shipments of less than 100 units. Therefore, the average production volume and lot size varies from very large in Type-1 (make-to-stock) MCP to 'close to one' in Type-5 (develop-to-order) MCP.

Bill-of-material (BOM) coefficient. Bill-of-Material (BOM) coefficient is defined as the ratio of the number of new components to be designed (for a specific order) to the total number of components in the product expressed as a percent. In customizing an automobile, for example, with 5000 total number of components and 50 new components, the BOM coefficient is 1%. Changes in the bill of materials or engineering can be quite expensive to the company.

When a product is designed and manufactured from scratch, the value of this variable can approach 100%. The number of new (added/modified) components manufactured for building a customized product can influence the cost of order fulfillment. For each and every order, some components could be unique. It can be generalized that the more the BOM coefficient the more the production cost. The BOM coefficient will be "close to zero" for Type 1 (make-to-stock) MCP systems and will increase as one progresses to other types of MC systems.

Cost of production changes. A change in the production system requires changes in the tooling and process, quality system, and management information system. While the requirements of several MCP systems can be met by general-purpose tools, there are some MC situations which generally require a new set of tools or modification of the existing tools.

In several MCP systems, and particularly in engineer-to-order MCP, changes in quality assurance systems (changes in hardware and software) can be very complicated and costly. The higher the magnitude of the changes the greater the additional cost of manufacturing and quality control.

In MCP systems, in addition to the product and the process databases, a customer database has to be maintained. Modifications to management information systems, changes in tooling and processes, and changes in quality systems are expected to be minimal in Type-1 (make-to-stock) MCP, and are generally very high in Type-5 (develop-to-order) MCP.

Learning rate. In mass production, a learning curve with a steep slope is possible because of continuous activity on a single repeating product. In contrast, the learning efforts in MCP systems can be time consuming and complex as each order may be executed only once. Therefore, learning rate varies from very high in the Type-1 (make-to-stock) MCP to very low in the Type-5 (develop-to-order) MCP.

Customer-order decoupling point (CODP). The Customer-Order Decoupling Point (CODP) refers to the point in the value chain at which a customer order drives or shapes the production activities. All the activities downstream of the CODP are driven by forecast-based planning rather than by customer orders (Wortmann, 1992). As shown in Figure 1, the value chain consists of marketing, development, production, and delivery. The chain starts with the customer and ends with the customer. Customers can be anonymous as in the case of a mass-produced and mass-customized products, or the customer can be a specific individual as in the case of make-to-order MCP. As can be easily seen in the value chain, marketing drives the development, and development drives the production, and production drives the delivery of products. The CODP is 'delivery' in Type-1 (make-to-stock) and Type-2 (assemble-to-order) MCP systems and it moves backwards to 'production,' 'development', and 'marketing' in the value chain in Type-3 (make-to-stock), Type-4 (engineer-to-order), and Type-5 (develop-to-order) MCP systems, respectively.

Manufacturing at the delivery end. In some MCP systems some manufacturing activity is required at the point of sales. The complexity of such a provision in terms of technology, human skills, and logistics should be worked out when developing an MC strategy. Extra manufacturing at the delivery end of the value chain might call for the installation of additional manufacturing facilities at the point of sales, and also training of sales personnel in those manufacturing operations. Obviously manufacturing at the delivery site doesn't exist in Type-1 (make-to-stock) MCP systems.

Personnel. In most MCP systems, special skills are required of the people employed to manage and operate these systems. Because it is neither possible nor desirable to automate all activities in MCP systems, the people are required to be highly skilled, multi-skilled craftsmen. Continuous improvement of the skills of the personnel an essential part of the MC philosophy. The teams working on individual orders must operate in a frictionless manner to make MC possible. At Bally Engineered Structures, Inc. the workers themselves are now responsible for planning and supervising their own work (Pine II and Pietrocini, 1993). The concept of virtual teams consisting of product designers, tool planners, and production engineers placed at different geographical locations is advocated by many experts. As in agile manufacturing, the objective here should be to combine the organization, technology, and people into an integrated whole.

Advanced technologies used. As shown in Figure 1, the implementation and operation of MCP systems are possible with both low and high technologies. The more the complexity of the technology involved in MC, the higher the cost of implementation and operation of the production system. Except in the production of mass-produced and mass-customized products, highly flexible general-purpose technologies will be required to operate MCP systems.

To operate an MC business, several technologies such as CAD, virtual reality, customer database, electronic kiosks, and multimedia communications may be required at the marketing stage for identifying a customer's requirement. For example, Motorola's pager sales representatives use laptop computers when they work with a customer. Once the design specifications are acquired from the customer, the pager sales representative hooks up the laptop to a telephone line and transmits the design specifications to the factory. In a very short time a bar code identifying the customer is created. From

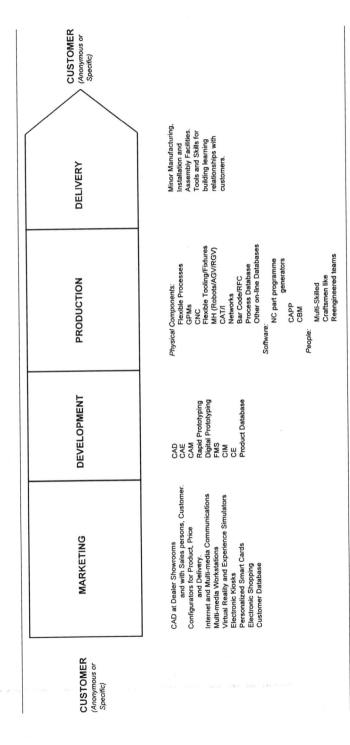

Figure 1. Value chain and its supporting components for mass customization.

CUSTOMER
(Anonymous or Specific)

MARKETING

DEVELOPMENT

PRODUCTION

DELIVERY

CUSTOMER
(Anonymous or Specific)

CAD at Dealer Showrooms
and with Sales persons, Customer.
Configurators for Product, Price
and Delivery.
Internet and Multi-media Communications
Multi-media Workstations
Virtual Reality and Experience Simulators
Electronic Kiosks
Personalized Smart Cards
Electronic Shopping
Customer Database

CAD
CAE
CAM
Rapid Prototyping
Digital Prototyping
FMS
CIM
CE
Product Database

Physical Components:
Flexible Processes
GPMs
CNC
Flexible Tooling/Fixtures
MH (Robots/AGV/RGV)
CAT/I
Networks
Bar Code/RFC
Process Database
Other on-line Databases

Software: NC part programme
generators

CAPP
CBM

People: Multi-Skilled
Craftsmen like
Reengineered teams

Minor Manufacturing,
Installation and
Assembly Facilities.
Tools and Skills for
building learning
relationships with
customers.

that point onwards, a flexible manufacturing system takes over to manufacture the customized pager.

Sales representatives from Anderson Windows use a software system on a workstation that features 50,000 possible window components to help customers design their own windows. Once the customer submits the desired window design, the system automatically generates error-free quotations and manufacturing specifications. At the development stage, technologies such as personalized smart cards, electronic kiosks, internet, CAD, CAE, configurators, and concurrent engineering design tools will be useful in reducing the design lead times. As listed in Table 1, some other technologies specifically required for mass customization are configurators, stereolithography, internet, multimedia workstations, personalized smart cards, virtual reality, groupware, multimedia communications, electronic data interchange (EDI), flexible fixturing, etc.

Production planning and control (PPC). Production Planning and Control (PPC) is based on demand forecasts for make-to-stock and assemble-to-order MCP. For example, Lego blocks are mass produced based on forecasts. Whereas in the other types of MCP systems, assembly, manufacturing, engineering, and development are driven by customer orders.

Order promising. Order promising in a make-to-stock MCP system is based on the inventory of the finished goods, and in an assembled-to-order MCP system, it is based on the inventory of modules out of which the products are assembled. In other MCP systems, it is based on their production system capabilities. For example, a company manufacturing industrial equipment must quickly produce equipment designs according to customer requirements, translate the design into a product, and deliver it. Meeting the due dates on the orders depends on the rapid response and flexible technologies of the production system.

Demand uncertainty. Demand uncertainty is handled by safety stocks in make-to-stock and assemble-to-order MCP systems. For example, in the Reebok's pump case, the safety stocks are calculated based on qualitative and quantitative forecasting models. In other MC schemes it is difficult, or almost impossible to forecast demand because there is no standard product for which forecasts can be prepared. Therefore the demand uncertainty in the make-to-order, engineer-to-order, and develop-to-order MCP systems is handled based on the capacity, capabilities, and the rapid responsiveness of the MCP systems.

Stages in MC implementation. For any company, the first phase in planing for MC is to assess whether to adopt the MC philosophy and its associated operational requirements. Joseph Pine discussed a method consisting of five stages for implementing mass customization (Pine II, 1993b). In this method a company in the first stage aims at customizing services around the existing standardized products. In the second stage, it tries to mass produce customized services or products that customers can easily adjust to their individual needs. In the third stage, the company adjusts its production strategy to provide point of delivery customization. At the fourth stage, the company tries to provide quick response production customization, and at the fifth stage it tries to modularize building-blocks to customize its end products and services.

Some large companies can effectively produce products for multiple markets. However, for such an operation to be successful, a company must separate and organize its manufacturing facilities to meet the needs of each product and then develop sales volumes that are large enough to make those manufacturing units competitive. For example, Lutron Electronics has over 11,000 products which can be delivered anywhere in the world within 48 hours. Lutron has achieved this kind of customization with four levels of customization. In level 1, the customer configures the product from modules. In level 2, cosmetic customization (of color or surface finish) is allowed. In level 3 customization, some product specific manufacturing is carried out. Level 4 customization is meant for totally new designs. The organization of the company must take all levels of customization into account.

Four distinct approaches to mass-customization are presented by Gilmore and Pine II (Gilmore and Pine II, 1997) that are helpful in redesigning a product, process, or business unit. These are named as collaborative, adaptive, cosmetic, and transparent customization strategies. In collaborative customization, the manufacturer conducts a dialogue with individual customers to help them articulate their needs, identify the product specifications, and make customized products for them. Adaptive customizers offer one standard product with embedded technology which users can alter themselves. Cosmetic customizers present a standard product differently to different customers, for instance, with changes in packaging, or with minor modification of some selected specifications. Transparent customizers provide individual customers with unique goods or services without letting them know explicitly that those products have been customized for them (Henricks, 1996).

The degree of collaboration between a manufacturer and a customer can vary in different types of MC situations. Hewlett-Packard Company implemented mass-customization effectively by adapting three organizational principles (Feitzinger and Lee, 1997). These three principles are, modularity in the product, modularity in the process, and supply network modification to provide customization at the last step in the value chain, that is, the task of differentiating a product for a specific customer is postponed until the latest possible point in the value chain.

Increase Modularity. Products should be designed employing the concept of modularity to achieve customization. Modular components can be configured into a wide variety of end products easily and inexpensively. Black & Decker provides an excellent example of how component standardization can yield low costs and greater variety. As a result of standardizing, Black & Decker found that it could produce an entire line of 122 basic tools—jigsaws, trimmers, circular saws, grinders, polishers, sanders, and so forth—from a relatively small set of standardized components (Meyer, 1997).

There are six types of product modularity: component-sharing modularity, component-swapping modularity, cut-to-fit modularity, mix modularity, bus modularity, and sectional modularity (Pine II, 1993a).

Key Concepts: Assemble-to-Order; Component Standardization; Customer Involvement; Customer Order Decoupling Point; Develop-to-Order; Engineer-to-Order; Make-to-Order; Make-to-Stock; Manufacturing at the Delivery End; Modularity Types.

Related Articles: Agile Manufacturing; Manufacturing Flexibility; Mass Customization.

REFERENCES

Anderson, D.A. (1998). *Agile Product Development for Mass Customization, Niche Markets, JIT, Built-to-Order and Flexible Manufacturing.* Irwin, Chicago, IL.

Davis, S.M. (1987). *Future Perfect.* Addison-Wesley, Reading, Massachusetts.

Elantora, E. (1992). "The Future Factory: Challenge for One-of-a-Kind Production." *International Journal of Production Economics,* 28, 131–142.

Feitzinger, E. and L.H. Lee (1997). "Mass Customization at Hewlett-Packard: The Power of Postponement." *Harvard Business Review,* 75, 116–121.

Gilmore, H.J. and J.B. Pine II (1997). "The Four Faces of Mass Customization." *Harvard Business Review,* 75, 91–101.

Handfield, R.B. (1995). *Re-engineering for Time-based Competition-Benchmarks and Best Practices for Production, R&D, and Purchasing.* Quorum Books, Westport, Connecticut.

Henricks, M. (1996). "Mass Appeal." *Entrepreneur,* 24, 68.

Hill, T. (1994). *Manufacturing Strategy, Text & Cases.* Richard D. Irwin Inc., Burr Ridge, Illinois.

ICM Conferences Inc. (1996). *Mass Customization, A Two Day Conference,* ICM Conferences Inc., Chicago, IL.

Kotha, S. (1995). "Mass Customization: Implementing the Emerging Paradigm for the Competitive Advantage." *Strategic Management Journal,* 16, 21–42.

Lau, R.S.M. (1995). "Mass Customization: The Next Industrial Revolution." *Industrial Management,* 37, 18–19.

Meyer, M.H. (1997). "Revitalize your Product Lines Through Continuous Platform Renewal." *Research Technology Management,* 40, 17–28.

Moffat, S. (1990). "Japan's New Personalized Production." *Fortune,* 122, 132–135.

Pine II, J.B. (1993a). *Mass Customization—The New Frontier in Business Competition.* Harvard Business School Press, Boston, MA.

Pine II, J.B. (1993b). "Mass Customizing Products and Services." *Planing Review,* 21, 6–13.

Pine II, J.B., D. Peppers, and M. Rogers (1995). "Do You Want to Keep Your Customers Forever?" *Harvard Business Review,* 73, 103–114.

Pine II, J.B. and T.W. Pietrocini (1993). "Standard Modules allow Mass Customization at Bally Engineered Structures." *Planning Review,* 21, 20–22.

Pine II, J.B., B. Victor, and A.C. Boynton (1993). "Making Mass Customization Work." *Harvard Business Review,* 71, 108–119.

Preiss, K. (1995). *Mass, Lean, and Agile as Static and Dynamic Systems.* Agility Forum, Bethlehem, PA.

Roos, D. (1995). *Agile/Lean: A Common Strategy for Success.* Agility Forum, Bethlehem, PA.

Sullivan, L. (1995). "Custom-made." *Forbes,* 156, 124–125.

Womack, J.P., D.T. Jones, and D. Roos (1990). *The Machine that Changed the World.* Rawson Associates, New York, NY.

Wortmann, J.C. (1992). "Production Management Systems for One-of-a-Kind Products." *Computers in Industry,* 19, 79–88.

25. INTEGRATED PRODUCT DEVELOPMENT: THE CASE OF WESTINGHOUSE ELECTRONIC SYSTEMS

JOHN W. KAMAUFF, JR.
ROBERT D. LANDEL
LARRY RICHARDS
University of Virginia, Charlottesville, VA

In the summer of 1993, Robert T. Barnes, general manager of the Manufacturing Operations Division for Westinghouse Electric Corporation's Electronic Systems (ES), was reconsidering the organization's approach to integrated product development (IPD). He knew IPD has exciting possibilities for overcoming chronic deficiencies in the company's traditional serial approach to product development, but he was concerned about its effectiveness in several recent projects. Since 1990, ES had invested significant resources in investigating and understanding IPD and had developed guidelines for program managers and IPD cross-functional teams. An integrated development approach had been adopted, to varying degrees, in such major developmental projects as transmit/receive modules, modular radar, and low-temperature cofired ceramics. In reviewing progress to date. Barnes was attempting to identify what aspects of the multifunctional team approach really worked well and, more importantly, he hoped, how to institutionalized an IPD approach within ES.

WESTINGHOUSE ELECTRIC CORPORATION

Westinghouse Electric Corporation is a world leader in the development of advanced products and services for government, industrial, and commercial applications for more

Source: This is adapted from the case developed by John W. Kamauff, Jr., Robert D. Landel, and Larry Richards of the University of Virginia, Charlottesville, Virginia. Copyright 1994 by the National Consortium for Technology in Business, c/o the Thomas Walter Center for Technology Management, Auburn University, Alabama. Revised 2000. Reprinted with permission.

Recipient of the "one of the best cases" award at the pre-conference workshop at the First National Conference on Business and Engineering Education, Auburn University, Auburn, Alabama, April 5, 1994.

than 100 years. In 1992, however, the corporation reported a net loss of over $1.2 billion, despite generating an operating profit of $750 million, on sales and operating revenues totaling nearly $8.5 billion.

WESTINGHOUSE ELECTRONIC SYSTEMS

Westinghouse ES, considered to be one of the corporation's core businesses, generated approximately 31 percent of the company's annual revenues. With sales of over $2.87 billion in 1992 and a backlog of nearly $4 billion, ES continued to be a world leader in airborne and ground-based surveillance radar and other high-technology defense systems. Although headquartered at Baltimore-Washington International Airport, ES had other facilities both nationally and internationally. The group's primary customers were large institutional buyers representing the U.S. government and international agencies; ES worked closely with the Department of Defense (DOD), Federal Aviation Administration, National Aeronautics and Space Administration, and the U.S. Postal Service, and with prime contractors and commercial airframe and ship manufacturers to "turn advanced technology into advanced products, rapidly and efficiently." The approximate 1993 distribution of customer orders is shown in Table 1. According to *Westinghouse Today: A Special Report* (1993):

ES has the resources to transform technical problems and market challenges into solutions our customers can depend on. Our strengths begin with our human resources: more than 14,600 highly skilled people, including nearly 5,000 engineers and scientist. They are equipped with the state-of-the-art research facilities and the tools needed to maintain our leadership in key technologies from sensors to software. Westinghouse can exploit technological advances as they develop.

As shown in Exhibit 1, ES had adopted a management structure oriented around four product families: Aerospace and Anti-Submarine Warfare (ASW); Command, Control Communications, and Intelligence; Information and Security Systems; and Integrated Logistics Support. The Design Engineering and manufacturing Operations Division (DEMOD) provided matrix support to the programs. Computer-integrated systems helped transform design engineering and manufacturing into a single, unified process. The ES flexible, distributed-manufacturing plants made it a low-cost producer of high-quality systems even in limited production quantities. Westinghouse also supported its manufactured products throughout the system's operational life cycles.

Table 1 1993 ES customer base

Customer	Percent (by $ volume)
Department of Defense	64%
Foreign military sales	6
Commercial	15
Government	8
Foreign	7

Exhibit 1. Westinghouse electronic systems (ES). (Source: Westinghouse).

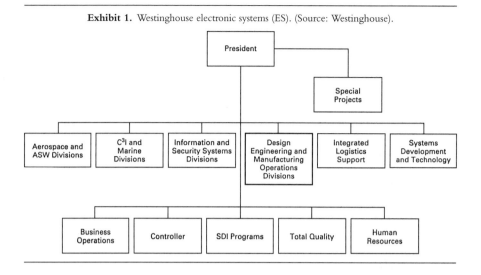

Overall, in 1993, ES was continuing its strategy to strengthen its commercial electronics operations while maintaining its defense market-share. Key contract awards in mobile satellite communications, transportation-management systems, and international air traffic control had positioned the group to compete for key infrastructure project since the 1990's. In conjunction with its continuing diversification into a wide spectrum of nondefense needs in commercial and civil markets, ES was also attempting to increase its global penetration by establishing strategic positions and joint ventures around the world.

DEMOD BACKGROUND

In addition to supporting its program management with a matrix organization, ES was one of the first U.S. companies to treat design engineering and manufacturing as a single, continuous practice by combining, in 1984, Design, Producibility, and Engineering Division and Manufacturing Operations Division (MOD) under a single general manager. The manager who spearheaded the move summarized his role as follows: "I'm the embodiment of concurrent engineering."

After nearly a decade of being under a single senior executive, in 1993, the DEMOD was structured as shown in Exhibit 2. Permanent functions in the key disciplines of product design, manufacturing, quality assurance (QA), logistics, human resources, and information systems retained authority and responsibility for the technical performance and professional standards of their functional units. Project management teams were created, as needed, to complete the work on specific programs.

R.T. Barnes, MOD general manager, cited the matrix organization as a competitive capability:

Exhibit 2. Design engineering and manufacturing operations divisions (DEMOD).
(Source: Westinghouse).

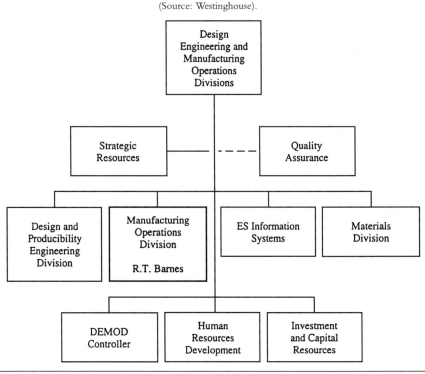

The ES matrix organization enables us to design and build our products with maximum efficiency. The matrix structure forms a network of functional and program responsibilities and allows us to apply our engineering and manufacturing resources equally to the common elements of every ES system, whether for defense or commercial use.

To ensure a smooth transition from development to manufacturing, ES had also established producibility and transition assurance center in 1988. This strategy has cut transition times between concept design and manufacturing up to 50 percent and more. Another innovative ES concept has cut its materials acquisition cycle time an astonishing 600 percent. ES has built an automated Material Acquisition Center (MAC) in 1982 to unify all the functions in acquiring, preparing, and distributing materials for manufacturing. The MAC's integrated information systems received and transferred manufacturing data throughout ES.

ES also developed a Manufacturing Systems and Technology Center (MS&TC) to develop the systems needed for manufacturing advanced products. For instance, ES engineers pioneered the computer-integrated manufacturing systems in use at their Electronic Assembly Plant (EAP). Located in College Station, Texas, this $55 million plant demonstrated the power of Westinghouse manufacturing technology. Its flexible manufacturing systems had reduced cycle times for printed wiring assemblies by 85 percent, while increasing first-time-through yields to a full 90 percent. In addition to EAP, ES operated specialized manufacturing centers in the Caribbean and Europe. Services ranged from cable fabrication to systems integration and test. This distributed manufacturing system enabled ES to build world-class products on a global scale.

Despite these accomplishments in integrated-design engineering and manufacturing, concerns remained that the existing organization did not adequately address the most difficult programmatic issues such as cost, schedule, and performance. According to Bill Newell, a supervisory engineer in Hybrid Engineering:

Our culture is driven by programs and, as a result, we are weak in the area of manufacturing collaboration. Operations is seen here as a stepchild, to simply support the programs. While we do accomplish some fantastic things in manufacturing, there is no vision that we could be a manufacturing-driven company.

Recognizing such concerns about the matrix organization's ability to support programs, ES began to investigate the concept of IPD.

INTEGRATED PRODUCT DEVELOPMENT (IPD)[1]

IPD is a process of using a systematic, structured approach to consider all stages of the product's life cycle during the initial design stage; product cost and performance are engineered to meet the customer's objectives. The basic principle is to integrate the design of a product with the design of its manufacturing, operation, support, and training processes; the goal is to achieve low-cost development, production, operations, and support within the shortest schedule and with robust quality of the products and services. In this context, IPD capitalizes on a systems perspective and structured program:

- Product and process alternatives are considered early to assure that the most cost-effective alternatives are chosen for further development.
- Using multidisciplinary teams ensures producibility and performance from system design to field support.
- Teams use quantifiable technical and management tools.

The IPD approach requires the simultaneous, rather than sequential, integrated development and qualification of all the elements of a total system (Exhibit 3). It focuses on establishing Integrated Product Teams at the "doing level" to ensure that all functional and special-interest groups are "integral contributors" rather than "monitors" in the process. For IPD to be[1] successful, the development process must change what people do and when they do it so that they actively participate by creating products that incrementally define the total system. IPD provides a structure to increase the emphasis on,

[1]Much of this section is based on *Results of the Aeronautical Systems Division Critical Process Team on Integrated Product Development* (WPAFB, OH: Aeronautical Systems Division, Air Force Systems Command, November 1990).

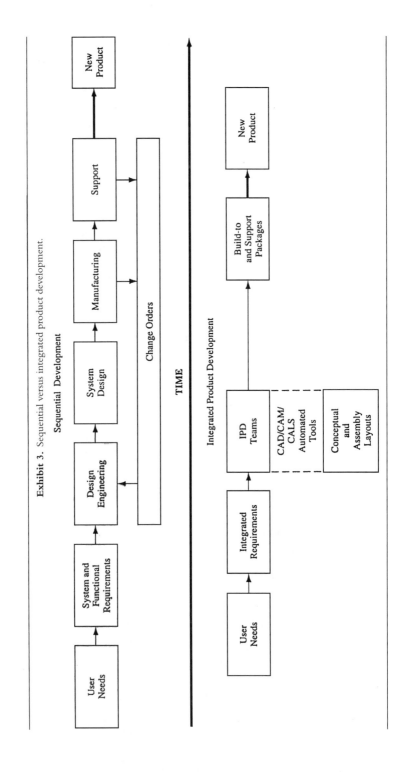

Exhibit 3. Sequential versus integrated product development.

and ownership of, the products and processes, improve horizontal communications, establish clear lines of responsibility, delegate authority, and institute clear interfaces with suppliers and customers.

THE IPD GROUP

The general manager of Systems Development and Technology cited the turbulent nature of the ES marketplace as a major driver for change:

> In our business, we need a flexible factory and a close interaction between design and manufacturing. In the 1950's, our manufacturing was characterized by a dictatorial supervisor who mandated what got done; now it's tough to distinguish between the design engineering and manufacturing people. When you combine that with IPD—which is simply good common sense—you get all the people together to design and build. For years (and even now), we have pushed the design, and our "macho factory," to not be outdone by design, got the job done. But it was hard work and often at the expense of the people involved.

Barnes's MOD was significantly affected by the effectiveness of product development planning. By the 1990's, there was great pressure to compress product development cycle times from the 36–48 months that had typified efforts in the late 1980's to 12–24 months. In addition, the products were experiencing ever increasing cost competition. In traditional serial development, when MOD was ready to produce, the product had reached a very-high-value-added state. Marketing and design had finished their inputs and had begun focusing on new projects. Any changes in product design at the production stage were costly in terms of time and money. The IPD concept appeared to be an excellent way to implement total quality management in MOD's core functions while avoiding cost overruns, missed schedules, and systems integration performance failures.

These factors spurred Barnes to initiate an IPD effort in the transmit/receive (T/R) module project, and IPD evolved with each succeeding development project. A dedicated IPD facilitation/support group was eventually formed in 1991 with a cadre of eight people:

> It's been an issue of survival . . . when you are #1 on the DOD list and the Berlin Wall comes down, you scramble. Industry won't stand by so that you can build a program. We were world class in DOD and "also-rans" in commercial applications. IPD was our way to get ahead.

This group was charged with facilitating IPD activities within the existing organization and implementing process control and optimization (especially in manufacturing operations). In addition, the group was asked to develop and implement IPD training initiatives. These goals alone provided significant technical and cultural challenges, but in the words of Jeff Tucker, IPD group leader, "The [technical] tools are easy, but the real challenge for us is in implementation of cultural change."

The size of the IPD group had fluctuated since its inception in 1991, but including the manager, it consisted of eight people in 1993 (two of whom were on leave attending graduate engineering programs). Over time, the group had evolved to serve two primary roles: (1) facilitator for process control and optimization, and (2) integrated product development liaison for key products.

In its role of product development liaison, the IPD group represented manufacturing and operations concerns to the design function. One group member saw the ideal IPD group tasks as evolving from these current roles:

> The IPD group would maintain its position as a technical support organization for process control, but the support would expand beyond the manufacturing division. In an ideal situation, the group's role as manufacturing liaison to design would be transferred to the relevant manufacturing group involved in each specific project. The IPD group would then act as a facilitator behind the scenes to help the process of IPD rather than do it themselves.

Some individuals within the IPD group were frustrated by the group's current position in the organization. Because the group resided *within* MOD, it faced difficulties in influencing the other functional divisions and truly integrating the various product development activities. Other divisions were frequently unaware of the existence of an IPD organization, and even when told, they often seemed uninterested in consulting IPD.

Expanding IPD

After considerable discussion, senior ES management decided that IPD offered enough benefits that the approach should be adopted in selected program situations. As with all major initiatives to modify organizational behavior, however, the new product development strategy found advocates and opponents, and many issues centered around the relatively new IPD group.

The IPD guide

To support the spread and acceptance of IPD, Barnes authorized the development of a user-friendly IPD guide that would solidify Westinghouse IPD concepts and also serve as a learning tool for program managers and the IPD teams that would consequently be formed for each project. This guide, published internally in 1992, was developed by a quality improvement team composed of seasoned department and program managers and based on their experiences. Members of the IPD group did not participate in the preparation of the guide but were given the opportunity to critique the results.

The guide included a commitment statement ("We are committed to the Integrated Product Development Process. We will assist the program manager and IPD team leader in understanding the use of this guide in their programs") and a sign-off by the general managers of all supporting matrix divisions, including MOD (Barnes), Systems Development and Engineering, Design and Producibility Engineering, and Quality Assurance.

The system team

The guide introduced the ES product development "wheels" (Exhibits 4 and 6), in which all functional areas were served equally by the team leader who occupied the

Exhibit 4. System team. (Source: Westinghouse).

center of the wheel. The guide's purpose was to lead users through the complete cycle of subsystem product development. It adhered to a typical DOD process flow, from contract award through first-piece production and did not specifically address preproposal or system level activities. It provided a list of IPD group members to help in IPD training and give "specialty guidance for the various disciplines."

The guide proposed a tiered approach. The system team (Exhibit 4) was responsible for the entire product system. It was led by the program manager who worked with the subsystem team leaders and support managers. The key task was the development and management of the comprehensive program plan (CPP):

The *program manager* is responsible for generating the CPP with appropriate division level general managers within one month after receipt of contract award. The CPP defines the product subsystems, the events necessary for the program to meet its requirements, and the support resources needed. The output of the CPP is required input for the IPD team leaders. The major components of the CPP are shown in Exhibit 5. The *program manager* also works with marketing and IPD team leaders to assure that the system (hardware/software) meets market needs.

The IDP team

The second-tier teams, the sub-system IPD teams, represented by the IPD wheel (Exhibit 6), would support the program launch team. Each IPD team would take responsibility for specific subsystems such as the power supply, antenna, transmitter, or receiver. These teams consisted of an *integrated team leader* and members of the following core disciplines, as appropriate, for each specific subsystem: system design,

Exhibit 5. Comprehensive program plan (CPP).

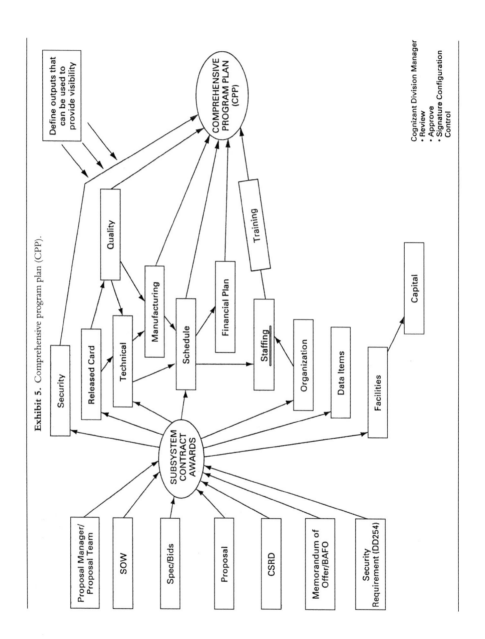

Exhibit 6. Integrated product development (IPD) team. (Source: Westinghouse).

product design, manufacturing, QA, test, materials, suppliers, and integrated logistics support. The guide spelled out the responsibilities of each team member as follows:

The *integrated team leader* is responsible for planning and managing the team activities to meet the requirements and goals for a segment of the contract/product; cost (both nonrecurring and recurring), schedule, technical goals and objectives; application of resources and tools; team effectiveness and implementation processes. The leader must also work with the program manager and marketing to assure that the subsystem meets market needs.

System design is responsible for systems analysis, system requirements, system hardware/software partitioning, system test and verification, documentation, and control of all system requirements.

Product design is responsible for specification and design of all operational hardware and software elements of the systems; incorporating quality, reliability, maintainability, producibility, and other specialty disciplines into the designs of all commodities in the system; documenting and controlling all design data; product design planning and managing; and engineering test equipment.

Manufacturing is responsible for manufacturing planning, product quality, production control, industrial engineering and process development, tool development and fabrication, assembly and test, manufacturing management, generating product cost data, and process source selection.

Quality assurance is responsible for quality flow-down and reporting; planning and managing quality systems; quality-related training; process and personnel certification; and product acceptance.

Test is responsible for test planning and process development, test hardware and software development, design and acquisition, test equipment and use documentation, test equipment maintenance, calibration, and configuration control.

Materials is responsible for purchased material quality, acquisition planning, requisition release, cost and schedule availability, supplier identification, requirements flow-down, subcontract negotiations, and subcontract management. *Materials* also participates in the design-to-cost activity to contribute to accurate target pricing.

Suppliers participate in an integrated product design activity. They extend in-house capabilities to provide product (material) in support of program requirements. They are responsible for requisite planning and operations execution to ensure timely delivery of compliant product at a reasonable price.

Integrated logistics support (ILS) is responsible for analysis, Life cycle Cost Analysis, spares provisioning, training, handbooks, field engineering and support, repair and maintenance services, ILS planning and management, and support and test equipment.

The IPD team was also to consult with other disciplines or, if conditions warranted, include them as full-time team members: internal and external customers, contract partners, management or technical specialists, analytical process facilitators and/or marketing. The roles of these groups were as follows:

Customers, both internal and external, are often considered to participate in making team decisions and tradeoffs. If a positive environment can be maintained on an ongoing basis, the IPD suggest that this relationship "be one of the most effective team arrangements." It is essential that contract *partners* are involved in each other's team activities. If the interface is major, it may be necessary that individual disciplines be represented by all partners. In addition, there will be times when *management or technical specialists* in specialty disciplines such as technology innovators, estimators, and management information systems developers will provide critical assistance to a specific team. Similarly, *analytic process facilitators* help to facilitate team problem-solving activities using such techniques as quality function deployment, Pareto analysis, and Taguchi methods (or designed experiments). Finally, *marketing* works with program management and IPD team leaders to ensure that future market options for the system are vigorously pursued.

IPD team assessments

The new IPD team concept was applied to many new product efforts, including transmit/receive (T/R) modules, modular radar (MODAR), and low-temperature cofired ceramics (LTCC). One of the issues that plagued a definitive analysis of the IPD efforts was that none of the programs had yet entered into full-scale production. Barnes was able to point to the establishment of the IPD group and the preparation of the IPD Guide as tangible evidence of applying the IPD approach at ES. He was also keenly aware that the lack of a clear, indisputable IPD success story contributed to varying views about the IPD process. Barnes believed each program had to be reviewed in detail, however, to get a clear understanding of the degree to which IPD could be successfully applied.

T/R modules

Radar system technology had changed substantially from the traditional mechanically steered antennas and the relatively inefficient passive phased arrays to active phased arrays. Active phased arrays are stationary antennas made up of up to several thousand radiating elements. Each element is directly attached to a T/R module. Each T/R module includes its own small transmitter and receiver, which are phase-adjusted so that (a) their transmissions combine in only one direction (a radiated beam); and (b) received signals are combined but from only one direction.

Since radar systems were ES's main products, and the T/R modules were the key determinants of the radar system performance and/or costs and ensure radar performance, ES designed and manufactured the T/R modules in-house. T/R modules were also used in electronic welfare systems. T/R modules had evolved into a generic commodity that could be sold to others, and these modules provided the stepping stones to enter new commercial markets.

Because one radar system could be thousands of T/R modules, the quantities needed for all ES systems could total millions of units per year. In addition, ES had decided to invest in developing advanced manufacturing methods to provide these low-cost, high-volume microwave assemblies to other radar companies. However, according to Barnes, who oversaw the effort: "this transition to high-volume commercial manufacturing (e.g., in million-size lots) has been a radical departure for us. IPD is the only way to do it; we have to integrate these technologies very early on."

The T/R module program. The T/R module program was thus the first major project to incorporate the new IPD concepts into the product development activities, although the formalized IPD integrated team activities, although the formalized IPD integrated team activities were not begun until 1990. Starting in 1997, Bill Newell, Manufacturing engineering Lead, and Dr. Ted Foster, the Microwave Engineering Manager, spent an average of five hours a day for four to five months defining the integration tasks. In this time, they formulated an umbrella developmental strategy for the project that was still being used in 1993. The strategy included commitments, initially, to such philosophies and methodologies as concurrent engineering, variability reduction, statistical process control (SPC), and subsequently the IPD approach. During 1988–89, integrated product/process development activities were investigated under the auspices of the U.S. Air Force T/R Module Manufacturing Technology program, and the IPD team concept was actually written into the F-22 radar proposal for the advanced tactical fighter, which was subsequently awarded in 1990 to the team of Westinghouse and TI.

The T/R module IPD team was composed of representatives from design, manufacturing, materials, factory test, components, quality, and variability reduction. Customer relations were handled on multiple levels: Top management interacted directly with the customer's top management. At a lower organizational level, the systems team translated specific product performance requirements into subsystem requirements, and the subsystem engineers worked directly with the customer's engineers to resolve problems that arose.

Realizing the importance of an integrated effort from the supplier to the customer, Westinghouse trained its suppliers in TQM techniques. Initially, some suppliers realized only a 5 percent yield, which was completely unsatisfactory for the aggressive cost goals required by the F-22 contract and set by the IPD team. The variability reduction training provided by ES, however, contributed to eventual soupier yields ranging from 95 percent to greater than 99 percent. The T/R module team also experimented with supplier relations by bringing a supplier into the design review. This approach had never before been tried but the results were significant. As the materials manager explained. "The supplier said the design would not work, and this resulted in a design change." Supplier relationships thus proved to be especially rewarding in the T/R module project.

Quality function deployment (QFD) proved to be a useful tool in the T/R modules. It was used, for example, in the translation of the housing or interface requirements into design guidelines. Team members used QFD to identify customer needs and determine the engineering tradeoffs between different customer requirements. Using the tool was not always easy, however, according to Robert Horner, the IPD group member who facilitated supplier partnerships: "QFD is a very good tool, but it is not much fun. Getting together for a consensus is really tough. We did try to include the customer in these meetings."

Design of experiments (DO) was also considered to be vital by the IPD teams to the project's success. Every process relied on a controlled experimental foundation Between 1990 and 1992, 30 DOEs were performed at an average cost of $30K each and resulted in at least a 10-fold reduction in defects in each of five critical assembly processes. Formal SPC was not used, but data for future use were collected and stored in a database. According to Horner, "SPC should not be necessary if the proper work is done up front with QFD and DOE."

Manufacturing analyses were also used where practical. No other T/R modules were available for direct comparison, but ES conducted comparative assessments of the individual components of the T/R modules. ES determined the technical needs for each purchased component, then went to the suppliers and asked how each planned to meet those needs. This method helped with the initial selection of suppliers.

Despite the apparent valuable contribution of IPD tools and techniques in some areas, their use was not universal. During 1990, ES embarked on an investigation of the process of benchmarking led by Steve Kramer, a senior manufacturing engineer and IPD group member. The initial results of his benchmarking efforts were disseminated throughout the IPD group but were not totally embraced by the team members.

Feelings about the matrix management support of IPD were mixed. Horner explained: "Matrix management is both a friend and a foe. It is good for sharing the wealth of knowledge between projects, but it is also a drain on getting things done." The major concern was that IPD team members frequently moved between projects, and the transitions took time and promoted inefficiency. Furthermore, team members were not evaluated by team leaders but by their functional supervisors. The concern was voiced that "an individual could do well on a project but not be properly rewarded."

On a more positive note, most team members believe that product-development duplication was reduced: "wheels are not being reinvented in every project." Newell commented: "If you asked me if it is working, I would have to say yes. We get incredible results while being significantly understaffed . . . upper management loses sight of how hard people are working."

Consistency of effectiveness proved to be a problem in the early IPD efforts. The IPD teams had different abilities and achieved different results. The F-22 operations manager, and a team member, attributed much of the disparity to 'the differing abilities of the leaders." Indeed, leadership was often seen as the critical component for a successful team. In the T/R module project, team leaders were chosen based on past performance in the "old management system" and were given little or no training in the new methods.

MODAR

The MODAR program was intended to provide low-cost predictive wind shear and improved weather detection in an expanded airborne radar market.

The key to the MODAR program was to develop a new product line by overcoming significant challenges in modularity, flexibility, and guaranteed high reliability in a short development time. The potential problems were exacerbated by the fact that limited nonrecurring development funds were available. In addition, MODAR was designated to be a dual-use product, built to commercial standards but applicable in both the military and commercial markets. With this product, ES was entering a highly competitive market where performance had to be cost-justified.

Launched in 1991, shortly after the T/R module program, MODAR adopted an IPD approach in order to meet a diverse set of objectives. For the most part, the goals were well defined and communicated throughout the teams. "Aggressive target pricing" was embraced as key to MODAR success. Extensive tradeoffs were made in establishing the design requirements and tolerances before 10 percent of the project was completed, and the baseline requirements did not change significantly for the remainder of the project.

The success of many Westinghouse projects often depended on the organization being able to learn new technologies and simultaneously learn to serve new markets. Unlike those projects, MODAR drew from a familiar technology (airborne fire-control radars). The MODAR program only had to extend an existing technology to a new, albeit commercial, customer base, which in this project eliminated a significant amount of uncertainty.

According to Rita Herlihy, a mechanical design engineer and IPD team member, the MODAR IPD team was composed primarily of design and manufacturing staff with quality and reliability staff playing minor roles. Prior to Jeff Tucker's assignment as IPD group leader in 1992, he served as a manufacturing engineer in the IPD group that facilitated the MODAR IPD process. While some people wanted to conduct general training in IPD tools for the entire MODAR staff, tucker believed that such training would waste valuable time because everyone would not be using all the tools and techniques. According to Bob Jelen, engineering manager for MODAR, "Jeff

Table 2 MODAR productivity tools

Quantitative tools and measures
QFD
Functional analysis
Failure mode and effect analysis
Value analysis engineering
Design-to-cost
Design of experiments
Design of assembly
Competitive cost comparisons
Structured problem solving Capability studies/SPC

Table 3 MODAR cycle-time reduction

	Typical program (months)	MODAR (months)
First processor hardware	10–12	5
First prototype system	14–18	$6\frac{1}{2}$
First flight	18–24	$8\frac{1}{2}$
First production delivery	24–30	14

brought a sense of reality to the program by introducing only the tools which were needed and could be used at the time."

The key elements introduced during MODAR were CPP (Comprehensive Program Plan), phased product build, quantitative tools/methods, cross-functional teams, team ownership/accountability, and training and development. In addition, the developers embraced proactive manufacturing participation, added purchasing coordinators to each IPD team, set "no purchase over cost" goals, established partnerships with suppliers, provided timely feedback on the status of cost goals, and used "prototype" hardware built by production facilities/personnel.

The MODAR project eventually used a variety of IPD tools and techniques, as shown in Table 2.

Notwithstanding the debate over the use of tools, the initial program results were dramatic, providing a reduction in the product development cycle time of over 50 percent when compared with previous efforts (see Table 3), which translated directly into reduced nonrecurring engineering costs. MOD was also able to automate 96 percent of the circuit card assemblies by designing to process and standardizing the criteria for component selection. The Advanced Manufacturing Technology Center, which was dedicated to high-volume production, increased process yields from 80 percent to 99 percent and, in one department, decreased defects by almost 80 percent in one year.

The IPD teams were credited with much of this success through their abilities to inculcate a holistic view of the product-development effort, obtain early manufacturing participation, and facilitate communication with the teams. Mike Fahey, a manufacturing engineering manager, stated "The MODAR project was the first time in my life that I had seen a design change before it went into manufacturing."

Despite these early successes, some observers noted that IPD teams could have performed better in some areas. Supplier nonrecurring engineering needed to be fully planned, for example, and the condensed schedule caused higher rework labor and material on the initial systems. The IPD teams met weekly to ensure that all groups were working toward the same goals and working well together. These meetings proved to be a source of frustration for some people, however. Herlihy explained those involved in MODAR noted a number of weak areas in the process. Accountability between task and schedule was criticized, for example. Jelen also noted a problem when the time came for the transition to production: "some of the design engineers simply did not want to let go of the product." Additionally, while communication within teams was strong, communication between IPD teams was weak.

The team approach was touted as a great idea, but the meetings often went off track. They would get bogged down in a few details which could have been solved by a few individuals. Major conflicts were resolved with a few key people. The decision-making process seemed to be hampered by the new team process. The emphasis on consensus decisions often acted as a detriment to decisive action. Furthermore, the team composition, especially in leadership position, was criticized as being poorly planned. According to Herlihy: "electrical engineers made up the majority of IPD leaders, but they did not really lead. I do not think they understood their roles; they were expecting to be in more of a technical position."

Barnes knew that team members were still struggling with team life cycle. Although the teams were technically intact in 1993, they had not really worked together since the assembly drawings had been created and delivered to manufacturing during the previous year. Herlihy believed that "the team should not fall apart after design is done. Members should still continue the contact."

Unlike in other IPD efforts, little conflict occurred between the functional matrix and IPD team leaders. Herlihy explained that she had devoted full time to the MODAR project during the early stages and used functional management as a consulting resource. Unfortunately by 1993 she had been assigned to two other jobs that demanded equal time and this caused problems.

In addition, the transition to production had not gone smoothly, some assembly and workmanship issues were noted in high-quality production, and Barnes had recently begun to encounter some significant problems:

On MODAR, we did great on the up-front planning, but then we unleashed our young people and told them to go do it. We lost the necessary discipline to get it done and did not ask enough critical questions throughout the process. We have, in essence, "empowered" our people without preparation. We failed to enforce the programmatic discipline. Everything went fine until we put it together ... from an overall systems viewpoint we failed to stay close enough to it. Our 25- to 40-year old workforce had not been prepared adequately for the tasks.

Exhibit 7. LTCC methodology. (Source: Westinghouse).

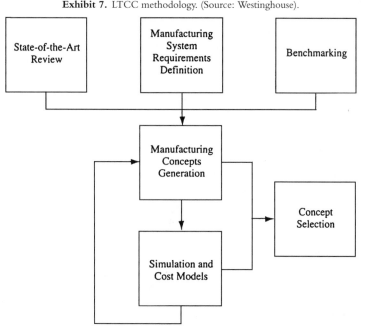

LTCC

Low-temperature cofired ceramics (LTCC) was a new technology that formed electronic elements or networks on a supporting substrate. It had emerged as a viable alternative to other packaging and interconnection methods because of its ability to integrate digital, analog, radio-frequency buried microwave and components in a hermetic, monolithic unit.

The LTCC development project followed the T/R module and MODAR IPD efforts and offered the greatest opportunity for introspection because it drew significantly on the learning from previous IPD efforts. Relying on the experiences of the past, the LTCC project used such tools as QFD, benchmarking, and cost modeling in the early stages to lower overall costs in the long run. To achieve the specific program goal of developing, demonstrating, and implementing a low-cost, high-throughput (250,000 square inches per week), and flexible LTCC manufacturing process, the IPD team developed a comprehensive methodology consisting of six tasks: state-of-the-art review, manufacturing-system requirements definition, benchmark study, process simulation and cost modeling, manufacturing-system concepts generation, and manufacturing system concept selection. The task interactions are shown in Exhibit 7. Recognizing that the six tasks were interdependent, the IPD team opted for close interaction between the parties responsible for each task. To facilitate daily communication, members were colocated in one office area.

As illustrated in Exhibit 7, state-of-the-art review, requirements definition, and benchmarking provided information for the generation of two potential LTCC *manufacturing-system concepts*. An iterative process involving the cost and simulation models, technical feasibility analysis, and risk analysis was used to determine the most effective low-cost, high-throughput, flexible LTCC manufacturing system.

The early results of the LTCC IPD teams had been mixed. Tim Parr, the manager of Printed Wiring Assembly Manufacturing Engineering and the LTCC team leader, told Barnes: "The LTCC project has been the most frustrating in my career at ES. The transition from design to manufacturing has been difficult, since design engineering has been reluctant to give up the design." Parr believed that an integrated team required that "team members have experience in both manufacturing and design, or team members from design and manufacturing need to understand their disciplines and be willing to hand off to the next discipline at the appropriate time." He believed the LTCC project faced a situation in which the good design engineers were unwilling to hand off to the somewhat inexperienced manufacturing engineers. Parr's concern with inexperienced manufacturing personnel was exacerbated by the fact that, although most of the team consisted of unquestionably dedicated industrial and manufacturing engineers, they had spent most of their time in ES labs prior the LTCC project.

Exhibit 7 (LTCC Methodology). Parr's views on IPD tools were varied. He said, for example, "I hate the term DOE . . . it's a buzz word . . . but I love concept." Parr was also a strong supporter of benchmarking when it was used judiciously. Although his team used QFD to identify the customers' needs and translate into specific product and process specifications, Parr remained skeptical: "I feel that QFD is grossly overrated. It consumes a great amount of time, and you must measure the benefits against the input. It may be appropriate in some cases and LTCC may be one of those cases but they try to use it everywhere. Sometimes it cannot be justified."

Barbara Mae, a senior manufacturing engineer, who led the QFD effort, was also suspicious of QFD's utility in the LTCC project:

I think QFD was unsuccessful in the LTCC effort. The houses [of quality] were too big and should have been scaled down. Furthermore, management kept pulling people off QFD; it was the lowest thing on everyone's priority list. QFD was further frustrated by management's misunderstanding that QFD could be accomplished in a short period of time. They simply did not understand that time between sessions was required for research.

QFD (quality function deployment) did further the LTCC effort. Different pockets of people contributed to LTCC development, and QFD brought these people together in one room. QFD helped form team relationships.

Training was also designated by Mae as a limiting factor in the LTCC application. Mae had been put in charge of several QFD efforts without any prior training, and although she had learned QFD (by self-education and experience), she desired some formal training. She believed the training needed to be accomplished at three levels: management, among QFD team members, and the QFD leader. "Management needs to understand the requirements and limitations of QFD. Team members and team leaders need training to understand the concept and their respective roles." One team member, Stephanie Caswell, a manufacturing engineer, also liked the concept of QFD but was having problems with the mechanics and had hoped for additional training.

Of all the IPD tools used in the LTCC project, benchmarking, which was facilitated by Steve Kramer, a senior manufacturing engineer in the IPD group, may have been the most appreciated by all participants. IPD team members and process owners (those who would eventually be responsible for implementing the manufacturing-system concept) participated in visits to investigate how other organizations approach problems similar to those encountered at ES. According to Caswell, one of the process owners, it was valuable to see how other people have solved similar problems. "You can learn a great deal just by looking around, but you have to see it. You cannot benchmark over the phone."

Cost control was an important element of the LTCC project. In its transformation from a government contractor to a commercial competitor, ES found that it had to pay much closer attention to the production costs of its products. The traditional military contract concentrated on higher performance at lower production quantities. The private sector spurred ES to begin an aggressive design-to-cost strategy to increase the cost viability of high-technology products which also carried over to the military sector.

The LTCC project used a process simulation and a spreadsheet cost model to assist with the design-to-cost effort. The ultimate goal was to stimulate the manufacturing of a product design. The model considered the technical difficulty of the design and the material/labor requirements and then calculated the manufacturing cost of each product. A single simulation took about three hours, however, and according to a manufacturing engineer. "When it takes that long to run a simulation, it really is not a useful tool anymore."

One potential weakness of the preproduction planning, which included QFD, benchmarking, and cost modeling, was noted by Caswell:

The study is primarily a paper study which aims to identify the optimal manufacturing facility. Unfortunately, Westinghouse faces incredible time pressures and is very reactionary in its response to this pressure. A result is that some equipment has been ordered prior to the identification of the optimal configuration. They try to get the study team involved, but in some cases, they simply could not wait for advice.

VIEWS FROM THE FUNCTIONAL AREAS

Barnes recognized that the existing matrix structure at ES had a significant impact on the effectiveness of IPD teams. Individuals reported to their respective teams and also to their functional leaders. The functional structure served as an important resource to all teams and team members. Each function had the dual responsibility of staying abreast of all developments in the functional area and also serving as technical support to the individuals who were members of project teams. The IPD efforts had a tremendous effect on all ES functions; instead of keeping all projects within the boundaries of the functional areas, functions were now expected to lend their members to project teams. The different functional leaders expressed different views of the effect of IPD on ES and their specific functions.

Marketing

The marketing department for ES consisted of 20 people who obtained additional support at the division level. Because of ES's historical focus on the military market, the marketing department concentrated more on strategic issues, such as which markets to pursue, than on day-to-day relationships with the customers. As Steve Winchell, marketing manager, explained: "Our engineers work with our customers only a daily basis. This is not bad, since they work with the customers' [the military] engineers." Winchell was clearly concerned that "this will not work as well in the commercial world."

Although marketing did not work closely with IPD teams on a daily basis (not designated on the IPD wheel of Exhibit 6), Winchell believed that his people were indeed a part of the teams.

In product developments, marketing must bring the initial idea. Marketing people are charged with looking for new opportunities and they do that to support the team effort. However, when they find a new opportunity, they are immediately looking for other opportunities. There simply is not enough time or enough people to work on the old one.

Winchell added that "most design and manufacturing engineers don't come in contact with marketing. Some don't even know we have a marketing group."

Manufacturing

In the 32 years in which he had been involved primarily in mechanical design and manufacturing engineering. Mike Fahey had seen many changes. Fahey viewed IPD as being much broader than concurrent engineering: "It is an integrated methodology." Despite his support for cross-functional team empowerment, Fahey did see some potential pitfalls in team management:

If you get a participative system that becomes inert because people are so participative, you are headed for trouble. You need a leader who will keep them on track, and he needs to be autocratic in this regard. He needs to say "[The task] will be done by this time." A participative manager must put pressure on people.

Fahey noted a major structural problem which interfered with the IPD process: "All budget money came through the functional management, and this fact weakened the power of the IPD team leaders." In his eyes, to be truly effective as a leader, the leader should be responsible for both team member evaluations and funding. The functional leaders could then concentrate on ensuring that their function was on the leading edge of its technology.

In terms of current acceptance of IPD, Fahey felt that much needed to be done. "I do not believe that there is a universal support of IPD. The only person who has given a blank check to IPD is Bob Barnes . . . I think you need to get a little autocratic to institutionalize this system. We need to say that this is the way DEMOD will do business."

Design engineering

According to Jim Redifer, general manager of the Design and Producibility Engineering Division: "A matrix organization leads to technical competence because mentoring is available, but the environment can be sheltered from the rest of the world. On the other hand, an IPD framework yields greater self-reliance and interplay with other functions, but people need to learn to ask for help when it is needed; it also requires a greater emphasis on interpersonal skills." In his opinion, IPD teams necessarily "make life more difficult for our functional managers—since they have to be able to reach out to a somewhat more spread out group of people—and somewhat easier on our program managers, but that's probably the way it should be."

Because they furnished 80–90 percent of the IPD team leaders, design had a unique perspective on the role of the team leader position. Team leaders had found the position to be extremely frustrating. According to Jerry Beard, manager, Signal Generation and Reception: "Vivian Armor, DEMOD Human Resources manager, conducted a survey of all team leaders. When asked if they would like to be a team leader again, only 1 of 14 responded positively." Beard felt that management had mismatched people to the job. "Although some did the job well, none were trained and no one knew how to train them. In the old days, we had a design leader who was primarily technical with technical responsibilities. The position now is more of a program manager and less of a technical role. People are falling by the wayside."

Beard felt that upper management wanted the functional organizations to be held responsible for product failures. This situation understandably created a reluctance on the part of functional managers to turn responsibility over to IPD teams. An example can be seen in the development of nonstandard parts. ES wished to reduce the number of parts it produced to reduce inventory and to facilitate automation. Upper management pressured the functions and required that functional management—not the teams—must approve all nonstandard parts. John McClure, Mechanical Design and Development manager, agreed and noted that "functional managers would often grab the team back since they didn't want the team to stumble."

Regarding functional managers, Redifer was not convinced: "You *can't* teach an old DOD dog new tricks!" In the first two years he headed Design Engineering, all of his department managers retired, died, or were laid off, so he was able to build a cadre of people committed to his vision of greater interaction between design and manufacturing. In general, he was not sure if you can train someone with the necessary expertise to run IPD teams: "People need to have an innate ability to lead."

ES has a dual track for management and technical promotion, but according to Redifer:

Functional people never viewed a move into program management as a desirable step in their vertical track careers. Now, because of flatter organizations and integrated efforts, we need much more lateral movement and adaptability to new environments, including a greater emphasis on interpersonal skills such as communicating to a scattered functional group. In any case, because of program autonomy, it's still difficult to get them to listen to you. IPD has been personally driven rather than implemented by management or a steering committee.

Systems design and development

Noel Longuemare, general manager, Systems Development and Technology, viewed the IPD process as consisting of potential tradeoffs in terms of innovation, a structured methodology, and risk management:

From our own experiences, defense engineers can very readily adapt to the commercial environment, but they have to understand the new ground rules. It's not an instant transition. Contrary to the military approach where there is a set of requirements, in the commercial world the number one issue is the price of the product as an independent variable; everything else is a dependent variable. We have to give these people a tool set that enables them to translate their actions into costs. We have shown that once a price is set it is possible to develop the performance accordingly. It is, however, much more difficult to invert the process.

In his opinion, the application of oversight and discipline within the DOD had gotten out of hand—no tradeoffs were allowed; in the commercial world, the customer who has to sell the product can readily make these tradeoffs. He hoped that IPD would help ES to adapt to this emerging environment. "We have found IPD to be successful beyond our expectations when we use it properly, so we are diligently pursuing ways to make it the norm. We have seen no disadvantages . . . it's a win-win."

Regarding tools, Longuemare believed the single largest failing was in providing enough depth in the requirements.

Understanding the relationships was important before we went off to do something. In the military, it always took them longer to get the contract written, so we were behind the curve to begin with; we marched off smartly and typically had to backtrack in order to match the changing requirements. Few were willing to invest the requisite time to do the up-front planning. One of the main advantages of the IPD process is that it is conducive to doing just that.

To do IPD, a fundamental importance is to have collocation for the core team. It is also important to give the team local authority to get the job done, particularly so that they do not have to fall back upon preexisting systems. At ES, one of the biggest mountains yet to be climbed is the development of the necessary systems to facilitate IPD work. Until recently, it has not been possible to bulk our costing for separate items. We had to segregate for different uses.

Quality assurance

Terry Hart, Quality Assurance manager, suggested that the IPD process had run into problems as teams began to cut corners.

Terry Hart tried to ensure that there was a flowdown of quality objectives to all teams so that they are all heading in the same direction: "The key is understanding what the mission of the team is. We need to understand the purpose of the team and staff it accordingly—it could be based on experience or a variety of other factors." Hart also contended that the problem of measuring engineering was still paramount: "How do you measure engineering? It's the most difficult area to measure . . . let's do it like we've always done it. We seem to learn the same lessons over time. Recently, we have tried to measure the profile of engineering change notices over time, but we are not convinced this will work."

Bob Glanville, DEMOD QA manager, supported the IPD effort, but in his opinion, but in his opinion, several changes were required before IPD would become an effective force in QA. He believed the organization had not yet realized the full benefits which IPD could provide. QA was the last group to see the product before it reached the customer, but Glanville believed quality awareness needed to move back through the organization. "We must do our quality work up front. This means meeting the customer requirements and minimizing variability." He was convinced that one part of a good IPD process was design and manufacturing working together to develop a process with a high yield. Unfortunately, the current situation was one in which "we measure a few processes, but manufacturing does not measure, or even understand, many processes."

Others in QA emphasized the importance of prioritizing customer requirements. Glanville added: "It is essential that we distinguish key functional requirements from other requirements. Often the customer does not differentiate key requirements. Hopefully, IPD teams can help us to prioritize what the customer really needs."

SUMMARY

Barnes knew the sentiment about IPD, both pro and con, was running high. Clearly, it was a topic of debate—one that generated significant emotion in the functional areas, the programs, and especially the increasing number of people who had actually been involved on teams. For example, many program personnel had began to focus on the difficulty of MODAR making the transition:

The initial shock of IPD was that we took our best artisans and then put them together under young technical leaders; this was something new, and the resistance was, "You want me to go and do this" The thing that has me concerned is that some of the dinosaurs are saying. "See, IPD does not work."

Longuemare echoed this belief:

Now we must have the greatest degree of personal interaction and quick decisions in our programs. That does not imply that there is not a discipline in the approach. There are some fundamental problems that remain. How do you create an environment that encourages innovation while maintaining discipline and configuration control while designing a product that can be built? No individual is necessarily best at all things . . . most of the time you cannot find people who can create and maintain discipline. We frequently take someone who is innovative and put them into a position that requires them to maintain discipline ... and often have to endure the consequences.

Barnes was an ardent champion of IPD and believed that the benefits could be captured only if all of ES embraced it. Therefore, he had to act quickly to demonstrate its viability for integrating product development before opinions opposed to IPD became entrenched.

X. THE REVOLUTION IN THE SUPPLY CHAIN

26. SUPPLIER PARTNERSHIPS AS STRATEGY

BRIAN LEAVY

Dublin City University, Dublin, Ireland

ABSTRACT

The partnership approach to supplier relations is now a very common feature in many industries. This is due in large part to the phenomenal success of the 'world class manufacturing' (Schonberger, 1985), and 'lean production' crusades (Womack et al., 1990; Womack and Jones, 1994) over the last two decades, which have drawn their inspiration from the proven competitive practices of leading Japanese companies (Dyer and Ouchi, 1993). One of the by-products of this success has been the ever-widening faith in the virtues of supplier partnering

The purpose of this article is to present a strategic perspective on supplier partnering to help companies to more critically assess the opportunities and risks associated with the adoption of the partnering approach. The article begins by contrasting the underlying rationale for supplier partnering with the more traditional arm's length approach to buyer-supplier relations. Most companies with aspirations to world class manufacturing standards are now adopting the partnering approach. The practices of the following companies are included in this article: **Amdahl; Apple; Bennetton; Digital Equipment Corporation; General Motors; Intel; Marks and Spencer; Toyota; Wal-Mart.**

A STRATEGIC PERSPECTIVE ON SUPPLIER-PARTNERING

To understand better the nature of the supplier partnering from a strategic perspective, it will be useful in the first instance to examine how it contrasts with the more traditional arm's length approach to buyer-supplier relations.

The traditional approach

The traditional approach to buyer-supplier relations is based on two main premises:

(i) *The relationship is best managed through the market mechanism.*
  • Suppliers compete with one another for the customers;
  • Buyers shop around for the best deal;
  • Repeat business is not guaranteed;
  • Buyers and suppliers communicate and interact with each other at arm's length;
  • The 'discipline of the marketplace' keeps them independently on their toes.

and

(ii) *The buyer and supplier see themselves as essentially competing with each other for margin.*
  • Both parties see it as a win-lose game;
  • The primary focus is on the division of profit margin;
  • The lion's share goes to the party with the most economic power.

Under this traditional arm's length perspective any company, at any stage in the industry chain, is seen to be in competition with its upstream suppliers and downstream buyers for profit margin (Porter, 1980). For example, the producer of soft-drink cans must compete for margin with the aluminum companies at one end of the supply chain and the soft drink companies at the other. All are ultimately dependent on the dollars of the soft drink consumers for their revenues, and the margin enjoyed by each of these players in the industry is seen to depend on its economic power in the overall market chain.

Within this perspective, company strategists are advised to try to reduce the power of their suppliers by maintaining multiple sources of supply, avoiding any uniqueness in the relationship that might make the cost of switching suppliers high, and searching for readily available substitute materials that would help to keep supplier prices in check. The ability to switch business easily among suppliers and pose a credible threat of bringing component production in-house can be used to keep the most aggressive and ambitious suppliers in line. Moreover, by avoiding long-term commitments and keeping its options continuously open, the buyer will always be able to move his business quickly and easily to those suppliers which, at any time, are the most efficient and technically advanced in their own sectors.

The strategic advice to suppliers in this perspective is the mirror equivalent. Suppliers are advised to keep their options open by not allowing themselves to become over-dependent on any one buyer. At the same time they are also advised to try to increase the switching costs for their customers by building elements of uniqueness into the relationship (in the form of product, service or quality distinctiveness, for example). Strong suppliers can keep their buyers in line, and their demands in check, by posing a credible threat of forward integration. They can also try to increase their influence in the overall market chain by directly developing a brand image for their components in the wider marketplace, as Intel has been able to do with its 'Intel Inside' campaign.

The partnership approach

The partnership approach to supplier relations, on the other hand, presents a very different picture, with its own set of strategic implications. Its main premises are:

(i) *The relationship is better managed through direct cooperative agreement rather than through the market mechanism.*
  - Buyers and suppliers seek long-term commitments from each other;
  - Repeat business is guaranteed;
  - Buyers and suppliers communicate with each other directly and share information, and learning/know-how;
  - The supplier is co-opted into the buyer's competitive strategy, and the discipline of the buyer's marketplace keeps the entire partnership on its toes;

and

(ii) *The buyer and supplier are essentially partners-in-profit.*
  - Both parties see it as a win–win (or lose–lose) game;
  - The primary focus is on the creation of profit margin;
  - The partners work together to ensure an equitable share of the spoils;
  - Trust replaces opportunism.

Under this approach, the buyer seeks to improve his competitiveness through the development of close cooperative relationships with a relatively small number of carefully selected suppliers. The emphasis is on the development of long-term partnerships. This strategy encourages a high degree of interdependence between the buyer and the supplier and it is pursued in the belief that it can offer very significant economies of cooperation, which can help to improve the profitability of both parties. The commercial benefits of such an approach are expected to come from the closer coordination of schedules, cooperation on product development and process improvement, and joint-action on cost reduction. Furthermore, the cooperative approach provides some insulation for the supplier from the full winds of competition in its own segment of the marketplace. It also offers the buyer many of the benefits of vertical integration (greater security of supply, and more direct control over cost and quality) without the associated investment or the risks that would normally be involved. This is the approach that has helped Toyota to become the world's most efficient manufacturer of cars, much to the benefit of the company and its suppliers (Womack et al., 1990). It is also the approach that has underpinned the successes of Marks and Spencer, sometimes called 'the manufacturer with no factories', and Benetton, 'the retailer with no shops'.

In short, under the partnership approach, the supplier's growth and profitability become less dependent on the competitive forces operating in his own segment of the market chain. Instead, the overall effect is to co-opt the supplier into the competitive strategy of the buyer, and the future growth and profitability of both partners become more closely tied to the evolution of the buyer's market and to his competitive position within it. The fortunes of both parties depend on the ability of the partnership to develop and fully exploit the potential economies of cooperation in ways that will strengthen and secure the growth and competitiveness of the buyer.

SOME RISKS AND IMPLICATIONS

While the partnership approach to buyer-supplier relations can offer many potential advantages to both buyer and supplier, there are a number of implications and limitations that company strategists in both buyer and supplier firms should keep in mind.

The rationale for partnership may change with industry evolution

To begin with, it is important to recognise that the underlying economic rationale can change with industry evolution, as it did in the case of the information technology industry over the last three decades.

In the 1970's the industry was dominated by a small number of highly integrated manufacturers, like IBM and Digital Equipment Corporation, which controlled the key technologies in both hardware and software. These companies sold very high margin products directly to a relatively small base of industrial and commercial customers. Manufacturing was very labour intensive, with little automation, few economies of scale, and little incentive to buy anything other than very basic components and raw materials. The whole independent sub-supply segment remained very underdeveloped in terms of range, scale, quality and efficiency as companies continued to find it more economical to produce their own printed circuit boards, manufacture their own semiconductors, processors and peripherals, integrate their own systems, write their own software and support their installed base with their own service operations.

The situation then changed dramatically during the 1980s, as the industry experienced the rapid growth that accompanied the increasing convergence of computing and telecommunications and the emergence of a whole new mass-market segment following the arrival of the personal computer. This explosion in the primary market has since allowed the sub-supply sector to grow to a size which now supports degrees of specialization and automation that would have been unthinkable less than two decades ago. Furthermore, many of the industry's strategic technologies have been migrating backwards into the component and subsystems levels, particularly in the printed circuit board and microprocessor areas. The overall effect of these developments has been to increase the opportunities and incentives for a whole variety of alliances and partnerships throughout the industry, as the range of value-adding activities has become too wide and complex for any individual company to be able to excel across the board.

The partnership approach is always a trade-off, never a panacea

Even when the industry context favors the emergence of supplier partnering, we should always remember that the supplier as partner approach involves a fundamental trade-off. It sacrifices some degree of market discipline for the benefits of closer cooperation, and may not always be the best option in all cases. For example, the partnering approach has been very effective in supporting Marks and Spencer, the leading British retailer, in its strategy of providing premium quality store-label products to middle income customers at very competitive prices (Montgomery, 1991). On the other hand, however, it is also evident that the traditional arm's length approach has been just as effective in underpinning the Wal-Mart Stores strategic mission of offering consumers branded goods at lowest prices (Huey, 1989).

Furthermore, while the security and stability of partnership can be beneficial and help to generate substantial economies of cooperation of the type discussed earlier, there is always a risk of complacency creeping in over time, particularly on the supplier side. The absence of a strong market mechanism may eventually blunt the inventiveness and enterprise of the supply sector.

The future of both partners is tied to the buyer's market

The fact that the partnership approach ties the fortunes of both parties much more closely together, and makes them dependent on what happens in the buyer's marketplace is perhaps its biggest implication. As long as the buyer's industry continues to grow, and the buyer's competitive position remains strong, then both parties will prosper. However, as the industry matures and the growth rate declines, the partners may find themselves facing zero-sum conditions that will really put a strain on the relationship, and test their commitment to each other to the full. The relationship will also come under pressure if the buyer's relative competitive position deteriorates, as suppliers of General Motors discovered a few years ago, when teams of GM "warriors" were sent into their operations to press for price reductions of 20% on previously agreed contracts in order to shave nearly $1 billion from their partner's overall components bill (Lorenz, 1993).

The evolution of the buyer's industry is influenced by many factors external to both partners, and the buyer's competitive position is dependent on many elements that have little direct connection with the quality and effectiveness of the buyer-supplier relationship, as the recent histories of such partnership-oriented companies as Digital Equipment Corporation, Amdahl and Apple can readily testify.

The partners may hitch their fortunes to the wrong star

A further danger for both buyer and supplier in the partnership approach is the risk that they may hitch their fortunes to the wrong star. Will the partners be able to keep pace with each other, and drive each other forward? Or will one partner eventually end up holding the other back?

For the supplier, the risk is that he commits a large portion of his capacity to a company that may be unable to hold or strengthen its competitive position and grow as the industry evolves. For example, Intel has a greater strategic stake in many of its buyer's markets than they have themselves. As the cost of developing each new generation of microprocessor continues to escalate by orders of magnitude (from $2M in 1980 to $1B by 2000), and ever greater volumes are needed to recoup the R&D investment and fill the high-technology fabrication plants to capacity, the company becomes more and more dependent on the ability of its own commercial customers to grow the downstream markets. This is why Andy Grove, Intel's CEO, believes that his company has to be the driving force in the industry, even if it makes its commercial partners uncomfortable and keeps them under pressure (Kirkpatrick, 1997).

The risk for the buyer, on the other hand, is that he invests heavily in a partnership where the supplier proves incapable of developing in line with him and ultimately undermines his competitive position. If this happens for either party, the process of unhitching can be expensive because of the switching costs involved. The more traditional perspective on the buyer-supplier relationship serves to remind both parties of such risks, and advises them to always keep some options open.

A partner may unintentionally mortgage its future

One of the major arguments offered by advocates of supplier-partnerships is that they offer firms enhanced opportunities to harness the complementary core competencies of an array of sophisticated suppliers in the creation of value for customers, and to enrich the development of their own competencies through the wider access to learning and new ideas that such relationships can provide. As Quinn and Hilmer (1994; 43) have argued, such strategic partnering can offer a company "the full utilization of external suppliers' investments, innovations and specialized professional capabilities that would be prohibitively expensive or even impossible to duplicate internally".

However, there is also a downside risk to this process. While partnerships may create more learning and innovation than would result from the more traditional arm's length approach, the benefits may not accrue equitably to both partners. As Hamel and Prahalad (1994) have pointed out, competing for the future is closely tied to competing for today's learning opportunities. When a company offers itself as an attractive supplier-partner of key subsystems to a number of major buyers in its industry, it can also be seen as borrowing the customers of its OEMs in order to increase its own production volume of these core components and accelerate its own learning in strategic technologies at the possible expense of its buyer partners. In this way, the hasty and uncritical adoption of the supplier-partner approach may result in the unintended transfer of critical learning opportunities from the buyer segment to the supply segment, which the buyer may regret. Something like this happened to IBM in its relations with Intel and Microsoft in personal computers. It also happened to General Electric in its relationship with Samsung in microwave ovens, and to Bullova in its relationship with Citizen in the watch industry.

CONCLUSION

Perhaps the essence of the partnership approach is best summed up in the old Japanese proverb as two partners "having the same bed, but different dreams". So while partnership clearly offers many strategic advantages, both buyers and suppliers would be well advised to keep the insights offered by the more traditional arm's length approach readily to mind. Buyer-supplier partnerships are rarely marriages among economic equals (Lyons et al., 1990). Furthermore, the partners remain separate commercial players with their own goals and aspirations, and the marriage, however close, will always be a conditional one, for better, for richer, and in health, for as long as it continues to make sound economic sense.

It is important, therefore, that buyers and suppliers continue to look out for their own interests, even when the partnering relationship is working well. Few buyer-supplier partnerships can ever be expected to develop to the point where the partners are fully prepared to sink or swim together in times of serious industry downturn or major strategic error. One of the enduring virtues of the traditional perspective on the buyer-supplier relationship is that it serves to remind each of the partners, particularly the economically more dependent one, be prepared for industry downturns or strategic error.

Key Concepts: Backward Integration; Economics of Co-operation; Forward Integration; Integrated Manufacturer; OEM; Supplier Partnering; Supplier Performance Management.

Related Articles: Developing a Supply Partner: The Case of Black & Decker and TEMIC Telefunken; Supply Chain Management: Competing Through Integration.

REFERENCES

Dyer, J.H. and W.G. Ouchi (1993). Japanese-style partnerships: giving companies a competitive edge, *Sloan Management Review*, Fall, 51–63.
Hamel, G. and C.K. Prahalad (1994). *Competing for the Future*, Boston Harvard Business School Press.
Huey, J. (1989). Wal-Mart, will it take over the world?, *Fortune*, January 30th, 56–62.
Kirkpatrick, D. (1997). Intel's amazing profit machine, *Fortune*, February 17th, 24–30.
Lorenz, A. (1993). Lopez to drive VW on a lean mixture, *The Sunday Times*, London, March 21st, Section 3, 9.
Lyons, T., A.R. Krachenberg, and J.W. Henke (1990). Mixed motive marriages: What's next for buyer-supplier relations?, *Sloan Management Review*, Spring, 29–36.
Mandel, M. (1997). "The new business cycle." *Business Week*, March 31st, 48–54.
Montgomery, C. (1991). Marks and Spencer, Ltd. (A), *Harvard Business School*, Case #9-391–089.
Porter, M.E. (1980). *Competitive Strategy*, New York: Free Press.
Quinn, J.B. and F.G. Hilmer (1994). Strategic outsourcing, *Sloan Management Review*, Summer, 43–55.
Womack, J.P. and D.T. Jones (1994). From lean production to the lean enterprise, *Harvard Business Review* 72, March–April, 93–103.
Womack, J.P., D.T. Jones, and D. Roos (1990). *The Machine that Changed the World*, New York: Rawson Associates.

27. SUPPLY CHAIN MANAGEMENT: COMPETING THROUGH INTEGRATION

STANLEY E. FAWCETT

Brigham Young University, Provo, Utah, USA

ABSTRACT

The primary objective of supply chain management is the elimination of barriers that inhibit communication and cooperation among different members of the entire supply chain. To eliminate these interorganizational barriers, managers must understand and manage the flow of goods and information from the initial source of raw materials all the way to the final customer. The terminology commonly used to describe supply chain management is, "the management of the value-added process from the suppliers' supplier to the customers' customer." Given the difficulties that most firms encounter in their efforts to mitigate the adverse effects of functional barriers that are entirely within the firm, the challenge of supply chain integration is daunting. To achieve synergies among supply chain members companies must find answers to many questions, including:

- Why should I be concerned about how somebody else does business?
- Can we really trust the supply chain members not to take advantage of us?
- How is our role going to change in the new, integrated supply chain environment?
- Who are the best partners to align our competitive efforts with?
- How many different supply chains can we work with effectively?

Most firms initially lack the answers to these questions, and struggle with the very notion of supply chain management (Elliff, 1996). Many view supply chain integration as a serious threat to independence. Some even view supply chain management as the latest effort of larger companies in the supply chain to squeeze smaller firms' contribution margin and autonomy than before. Despite these serious reservations, and the many unanswered questions, today's changing competitive environment has left most managers feeling that they have no other legitimate options than to participate in integrated supply chain management programs. The fact that key customers request participation while serious competitors are willing to enter into such integrated channel alliances provides strong motivation for adopting a supply chain management perspective. Besides, the competitive improvements that emerge from well-designed and carefully executed supply chain integration are attractive, and create considerable motivation in their own right. The practices of the following companies are mentioned here: **Home Depot; Proctor and Gamble; Toys-R-Us; Wal-Mart.**

THE SUPPLY CHAIN

The quest to achieve higher levels of customer satisfaction is driving supply chain integration initiatives. To better meet the needs of customers and to increase organization effectiveness, companies have for many years restructured and reorganized their operations. Most of these efforts have failed to achieve their objectives of sustainable competitive advantage and customer loyalty. Part of the problem is that ever higher levels of service and value are needed to retain customers (Jones et al., 1995). Indeed, managers at world-class firms now believe that, to compete effectively, their firms must develop value-added processes that deliver innovative, high-quality, low-cost products on-time, with shorter cycle times, and greater responsiveness than before. Yet, even as all-around superior performance is pursued, managers have come to realize that their firms may lack the resources and the competencies required for success. This realization has led managers to look beyond their firms' organizational boundaries to consider how the resources of their suppliers and customers can be utilized to create the exceptional value that is required for long-term sustainable advantage.

The endeavors to align objectives and integrate resources across organizational boundaries are known as supply chain management initiatives. The typical supply chain involves the firm various tiers of materials suppliers, service providers, and one or more levels of customers (see Figure 1). In the past, materials suppliers and service providers have been managed differently, frequently by different functional areas within the firm. For Example, materials suppliers have been managed by purchasing and production while service providers such as distributors and transportation providers have been managed by logistics, marketing, and at times purchasing. For superior operation, a firm must manage both types of suppliers in a coordinated, seamless manner.

The essence of supply chain management is for the firm to focus on doing exceptionally well a few things for which the firm has unique skills and advantages. Non-core activities and processes are then obtained from firms that possess superior capabilities

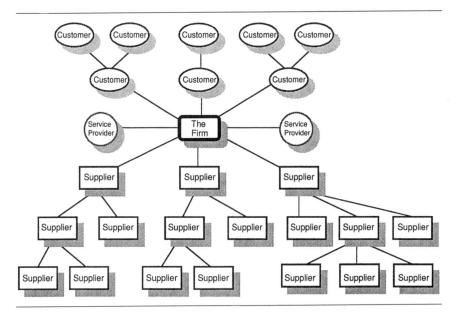

Figure 1. A simplified supply chain.

in those areas, regardless of the firms' position in the supply chain. Close relationships are formed to assure outstanding and seamless performance levels. In effect, "teams" of firms are formed to create the very best product/service offerings possible. As with other "teams," the most successful supply chain teams are those that not only have the best players but that have established true chemistry—a common understanding of supply chain success factors, an understanding of individual roles, an ability to work together, and a willingness to adjust and adapt in order to create superior value for the customers. These allied teams of companies form a supply chain.

HISTORY

The seeds of today's approach to supply chain management in the U.S. were planted in the early 1980s as firms began to seriously consider the threat of Japanese rivals and the competitive implications of just-in-time manufacturing. After several years of focusing on inventory reduction as the key to JIT and improved competitiveness, managers saw the broader aspects of JIT. One of the techniques that emerged from JIT is quite different from traditional management practice, it is the use of long-term partnership relationships with suppliers. Until the advent of Japanese buyer/supplier relationships, U.S. firms had emphasized the practice of multiple sourcing so that suppliers could be leveraged against each other to obtain the lowest possible price on purchased inputs. However, Japanese rivals appeared to achieve significant benefits from supplier certification/development and sole sourcing. Thus, U.S. firms began to reduce their supply bases and emphasize closer buyer/supplier relationships.

Even as greater emphasis was being placed on the strategic management of suppliers, the competitive environment was changing such that channel power began to shift. That is, Channel power, which had long resided with dominant manufacturers, began to dissipate because globalization in recent decades increased dramatically the number of competitors in the U.S. and world markets.

As a result, both retail and industrial customers had greater access to an array of competitive offerings. Channel power therefore began to shift toward the customer from the manufacturer. Moreover, "category killers" like Wal-Mart, Home Depot, and Toys-R-Us quickly established dominance in their respective industries. Their sheer size created the leverage necessary to further shift channel power in a way that put serious pressure on major manufacturers to better manage customer relationships.

Today, power is increasingly concentrated closer to the final consumer, greatly altering the way leading manufacturers operate, and forcing them to pay much greater attention to delivery, timeliness, and cost. For example, a custom-designed delivery system that uses cross-docking, a satellite communications system, and a private trucking fleet enables Wal-Mart to keep its shelves consistently well stocked with low-priced products. Further up the channel, Proctor and Gambles' status as a key Wal-Mart supplier required shorter delivery lead times and a generally much higher level of logistics service than Proctor and Gamble was used to.

The emphasis on closer relationships with both upstream and downstream channel members naturally led to an increasing reliance on strategic alliances. The bottom-line was that only by working as "partners in profit" could firms develop the efficient and effective processes required to produce the high-quality, low-cost products that world consumers required (Schonberger, 1986). This pressure and sentiment continues to persist today, and is exemplified by the adoption of JIT II and customer rationalization strategies by many leading organizations. First, JIT II represents substantial integration of buyers and suppliers in an effort to further reduce inefficiencies and lead times. In JIT II, managers from key suppliers work on-site at the buyer's facilities in order to closely manage important inventories. Second, customer rationalization involves the classification of customers into key accounts and less important customers. Tailored products and services are then offered to those customers that are viewed as important to long-term success. Efforts to customize product/service offerings place tremendous emphasis on information sharing and process integration.

Other circumstances have also led to a greater emphasis on extended supply chain alliances. For example, the drive toward global operations has meant that companies need alliance partners to extend their reach into global markets and technologies. Further, in a global marketplace, companies no longer compete against each other exclusively; rather they and their alliance partners compete against other alliances of global partners. In some instances, fierce competitors in one market find themselves working together in other markets. Managing such complex relationships requires ever-greater flexibility and a constant awareness of the competitive environment. Thus, an emerging key to competitive success is to put together the best team of alliance partners possible, regardless of geographic or channel position. After all, the supply chain is only as competitive as its weakest link.

To summarize, "As the economy changes, as competition becomes more global, it's no longer company vs. company but supply chain vs. supply chain"(Henkoff, 1994). Thus, meeting the imperatives of today's global marketplace requires that firms and their managers adopt a supply chain management mentality.

IMPLEMENTATION

The foundation of effective supply chain management is put into place by top management—only top management can provide the direction and the resources needed to build strong supply chains that possess distinctive capabilities (Stalk et al., 1992). Without support from the highest managerial levels, the right people within the firm and across firm boundaries would never come together in a cooperative manner. Without their support the mechanisms needed for successful integration cannot be established. In fact, managerial support is critical to the implementation and management of each of the four primary integrative mechanisms: cross-boundary integration via process change, information systems support, performance measurement, and alliance management systems. Each of these implementation facilitators will be briefly discussed below.

(1) Process Change: Process change has become a core element of competitive strategies of most firms. The focus of many process change activities is on the integration of value-added activities that occur within the firm and throughout the supply chain. Heyer and Lee (1992) noted that greater coordination among operating departments is a critical outcome of process change. Greater supply chain integration has also been identified as an important outcome of process change (Lee and Billington, 1992). Ultimately, most firms believe that some degree of process integration is a prerequisite to competitive success. However, making the transition from reactive, cost-driven relationships to proactive, customer-oriented cooperation requires strong, sustained and targeted emphasis (Hammer, 1990). Several characteristics are identified as important to supply-chain integration initiatives:

- An increased emphasis on customer input in strategic planning. It is not uncommon today for senior executives to spend upwards of 20% of their time working directly with customers.
- A high degree of consistency among interdepartmental and interorganizational operating goals (St. John and Young, 1991).
- A willingness to share information across boundaries. Interestingly, while many firms exhibit a high degree of willingness to share information, they often lack the technological ability to effectively share information. Bowersox et al. (1995) describe this technology compatibility as connectivity.
- The co-locating of employees among supply chain members.
- A greater focus on formal rules and procedures to guide process integration.

A tremendous amount of rhetoric exists regarding these integrative characteristics; yet companies are slow to implement them (Bleakley, 1995).

(2) Information Capability: Information systems play an important role in supply chain management because they link the diverse and often geographically dispersed members

of the supply chain. Information is substituted for inventory throughout the supply chain, and is the key to postponement, continuous replenishment, and other time-based competitive strategies (McGrath and Hoole, 1992; Bleakley, 1995). Recent research has shown a wide disparity in the effective use of information system capabilities (Fawcett and Clinton, 1996). Even so, the use of electronic data interchange has become the standard operating procedure for world-class firms. Leading firms have also been more successful at integrating their information applications. Despite this progress in building enhanced information capabilities, much work remains. In fact, while a large majority of leading firms have made significant investments in new information technologies in an effort to keep pace with customer demands and competitive threats, they continue to believe that their information systems are inadequate. Part of the challenge is that information demands in terms of accuracy and timeliness are escalating rapidly in today's information driven and time sensitive marketplace. Thus, continued emphasis and investment is required to develop information systems that can satisfactorily meet the requirements of successful supply chain integration.

(3) *Performance Measurement:* For the supply chain to establish the right strategic orientation, promote integration among the channel members, and achieve the necessary continuous process renewal, an appropriate performance measurement system must be put in place. Performance measurement's impact on supply chain integration is pervasive since it: 1) provides insight into the real needs of important customers; 2) yields understanding regarding the value-added capability of supply chain members; 3) influences behavior throughout the supply chain; and 4) provides information regarding the results of supply chain activities (Clinton et al., 1996; Kaplan, 1991). In effect, performance measures direct the design of the supply chain as well as assist in monitoring the integration and day-to-day management of supply chain activities.

In recent years, almost all firms have placed substantial emphasis on improving their measurement systems. Leading firms in particular have expended great efforts to develop measurement systems as a strategic facilitator and supply chain integrator. These firms have truly recognized that without proper alignment among key measures, supply chain activities will not yield the desired product quality, service and competitive advantage. As a result, world-class firms are seeking higher levels of customer input in the measurement of performance, while at the same time using techniques such as performance scorecards to better align supply chain members. Total costing and activity-based costing are also popular integrative tools. Overall, an increasing number of today's firms appear to recognize the importance of performance measurement and have actively sought to enhance their ability to use measurement to direct and mold their supply chain integration efforts.

(4) *Alliance Management:* Alliance management is the final issue that has received considerable attention in recent years as a key facilitator of supply chain integration. Indeed, alliance management is the essence of supply chain integration. However, in supply chain management the traditional dyadic management of alliances must be successfully extended throughout the entire supply chain. This extension requires the application of best alliance practice and technique. Four practices common to supply

chain alliances are: (1) the use of written contracts; (2) the use of clear guidelines and procedures for selecting alliance partners; (3) the use of clear guidelines and procedures for monitoring alliances: and (4) the sharing of risks and rewards.

Recent research has found that establishing alliances on the principle of shared rewards and risks is perhaps the single most important key to success (Schonberger, 1986; Bowersox et al., 1995). World-class firms are particularly adept at sharing both risks and rewards that emerge from collaboration. These high-performing firms also tend to be active in establishing formal guidelines and procedures for creating and managing alliances. Interestingly, recent research has found that techniques that promote the formalization of alliance management remain relatively little used (Fawcett and Clinton, 1996). This finding suggests that while alliances have been touted as vital for competitive success, most firms have not determined the best way to structure and manage important supply chain relationships.

TECHNOLOGY NEEDS

Currently conceived supply chain management could not exist without recent technological advances, especially in the areas of computing power and communications. The role of information systems was outlined above. These information systems include many of the following: EDI, barcoding, satellite tracking, data warehouses, and a variety of input/output devices. Together, these technologies allow for accurate information to be collected, stored, processed, and exchanged efficiently and in a timely manner. Without adequate information exchange, supply chain coordination could not occur. It is important to note that the ability and the willingness to exchange information are not the same thing. For many firms, investments in technology have not been matched by a change in mindset to freer exchange of information that is considered to be proprietary.

BENEFITS

In theory, supply chain management enhances communication and coordination among the different members of the entire supply chain or distribution channel. Improvements in coordination lead to more efficient materials management—total supply chain inventory is reduced, cycle times are shortened, and the supply chain is more flexible and responsive. Better communications play an important role in developing innovative products and services, and reduce concept-to-market leadtimes. Role shifting has become an important element of supply chain management since it places value-added activities with the team member best suited to perform a given role based on competency rather than tradition. Of course, role shifting also represents a threat to any firm that does not possess a real and valued competency—such a firm would likely lose its role in the supply chain. When done appropriately, role shifting enhances the entire supply chain's performance.

Most firms struggle with the implementation of supply chain management because of numerous, inherent information-related intricacies, difficulties encountered in modifying performance measures, and obstacles to developing working relationships within and across firm boundaries. However, some firms have aggressively and judiciously implemented supply chain integration initiatives with dramatic success. At such leading firms, total cycle time reductions of 80 percent have been achieved, inventory has been cut in half, and the cost of purchased materials and services reduced by 10 percent or more. Perhaps the most important fact is that these improvements have enhanced customer service (Elliff, 1996).

CONCLUSION

A fundamental challenge to understanding and achieving effective supply chain integration is that the term is used so differently by different individuals. In many cases, the term supply chain management is used for either simple dyadic relationships between a firm and its supplier or for a slightly more complex three-company relationship that extends forward to include the firm's customers. This frequently used, but narrow, approach to supply chain management has led to a relatively meager understanding of how entire supply chains with their multiple levels and diverse participants should be structured and managed. Supply chain management is one of the most talked about competitive strategies of the 1990s.

Key Concepts: Alliance Management; Continuous Replenishment; Customer Rationalization; Functional Barriers; Integrated Supply Chain; Japanese Buyer/Supplier Relationship; Role Shifting; Strategic Management Of Suppliers; Supplier Development; Total Costing.

Related Articles: Developing a Supply Partner: The Case of Black & Decker and TEMIC Telefunken; Supplier Partnerships as Strategy.

REFERENCES

Bleakley, F. (1995). "Strange Bedfellows." *Wall Street Journal*, (January 13), A1, A6.

Bowersox, D., R. Calantone, S. Clinton, D. Closs, M. Cooper, C. Droge, S. Fawcett, R. Frankel, D. Frayer, E. Morash, L, Rinehart, and J. Schmitz (1995). *World Class Logistics: The Challenge of Managing Continuous Change*. Council of Logistics Mgmt, Oak Brook, IL.

Clinton, S.R., D.J. Closs, M.B. Cooper, and S.E. Fawcett (1996). "New Dimensions of World Class Logistics Performance." In *Annual Conference Proceedings Council of Logistics Management*, (21–33).

Elliff, S. (1996). "Supply Chain Management—New Frontier." *Traffic World* (October 21), 55.

Fawcett, S. and S. Clinton (1996). "Enhancing Logistic Performance to Improve the Competitiveness of Manufacturing Organizations." *Production and Inventory Management Journal*, 37 (1), 40–46.

Hammer, M. (1990). "Reengineering Work: Don't Automate, Obliterate." *Harvard Business Review* (July–August), 104–131.

Henkoff, R. (1994). "Delivering the Goods." *Fortune* (November 28), 64–78.

Heyer, S. and R. Lee (1992). "Rewiring the Corporation." *Business Horizons* (May–June), 13–22.

Jones, T.O. and W.E. Sasser, Jr. (1995). "Why Satisfied Customers Defect." *Harvard Business Review* (November–December), 88–99.

Kaplan, R.S. (1991). "New Systems for Measurement and Control." *The Engineering Economist*, 36 (3), 201–218.

Lee, H. L. and C. Billington (1992). "Managing Supply Chain Inventory: Pitfalls and Opportunities." *Sloan Management Review* (Spring), 65–73.

McGrath, M. and R. Hoole (1992). "Manufacturing's New Economies of Scale." *Harvard Business Review* (May–June), 94–102.

Schonberger, R.J. (1986). *World Class Manufacturing*. The Free Press, New York.

St. John, C.H. and S.T. Young (1991). "The Strategic Consistency Between Purchasing and Production." *International Journal of Purchasing and Materials Management* (Spring), 15–20.

Stalk, G., P. Evans, and L.E. Schulman (1992). "Competing on Capabilities: The New Rules of Corporate Strategy." *Harvard Business Review*, 70 (2), 57–69.

Wisner, J.D. and S.E. Fawcett (1991). "Linking Firm Strategy to Operating Decisions Through Performance Measurement." *Production and Inventory Management Journal*, 32 (3), 5–11.

28. DEVELOPING A SUPPLY PARTNER: THE CASE OF BLACK & DECKER AND TEMIC TELEFUNKEN (1992–1994)

TIMOTHY W. EDLUND
GEE-IN GOO

Morgan State University, Baltimore, MD, USA

In 1987, Rosemary Smith, a representative of AEG (Germany) in the U.S., contracted Black & Decker (B&D) to solicit business. She talked to Bob Wall, manager of corporate commodities purchasing, at B&D's Towson, Maryland, corporate headquarters. She communicated directly with TEMIC regarding several possible programs at B&D, which eventually resulted in the program described below.

Seizing on the opportunity reported by Rosemary smith, Klaver, Director, TEMIC, joined her in a 1987 call on Bob Wall at B&D. Serious discussions resulted about providing TEMIC's developing technology and engineering capabilities to B&D, which Klaver saw as a useful way to enter the strategic U.S. market much earlier than might have otherwise been practical. As these discussions developed, both firms slowly came to realize that they could enter into a strategic partnership, from which both might benefit.

The Black & Decker Company (B&D) was developing a new coffeemaker, to fit under kitchen cabinets. This drip coffeemaker would come in three models; two would use sophisticated electronic controls, and a third would use a simpler manual control. TEMIC Telefunken (hereafter TEMIC, short for TElefunken MICroprocessor) of Ingolstadt, Bavaria, Germany, received B&D's letter of intent (LOI) contracting for

The authors thank TEMIC Telefunken and the Black & Decker Company for their enthusiastic cooperation.

Source: This is adapted from cases developed by timothy W. Edlund and Gee-In Goo of Morgan State University. Copyright 1994 by the National Consortium for Technology in Business, c/o the Thomas Walter for Technology Management, Auburn University, Alabama. Revised 1999.

designing and making the electronic control module for these two products. 400,000 of each model (and 400,000 of the simpler, manual coffeemaker) were scheduled for 1994; mass assembly of the complete coffeemaker was scheduled to begin in February 1994, 13 months away, in Asheboro, North Carolina. By then a great deal had to happen. In early 1993 Johannes Werther (Sales and Marketing) and Michael Zimgibl (Manager of Engineering) discussed how to organize the project team and make other arrangements to ensure that TEMIC completed its commitments to B&D on time with desired quality, function, and cost.

B&D had scheduled initial shipments of the new coffeemaker for June 1994 to stock merchants and fill distribution pipelines for the Christmas 1994, a major selling season. TEMIC had to be sure it did not cause its partner to miss that crucial selling season.

CRITICAL UNIT FUNCTIONS

The outside of the coffeemaker was a rectangular block, with the carafe and the filter funnel on the left; the control panel was to the lower right, below the water supply. See Exhibits 1 and 2. Therefore, the face of the electronic control panel had to completely seal off electronic components to protect them from water spills and vapor. The controls would be miniaturized state-of-the-art controls. It was required that there be no gap around the control panel, or any of the joints between the panel and control buttons, or past the clear plastic inserts which let the LEDs (light emitting diodes) shine through. A tactile feel to the buttons was required; that is, the user must know that activation had happened.

In the unit, the water reservoir was removable for filling, and would plug into the heating system. There was a valve to let the water into the heating mechanism when the reservoir was inserted. The whole assembly was designed to be mounted under kitchen cabinets on special brackets.

TEMIC TELEFUNKEN

In 1985, the present firm was in the "board stuffing" business. This meant that they operated as a subcontractor, assembling parts onto printed circuit boards (PCBs) in accordance with customer designs and specifications. This business was essentially a commodity business. Competition was on price alone; the firm added no design value. What became TEMIC was a typical German-oriented company, focused on manufacturing quality, and with little focus upon marketing or volume production. There was little awareness of the needs of consumer products, no orientation to US companies, which was also true for their parent firm, AEG. TEMIC's exports were in the range of 10 to 15 percent, all to other European counties; the balance of their business was done with German firms.

In 1987 there was a change in management. Mr. Klaver, previously with the Dutch firm Philips, was appointed director. He had been in Brussels about 8 years, and previously had spent 10 years in the Far East. A Netherlander, Mr. Klaver was used to thinking in terms of global business. Appointed by Telefunken to take charge of the business unit in Ingolstadt, he began the process of creating the present business

Exhibit 1. External view, under-the-counter coffeemaker. (Source: Black & Decker preliminary owners' manual for SDC series spacemaker optima™ coffeemaker. Spacemaker is a trademark of general electric company (USA).

unit. Essentially a subassembly firm then, now it's a microsystems development and manufacturing organization. He saw that it was necessary to obtain business from America. Business growth in Germany was declining. Germany had very high labor rates, so it became necessary to build a manufacturing organization outside of Europe, to permit manufacturing in the Far East or in Latin America.

Exhibit 2. View of controls and internal arrangement, under-the-counter coffeemaker. (Source: Black & Decker preliminary owners' manual for SDC series spacemaker optima™ coffeemaker. Spacemaker is a trademark of general electric company (USA).

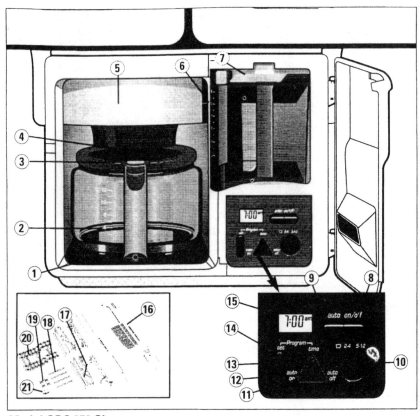

Model ODC 350 Shown

1. "Keeps Hot" Plate
2. Carafe with Cup Marks (12 Cups)
3. Flip-Up, Removable Carafe Lid
4. Sneak-A-Cup™ Interrupt Drip Stop (Under Basket)
5. Filter Basket
6. Water Reservoir with Water Level Markings
7. Reservoir Cover with Thumb Latch
8. On/Off Button/Light
9. Activate Auto Button/Light
10. Cup Select Dial
11. Adjustable Auto Off Program Button

12. Auto On Program Button
13. Time Program Button
14. Up and Down Set Button
15. Digital Clock Display Window
16. Mounting Template
17. 2 Mounting Brackets with Tabs
18. 4 Screws
19. 4 Nuts
20. Set of Spacers/4 Washers
21. Cord Hook

Items 17-21 are packed in the Water Reservoir. Be sure to remove them before mounting or operating the unit.

In 1993, total revenues of the Telefunken semiconductor businesses was about DM 1.1 billion (deutsche marks[1]). Total Telefunken revenues were about DM 3.1 billion. For about two years, Telefunken results had been fully integrated into their parent firm, DaimlerBenz, whose revenues were DM 98.5 billion in 1992. Organization and ownership changes affecting TEMIC are described in succeeding sections of this case.

Klaver believed it necessary to completely redesign the business to be competitive and profitable in the new markets. Together with improved products, including design capabilities to create them, production capabilities in the Far East had to be started to be competitive, to first provide for European business. Subsequent moves into the Americas would be possible, for both sales and production. TEMIC's business was 85–90 percent in the German market. There were very few people in the company who were able to speak English or other languages.

BLACK & DECKER[2]

Founded in 1910, Black & Decker (B&D) manufactured a wide range of corded and cordless portable power tools, small household appliances, other household products, and other products. In 1992, household products accounted for 14% of B&D's $4.8 billion revenue. (Exhibit 3). B&D was the world's largest and oldest power tool maker; and its "Household Products business[was]a major global competitor and North American market leader in the small appliance industry." B&D complemented its power tool business by purchasing GE's small appliance division in 1984. This provided a wide range of small appliance products (irons, mixers, etc.) to which was added B&D's popular line of Dustbuster© vacuums and other battery-driven appliances.

In 1985 Nolan Archibald, formerly a marketing and operations specialist with Beatrice Foods, was elected president and began reorganizing B&D. He organized plants around motor sizes and reduced product variations and streamlined manufacturing processes. B&D's restructuring drew attention. They developed an approach to simultaneous engineering, in which manufacturing engineering met regularly with product engineering, an approach aimed at reducing lead times and costs. B&D's factory in Spennymoor County Durham, England, was called one of six superplants in the United Kingdom. Archibald's cut-and-build concept drew compliments as an effective approach to cost cutting and firm rebuilding.

Major Changes in Product Requirements: Before the redesigned coffeemaker, the normal B&D household product used electronics built around stuffed PCBs, which took considerable space considering what functions were accomplished. Almost any subcontractor could build these, and there was substantial pressure on price. Use of microprocessors and miniaturization promised to change all that, providing both enhanced function and reduced size and cost. Not every firm could design and build these, nor was every customer ready to utilize such capabilities.

---

[1] In 1993, conversion ratios were approximately $1.00 = DM 1.65 and DM 1.0 = $0.60.
[2] Adapted and updated from "the Black & Decker Corporation (A) & (B)," by Timothy W. Edlund and Sandra J. Lewis. Copyright © 1991 by Timothy W. Edlund. Used and adapted with Permission.

Exhibit 3 Black and Decker 1992 revenues by product group within business segments

(Millions of Dollars)	Year ended December 31, 1992	
	Amount	%
Consumer and Home Improvement Products:		
Power tools	$1,175	25%
Accessories	674	4
Household products	341	7
Security hardware	515	11
Outdoor products	312	6
Plumbing products	173	4
Product service	189	4
Total consumer and home improvement products	$3,379	71%
Commercial and industrial products		
Fastening systems	$ 384	8%
Other commercial and industrial products	283	6
Total commercial and industrial products	$ 667	14%
Information Systems and services	$ 734	15%
Total consolidated revenue	$4,780	100%

Narrative description of the business (partial): Household products include a variety of both corded and cordless cleaning and lighting products, and a full line of small home appliances, including irons, mixers, food processors and choppers, can openers, blenders, coffeemakers, kettles, toasters, toaster ovens, waffle bakers, knives, and smoke alarms.

Source: Form 10-K for the fiscal year ended December 31, 1992, pp. 4–5.

CHANGES IN TEMIC

In 1988, Johannes Werther was hired to direct the sales and marketing effort for this new business. Previously he had been employed by Westinghouse Electronic Co. & General Signal, and had wide experience in European business. About this same time Michael Zirngibl was engaged as manager of engineering.

Determined to change the culture at the Ingolstadt business unit headquarters, TEMIC started emphasizing the need to know the English language. Special English lesson programs were provided. English became the official language of the firm, although German was still used for traditional German customers. New employees either knew English, learned it, or found other employment. New, younger, bilingual design engineers were hired for Ingolstadt, and similar industrial and production engineering capabilities were also added.

Also in 1988, TEMIC obtained their first piece of business from B&D. They contracted to produce the electronic module for a new electric iron, of rather basic design. TEMIC already had a semiconductor factory in Manila. Alongside the old plant a new plant was erected to produce the new product line. It was air conditioned and contained subassembly facilities and other necessary equipment. An initial workforce of 200 people was hired, which swelled to 1,500 by mid-1993. Although the module for the iron was a board stuffing project, TEMIC considered this project the first step in training their people to learn how to do subcontracting well.

It represented a way to get started in consumer-type business; including delivery on tight timetables, testing, signoff of the product, and the like. Up to then nearly all

TEMIC's business had been industrial. Time pressure became much more important. It was absolutely necessary to learn to figure in parts worth a cent, not parts worth a deutsche mark. They entered intense negotiations on pricing, the consumer product kind of tough pricing, with their purchasing people to show them how important it was to be accurate in pricing in this kind of business, considering that 1 or 2 million pieces would be delivered. They transmitted this kind of price consciousness to the factory. Managing material pricing was crucial. It was very important whether cost was 36 cents or 36.1 cents per piece. The transfer to the Far East, under absolute price consciousness, with more engineering, and higher technology, provided a focus on these three positive attributes. These were important in the developing partnership with Black & Decker.

TEMIC considered that they "bought" the B&D program then. There were lots of programs in production, including two with B&D: first the iron and then a new toaster, which was also a relatively simple project, having no part of the electronic module visible on the outside of the toaster. For a while there were no profits. Delivery of the first parts from the new factory was made at the end of 1988; it was believed that two years were needed to learn how to do this business.

As a simple example for their personnel, they made the comparison, of how important 10,000 pieces of an industrial product having one dollar unit profit were, compared to a 1-million-piece consumer product run, yielding a dime profit for each unit. This was a simple calculation, and demonstrated the value of consumer-type products.

NEW OWNERSHIP OF TEMIC

In 1993, Daimler-Benz acquired the organization. When the reorganization was complete, Telefunken was owned 50 percent by AEG and 50 percent by Deutsche Aerospace, which were in turn both owned by Daimler-Benz. Major decisions had to up through all these levels. Moreover, under German law each organization had two boards; a managing board and a supervisory board, both of which had to consider and approve each decisions. By law, labor was represented, and could delay any project labor felt was against their interests.

Moreover, Mr. Klaver left the organization in 1991. As a replacement, TEMIC got what seemed to be a typical Daimler-Benz-oriented management, including their new director, Mr. Keller. "They couldn't understand what we [TEMIC] were doing with Black & Decker, doing low-unit-profit, high-volume business." In Europe, and particularly in Germany, B&D did not have a high-quality, high-reputation status. B&D was selling drills[3] and consumer products of perceived lower quality, at lower prices, and management was not particularly interested in doing business with firms like B&D.

Werther set about to convince the new director, Mr. Keller, of the value of this type of business. Together with Mr. Kohl, the new manager in Manila, he showed that

---

[3] AEG Telefunken also made a line of power tools, although this line was considered to have higher quality and competed more in the professional market, composed of construction workers and others who depended on their tools for their living.

consumer business was a stable business, which permitted training people to handle increasingly complex work, both in design and in production. The plan was to keep both types of work, using the same plants and design staff. Very sophisticated industrial/military and consumer products should be in the same plants. It was absolutely necessary to keep this type (volume consumer products) of production in the plant to train people in the program, so they would learn how to do more sophisticated manufacturing. Werther stated that it was strategically important to keep this business, because it would teach how to do profitable business, which in turn would provide substantial competitive advantage for industrial and military contracting.

Rosemary smith of TEMIC, resident in Towson, Maryland, was now concentrating solely on B&D business. B&D was increasingly interested in developing new programs with TEMIC. Because of the economic slowdown in Germany and Europe, in 1991, people in Germany were very open to new programs. Although the B&D name had not earned much credence in Germany yet, they argued that B&D could be a very strategic customer. They showed information and independent articles about B&D, to provide awareness of the U.S. market size and B&D's share of the U.S. market. They emphasized the high volume, the stable business, the acceptable level of profits, and the restructuring going on at B&D.

STRATEGIC PARTNERSHIP TAKES SHAPE

Mr. Keller set up meetings between TEMIC and B&D. He prepared a strategic analysis, which considered the actions of B&D: its future in high-quality, high-volume, high-design products; its careful selection and qualifying all important suppliers; and its efforts to be at the very high end of technology. It was pointed out that U.S. technology was the best on circuit boards and other electronic components. Senior management in TEMIC began to accept the strategy of partnering with B&D after a series of meetings, both internal and with B&D. The message became clear that B&D tries to be in strategic partnership relationships with crucial manufacturer/suppliers with the intent to share all pertinent technologies. Mr. Keller accepted that these ideas were valid and correct. It was projected that $30 to $40 million in business would develop in the future for TEMIC through the partnership.

Next, TEMIC planned to start an operation in Mexico. They already had contracts with Volkswagen Brazil, providing them with ignition and motor management systems for vehicles assembled there. A Mexican plant, eliminating many subcontractors, serving VW operations in Brazil and Mexico, combined with B&D work, was proposed and endorsed by Mr. Keller. He agreed that it made sense to start up production on consumer products with high technology in Mexico, taking a year or so, along the lines of Manila. They would make products in consumer electronics first, particularly the existing B&D products, then automotive, then others.

They proposed spending $10,000,000 to start a Mexican operation, relying on only two customers: VW and B&D. It took about 10 months to get approval from Daimler-Benz. A new building was started, about 60 kilometers from Mexico City to house this production. Legally, ownership was vested in AEG Mexico, which was already licensed to do business in Mexico. Under construction before NAFTA legislation passed, the plant was expected to start at the beginning of 1994 and would train on the older products before beginning assembly of the new coffeemaker module.

THE LETTER OF INTENT (LOI) FOR THE COFFEEMAKER

Michael Zirngibl visited B&D in December 1992 and discussed requirements for the new coffeemaker, concentrating on requirements for the electronic control. He showed a product TEMIC was building for Carrier Corporation USA: a remote control for air conditioning units. It was a two-way infrared communication and control system for major units, permitting no displays at the air conditioning unit. He contrasted this with products made in 1985, the board stuffing technology used in the two products being made for B&D, and compared those with the next generation technology product, having chips on silicon, using 85 to 90 percent full automatic insertion of components, full quality control, and closer mechanical relationships. The control for Carrier also had rubber buttons, backed with carbon to complete circuits when pressed, with backlighting provided by LCDs (liquid crystal displays) showing through clear inserts in the control panel. B&D expressed great interest, indicating that the Carrier remote control incorporated about 80 percent of the features they wanted in the new coffeemaker. Most of the technical problems in the coffeemaker control had already been solved by TEMIC for other programs, it was the combination that would be different for B&D.

B&D was in a hurry, Zirngibl was asked how the product should be designed and built, and what the cost would be. He borrowed a desk at B&D's Towson headquarters, sketched out his ideas and developed costs, taking two days. His proposal was presented on 15 December 1992; TEMIC received the letter of intent on 7 January 1993.

LOGISTICS

As Werther and Zirngibl conferred, they knew several things that served to complicate their planning. So far, most negotiations had occurred at B&D's headquarters in Towson, Maryland. But the Household Products group was based in Shelton, Connecticut, five hours drive north. Black & Decker would assemble the electronic module into the coffeemakers at Asheboro, North Carolina, a day's drive south of Towson. Both locations were remnants from the time that General electric (USA) owned the household products business. Design and marketing would be done at Shelton, as would major initial purchasing; some purchasing including purchase of plastic mold tooling would be done at Asheboro. This was critical because the front panel designed and made by TEMIC must seal perfectly into B&D-designed and-specified housings, permitting no leaks of water, whether liquid or vapor. Moreover, much of the production engineering would be done at Asheboro; this was important because assembly tooling would have to be fully compatible with the electronic module, both for quality and for cost.

TEMIC would be communicating over long distances with others beside B&D; the module would be made in Mexico, in a still inexperienced plant. TEMIC's own supplier network stretched around the world; important components would probably be made far from Germany, Manila, or Mexico.

THE CHALLENGE OF THE NEW PARTNERSHIP

On July 13, 1993, after a joint meeting of B&D and TEMIC personnel at B&D's Household products Headquarters in Shelton, Connecticut, Esko J. Nopanen of B&D's Sourced Products Engineering and B&D's program manger for this project, commented:

For almost 25 years, I've been involved in buying products from Far East suppliers, first with GE, and then with B&D, since they bought us. I've been involved in specifying and identifying what the product should be and which suppliers will make it, private label it, and send to us. I've got experience working with outside suppliers and with the communications needed to make it work for a long time.

For this project TEMIC is a supplier in a way, but not the traditional kind of supplier. They were selected to be a partner rather than a supplier, which puts a somewhat different perspective into the communications that go out. In a lot of ways a partnership evolves after having a supplier-customer relationship. It matures into a partnership. You already know each other very well. You may change attitudes a little because now you're partners; here may be less checking upon each other, accepting just the word of each other. With TEMIC it was a little different; we didn't have a supplier-customer relationship first. At the very first we agreed that this will be a partnership, and then we went into working within that partnership. It's taken both us a little while to say, what does that really mean? It would have been easier in several respects to have a customer-supplier relationship first.

Deidre Elloian, director of electronics engineering, had other concerns about the partnership. She said:

I'm puzzled why TEMIC is putting so much effort into this project. We only do a few hundred million [dollars] here; they're a multibillion dollar outfit and most of our products won't need this degree of electronic sophistication. Of course, there's more volume in our tools division, but I'm not sure about the application there. What's in it for TEMIC?

Also, to some extent, we're giving up the ability to do our own electronic work. About all we'll be doing is second-guessing their work, and, if the partnership works out, we shouldn't have to do much of that. Of course, trying to build the capability they already have would be very difficult, there's not enough of that kind of work to attract and keep such an engineering team together.

They both agreed that the meeting went well and that difficulties were being resolved on a timely basis. At this point, the focus was upon completing the project successfully. That involved the electronic module, the mechanical parts, the various connections, and how well everything fit together. The first volume production was scheduled for February 1994.

QUALITY AND MILESTONES

Convened promptly at 10:00 A.M., the meeting on July 13, 1993 quickly progressed to discussing a number of technical details. Among the Black & Decker personnel present were Nopanen, Elloian, Duc Tran (senior electronic design engineer), Paul Donoski (manager, Beverage Products Development), Julian Watt (senior electrical engineer), and Jon Rayner (director, Corporate Purchasing). TEMIC was represented by Werther, Zirngibl, Smith, and Blank (an engineer and the project manager for the B&D project).

One intense discussion focused on the sample panel brought here to show what the current design was like. It was pointed out that the lighted panel had to be visible not only in the dark, but also in a fully lighted kitchen.

TEMIC submitted a proposed schedule for completion of various milestones in the program. This was compared to B&D's desired schedule. B&D asked that two dates be improved by seven days. TEMIC would meet those requests, Werther replied without hesitation, as Zirngibl smiled.

Following several technical discussions, Nopanen inquired about TEMIC milestone release points. Werther described these as follows: QB (an abbreviation of the German words for quality control) was the general descriptive term, resulting in the following checkpoints:

QB 1: release indicated that the concept is good, design could proceed.
QB 2: release for engineering builds.
QB 3: release for preseries production (trial/pilot lots).
QB 4: released for series production.

Approval was required from purchasing, sales, and production engineering at every step, including everyone concerned. The QB 4 release required approval by every department, certifying that all work had been done; it was designed to generate cooperation to solve problems and to complete the project under the responsibility of the project manager, who also is responsible to see that everyone has done their job.

Nopanen responded that it was very similar to the B&D procedure, except that different names were used. He added that B&D would now be able to understand these terms when used in TEMIC releases. But Werther responded that all communication to B&D would use the B&D terms, to avoid any possible confusion. Moreover, these would be used internally at TEMIC along with the QB designations, so that all project personnel would be fully familiar with them. B&D's quality posters in Exhibit 4 reveal the company's quality related values and principles.

After resolution of all pending technical problems, the next item of business was a slide show presented by Werther explaining TEMIC's relationship to its parent organizations, its own internal organization, and the various types of work they were qualified to do.

After the meeting the comment was made to Elloian that it was interesting to see that TEMIC and B&D personnel were intermingled around the conference table. She responded with a chuckle: "That didn't just happen; we made sure it would."

Exhibit 4. Black & Decker quality posters.

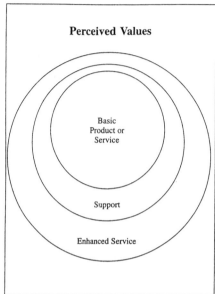

Perceived Values

Basic
Product or
Service

Support

Enhanced Service

Quality: The Basic Principles

1. Focus on the work process, issue, or behavior, not the person.

2. Maintain the self-confidence and self-esteem of others.

3. Maintain strong partnerships with your internal and external customers and suppliers.

4. Take initiative to improve work processes and partnerships.

5. Lead by example.

QUALITY POLICY

WE are dedicated to exceeding the expectations of our internal and external customers with uncompromising integrity.

In a participative environment of continuous improvement, our commitment is to satisfy our customers by providing error-free products and services every time.

QUALITY is the most important element of our business.

CORE VALUES

• INTEGRITY means we have complete and uncompromising honesty in the conduct of our business. Individually and collectively we do what we say. We make appropriate commitments and we honor them.

• INNOVATION means we will strive to delight our customers by maintaining an environment which stimulates new and creative approaches to all aspects of our business.

• EXCELLENCE means we will excel in all areas of our business by setting goals and standards which foster continuous improvement in our people, products, and services.

• TEAMWORK means we will create a participative environment in which each employee is valued as an important contributor, as an asset or resource of the business. We will have an atmosphere that attracts, retains, and develops excellent and committed people who want to share in the success of the company.

TEMIC'S APPROACH TO NEW PRODUCT LAUNCH

In July 1993 Johannes Werther described some of the internal workings of TEMIC, including its quality system. The following material is condensed from that conversation.

Given the changes in the product and quality focus at Black & Decker, we saw ourselves becoming the supplier of microprocessor-based electronics for their high-end products. We are a qualified supplier for General Motors. We furnish 1.5 million assemblies per year of their ABS (anti-lock brake system) systems, at $100 each, priced from Manila. When we make them in Mexico, we'll be at or close to that price. Our production technology and engineering capability are oriented to production line requirements. We build to automotive rules, using the same equipment and quality. Even for consumer products we produce with very low ppm (parts per million) defects, and guarantee very fine quality in all our products.

You may remember that, before Bob Wall (of B&D) started the program with us, the usual B&D electronic product was a stuffed PCB (printed circuit board). It was bulky and relatively crude.

Our design was done by Michael Zirngibl's staff. We developed the program plan, then did the engineering work, and did our part of the hardware design. We established our milestones, built them into the program plans. Normally our scheduled completion dates were two weeks earlier than we told B&D they would be.

We intensified our English language training program; all of our people have to speak English. By the way, we found that helps us in France, too. You can't get business in France speaking German; no one can speak French well enough to please a Frenchman, but it's okay to speak English there!

The design was finalized; we built our first prototypes on PCB boards (showing how they would function), prepared an overview discussing whether the concept and program were working, and put together the breadboard (an early type of prototype model to demonstrate function) by late April or May (1993). They visited us in May. The basic concept was released for further development. This basic concept was released for further development. This was very fast work; we Germans take a lot of holidays, but we work very hard, too.

There have been some small bumps so far. We're in daily communication with Rosemary Smith, and with the local TEMIC engineer in Shelton. The plan is, we send an engineer over there for six weeks. He has an office at B&D, Shelton, works with their design staff, and communicates with us. He will return for two weeks to confer with the rest of the engineering staff here in Ingolstadt. Then another engineer goes to Shelton for six weeks, returns for two, and is followed by a third, after which the cycle starts over again. The aim is to have essentially interchangeable people, who have knowledge of the department and specific knowledge of the program. Each engineer spends about one-quarter of his time in the States. We have an apartment; he can take his family. Last week Mr. Blank, one of our younger guys, went over and will be there until September.

As soon as this partnership idea arose, we discussed it with Mr. Keller, our director. We decided it was vital to get into a very close relationship with B&D. In the beginning, we decided that if we would do this at all, we want to do it right. From a business point of view, a partnership was necessary. That was our position from the beginning. It may also have helped convince B&D that we were serious about building a partnership.

After prototypes and the engineering build, the electronic module is tested in the finished product; we do this together. We did it some weeks ago for the toaster, using the international B&D Procedure. The Engineering build 1 and 2 for the coffeemaker module are planned; if necessary, number 3, 4, and so on, will be scheduled. We go to preseries production next, first 1,000 pieces, then 2,000. February 1994 is scheduled for mass production: 400,000 per year [those coffeemakers having the electronic control]. We have 10 months to make sure everything is working; their procedures and ours are reasonably congruent.

We are now between stages QB 2 & QB 3. Our internal release dates are two weeks ahead of B&D. We always schedule earlier than our customer's wanted dates.

Mr. Blank, one of Michael's engineers, is responsible for our B&D programs. He is the engineer now in the States. Also on the project business team are Mr. Bartoff (one of my sales engineers), Michael, and myself. The marketing and engineering group people are the business team. It is a part-time assignment for everyone.

Our business team involves engineering and production, quality and logistics (components availability after purchasing), purchasing, Mexico production, and project responsibility. The QB procedure is totally separated from customers' outside information. Logistics ensures that everything is shipped on time. We have our plan and then we provide the customer with a plan. We meet our standards, qualify under our automotive standards, which are tougher than customers' standards. We prequalify everything to our standards. We do live tests and mechanical tests, even if the customer doesn't want to have that. We do it because we have to sign that the product is acceptable to us, even though those requirements may be 20 percent higher than the customer's requirements. It doesn't matter whether the customer specifies this; we produce according to this rule. This is clearly supported by all of us, from our top management down.

Again, it doesn't matter whether the customer wants that quality standard; we produce according to our requirements. Therefore, if in six months the customer has a problem in the field, we can show that we have tested it 100 percent according to our rules, even though he hasn't seen those tests. [This is 100 percent testing in planning and in qualification, not 100 percent testing of all series production parts.] We include drop testing, shock testing, durability testing, whatever is necessary according to design and intended use, we do all this even though not asked. We follow the Ford standard; that's what Ford specifies for all their products, even those produced in the United States. It has higher standards than the ISO 9000 standard.

We communicate to the customer that we should be more involved in the customer's activities. Communication does not always happen. Our people are busy doing what they have to do, the B&D engineers get worried, they haven't heard anything for two or three weeks; it's in-between milestone times. The residential engineer is stationed in Shelton to solve those concerns, so that B&D can know "the Germans are working on it." They will come to know him and believe they can trust him. There are regular conference calls within the TEMIC engineering group so that the residential engineer and the sales engineer (Rosemary Smith) can be fully informed about what's going on.

We also have ongoing communication with our management, up to the boards of directors every two or three months, much more frequently with our own top management, and also with the B&D board. We provide communication lines from engineers and sales group up to our vice president. Moreover, everyone is talking to his equivalent in B&D; engineer to engineer, production to production, sales to purchasing, and with our purchasing director in Mexico. That director knows B&D; he set up our quality procedure while here in Ingolstadt. He wrote the rules; he knows *how to follow them*.

CHALLENGES IN DEVELOPING A SUPPLY PARTNER

It was February 28, 1994. Esko J. Nopanen, B&D program manager for the new coffeemaker project, described the current status of the project and the developing partnership with TEMIC. The following material is condensed from that conversation.

Communication and Expectations

As you know, this partnership was started without first having a supplier-customer relationship. Although I've had over 20 years experience with sourcing our products from many vendors, this has been different. It was already understood that TEMIC would provide the electronics and the control panel; they [management] said TEMIC will be the expert, we won't specify our requirements very closely to them. We told them in general what the control module has to do and they'll tell us how it's going to work, and exactly what the customer interface will be. We relied on TEMIC to provide a design that will satisfy our customers.

We encountered difficulties in that we tended to sit back too much and let things happen. We should have had more involvement to specify what was needed more fully in the beginning. Yes, we did the iron and toaster with TEMIC earlier, but those involved internal parts only. Those controls were not visible to the customer. On the coffeemaker, the panel and the buttons, the parts the customer touches, are TEMIC supplied. They showed us a remote control they did for Carrier; that background said they're capable of doing the whole design. [They did the whole remote control; it didn't have to fit into anything else.] With both sides learning what the strengths and weaknesses of each firm were, in November [1993] we had the first engineering build of what was to fit into our coffeemaker.

Technical Problems

There were problems of function, fit, and cosmetics. All three! Some problems showed up in the customer interface, involving the feel of pushing the buttons. There were some cosmetic problems in getting exact color matches on the external parts.

Just a week or two ago, reliability problems started showing up in the electronics. Water was getting where it's not supposed to be. The display window was fogging up, just where we didn't want water to be. As you know, the controls are below the water area.

Duc Tran found that the solenoid we furnished engaging the heater would sometimes cause a reset of the clock timer back to 12:00. Since the clock controls the timed start of the brewing cycle, modifications were needed. He came up with a solution; TEMIC didn't agree with his solution; but together we worked out something else that addresses that problem.

We felt we needed better tactile feedback from pushing the buttons. The bezel is made in Spain; both the tooling and the plastic molding are done in Spain. The rubber parts come from Taiwan. All these parts are assembled into TEMIC's product at their plant in Mexico. They have a very long supply line; many of their components are sourced worldwide. That caused problems getting everything to the final assembly point in Mexico on a timely basis.

Much of the integration of the parts was done in Shelton [Connecticut] at Black & Decker. One interface that had to be done by B&D was the panel's fit into our product. That interface had to be defined very closely. At first we said we don't care very much about this; we said it should be TEMIC's concern. But we had to develop that fit. The panel is a snap fit into our housing from the back, which is then covered by the back panel.

We have Underwriters Laboratory testing of all our products, which includes testing for structural integrity. They have a metal ball on a string, which they let crash into various parts of the coffeemaker. When directed at the panel, the holding latches weren't strong enough to keep the panel in place. They had to be strengthened.

When it came to the heater, the one that boils the water, the more heat, the quicker the boiling, the quicker the pot of coffee is delivered. The industry standard has been one minute per [5-ounce] cup. We wanted to do better than that. We specified a heater with higher watt density to heat faster. In our first design, we found that the heater would not survive. After just a few brewing cycles, it burned out, just like a light bulb. We had to downgrade the watt density [to the industry standard]. There was not enough room to put in a larger heating element. We had to lower the watt density, so the heater would survive. Even though the goal was to shorten the brewing cycle, we had to relax that objective and build in reliability.

Cost and supplier Network

The original TEMIC pricing was done in Towson. We know and agree that, as the design evolves, cost changes, and so should price. We kept asking TEMIC, what is your cost estimate now? They kept replying, no change. Very late in the game, our marketing department set the actual market price for the product. Then TEMIC came back and says, now we really know what the cost is. We responded, it's a little late. We were asking for this kind of update earlier. So TEMIC said they'd stay with their earlier estimates.

This puts engineering under pressure. Almost always, actual costs tend to be higher than first estimates. The pressure on TEMIC is showing up. They're finding quite a few alternate suppliers for components. In the early stages, the samples had a certain supplier network already in place which provided components, and as time has gone on they would like to use alternate components instead. This has placed pressure on B&D, because component qualification cycles are long. If suppliers for key components are changed, we have to requalify them, because B&D is concerned with the entire product. Minor components (like screws) are not important; but if TEMIC wants to use a different relay from a different maker, from what we know about relays, each specific relay has to be evaluated. We would have to insist on the original relay until qualification is completed, and then make a running change. If we qualify a second relay, it will be substitutable for the original one.

Schedule Issues

Concerning our delivery schedule, through last May (1993) the expectation was that introduction would be earlier than June 1994, that we would be in production in February 1994. Last summer, we realized that was overoptimistic. The introduction points for these products are twice yearly. January for Spring/Summer, June for the Christmas sales period.

The first production lot is scheduled for the end of March 1994. The TEMIC [revised] schedule said they would provide control modules to support that date. There is one model that is not associated with TEMIC, the one with the manual controls only. We can start on that, and fit the other models in as the electronics arrive. The only difference is the control module; mechanically the rest of three models are identical. TEMIC may be able to meet the start date; the latest indication is "no." Although there was some slack built in, it wasn't enough.

One thing that got in the way was time differences. Ingolstadt is six hours ahead of us, and seven or eight ahead of Mexico. We're not working at the same time, so we tend to lose a day or two on every communication. Now TEMIC is starting an engineering center, somewhere in Texas. We'll be able to communicate better, just because of the time. Our deliveries to our customers must start taking place in the June time period. So, if we start production by the first of April, we can build up some stock before we start deliveries of all three models by June 1994.

It is notable in this case that, in spite of much effort, the original schedule could not be met. One should reflect on why this happened.

XI. COMPETING GLOBALLY

29. INTERNATIONAL MANUFACTURING

ARNOUD DE MEYER

INSEAD, Fontainebleau, France

ABSTRACT

The globalisation of the economic activity has dramatically increased the extent to which manufacturing is carried out and managed on an international scale. International manufacturing poses some specific challenges. These challenges are listed below:

1. What determines the architecture of the plant network: how many plants does one need? What kind of charter should the different plants have? What should be their capacity?
2. Where should one build the factories?
3. How does one determine the contributions that a specific plant must make to the overall network and the business organisation to which it belongs? In particular what should be its contribution in terms of product portfolio and how should it contribute to the knowledge creation of the organisation?
4. How does one manage the flows in the network of plants: networks are characterised by flows of people, knowledge, goods and capital. How should these be managed?
5. How does one manage the dynamics of the network of plants?

The practices of the following companies are included in this article:
Hewlett-Packard; Renault; Samsung; Toyo Ink.

THE ARCHITECTURE OF THE GLOBAL NETWORK

Multiple plants are integrated in an international network of subsidiaries of the company. One legitimate question is whether the different plants should be clones of the original 'mother' plant, or whether international plants have a different role to play. The case for the idea that all plants should be copies of the original plant is not without merit. Indeed the original plant is probably successful, since the company has been able to grow and internationalize. Repeating that success in another part of the world may make sense. Yet, the idea of standardisation of all plants according to some original blueprint is intuitively not attractive, because it negates the need for local responsiveness.

A useful model of the architecture of networks, and the roles that plants can play in the overall network, is the one proposed by Ferdows (1997). He distinguished plants along two dimensions: the primary strategic reason for the site, and the depth of technical activities on the site.

The primary strategic reasons being the access to market, the cost of production input factors, and the access to technological ideas. Though many authors have proposed different strategic reasons for manufacturing abroad, the categories offered by Ferdows are general enough to include most other proposals. The one strategic reason that deserves to be added to his list is the need to learn about international management. But, this can be partially included in the access to markets or technological resources. It could be a fairly short term reason. Ferdows himself mentions two other reasons that may occur, but are not sufficiently important to be included in his model: pre-emption of the competition, and the control and amortisation of technological assets through economies of scale.

The second dimension, that is, the extent of technical activities at the particular site, refers to the value added activities in the plant. Ferdows notes that the depth of technical activities increase progressively from simple production and assembly, to planning, quality management, process adaptation and development, to product adaptation and development. Based on these two dimensions his map of the network of plants describes six types of plants (Figure 1):

- *Offshore plants* with a limited extent of activities and located in an area of low cost production.
- *Source plants* located in an area where the cost of production factor is still low, but have a charter to source a larger region of the world by serving as a focal point for a specific component, product or process.
- *Server plants* serve a specific market with a relatively low level of value addition.
- *Contributor plants* serve a particular market which has a strong local industrial network.
- *Outpost plants* tap into local technological resources, and their main function is to gather and process information.
- *Lead plants* tap into local technological resources and are also major producers and diffusers of ideas and information.

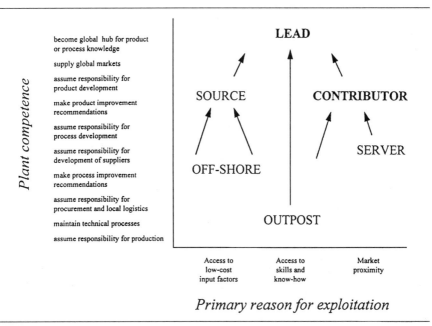

Figure 1. Strategic role of the plant: Ferdows' model.

This architecture as outlined by Ferdows enables us to propose a few assertions about good and less than good performing architectures. Ferdows claims that superior manufacturers have a larger proportion of their global factories in the upper part of Figure 1: source, contributor and lead plants. These plants provide their companies with a formidable strategic advantage. As a corollary, it can be argued that better performing networks are characterised as follows (De Meyer and Vereecke, 1996):

- Better performing networks have a higher degree of variety in the portfolio of plants; in other words they have less clones, but adjust each plant's charter to its specific strategic objective.
- Plants on the different locations in Figure 1 require different management systems and performance metrics. For example, an offshore plant has to be measured on its cost efficiency, a source plant on its market responsiveness, and a lead plant on producing and diffusing ideas throughout the network. (Contrast this with the prevailing practice in many large multinationals to use standardised management accounting systems and performance measures.) Since objectives and measurement systems differ, management of the various types of plants must be different too.
- Plants on the upper end of the map entail more assets and thus more power and status in the organisation. Therefore, plant managers have a tendency to move their plants up in Ferdows' map. Plants on the upper end are more difficult to close down: the

impact on the local business community of the closure of a contributor plant can be quite high, and may be severely hindered by social or political influences. The case of the closure in 1997 of the Renault plant in Belgium, which was a clear example of a contributor plant, illustrates this.

For over 75 years, Renault had an assembly plant for private cars in Belgium. This plant was originally created to serve the local and northern European market. Over the years it had developed into one of the most productive plants in the Renault network, and was heavily integrated into the local network of suppliers. The plant was seen as an essential element in the industrial fabric of Vilvoorde, the city in which it was located. However the rising labour costs in Belgium were not sufficiently compensated by the high level of productivity, and when Renault was forced to rationalise its worldwide production capacity, this plant was one of the first to be closed. The closure caused a lot of political and social tensions, not only because 3000 people lost their jobs, but because a whole region lost its key industrial player. One of the consequences of this friction was that over a period the Belgian population bouycotted all Renault products.

Having most plants in the upper end of the map may be a characteristic of superior companies according to Ferdows, but they also limit the flexibility to manage the dynamics of the plant architecture. The activities of local managers to move the plant up in the map will require entrepreneurial behaviour, which may well benefit the company. The case of Hewlett-Packard in Singapore (Thill and Leonard-Barton, 1994) illustrates this entrepreneurial behaviour of local management to move from electronic assembly to process development, and product adaptation to product development for the Asia Pacific market. This provided the company with a breakthrough in the Japanese market.

PLANT LOCATION DECISIONS

This is probably the most extensively researched area in international manufacturing. Both economist and students of manufacturing have contributed to this literature. Traditionally, the location problem has been mathematically modelled, originally to optimise locations with minimal transport distances between supply, production, distribution and consumption. In more recent work, there is more interest in the optimization of complex production and distribution cost functions. Some of these models are discussed in Brandeau and Chiu (1989).

An equally large stream of publications emphasises the more qualitative aspects of site location decisions. Subjective location decisions based on the preferences of top management, political risk analysis of a particular country, the link with other functions of the company, the example set by competitors, the competitive move against a competitor, have been used extensively. Anecdotal evidence says that site selection based on the availability of a golf course, or more relevant perhaps, the living conditions—housing, schools, safety, etc. for expatriates. They emphasise and document the point that site location is not merely a rational optimisation problem.

There are two important issues concerning this location problem. First, the choice is usually a multi-stage problem: after having determined the strategic reason for creating a new site, one chooses a region or country, and then a site within that region or country. The first two steps should be determined with the company's long term strategy as the most important guiding factor. The third step is often influenced by short term perspectives, e.g. tax optimisation or personal preferences. This is most probably not so bad as long as the first two steps are decided with the long term interests of the company as the major criterion.

Secondly, most of the location models do not capture well the dynamics of a site during selection. The advantage of a site (in terms of market attractiveness or cost advantages) evolves over time. Some advantages may disappear, new advantages may arise. A successful plant will be one that can adapt itself to these changing conditions. Site selection processes should consider the evolution of both the strategy of the company and the characteristics of the site.

THE CONTRIBUTION OF A PLANT TO THE NETWORK

Within the network each plant plays a significant and different role. In addition to the Ferdows model, two other models are useful to interpret the contribution of a plant to a network of plants.

Hayes and Schmenner (1978) describe two extremes of plant type: the product focused and the process focused plant. In the product focused plant, each plant produces an end product for a particular market. In the process focused plant, the plant focuses on a particular process, and produces components or intermediary goods for a range of products and markets. Later on, Schmenner (1982) recogmozed other types of focii: general purpose plants, and market oriented plants.

Vereecke (1997) has empirically tested the idea of focus and finds that focus improves performance relative to their objectives. In particular, product and market focused plants do better than unfocused complex organisations. In the same study she proposes four types of plants, which differ in terms of their role, focus, and coordination mechanisms tying them to the headquarters and to other plants in the network:

- *Isolated plants* which communicate little or not at all with the rest of the network and do not exchange innovative knowledge throughout the network. In many cases these isolated plants are also physically isolated and they do not depend on other plants for components or intermediary goods.
- *Blueprint receivers* are plants which get the innovative ideas from other plants, and which implement them. There is little other communication, and usually the managerial capabilities of these plants are relatively less developed.
- *Integrated plants* are plants which have a lot of interaction (both in terms of information exchange and supply of goods) with the other plants in the network. The integration is often a consequence of the long history of these plants.
- *True network players* are plants with a high level of capabilities and play a proactive role in the management of the network. They share innovative knowledge and are highly integrated in the flow of goods and information.

Isolated plants and blueprint receivers play a limited strategic role and are vulnerable in the medium term. They do not contribute to the strategic development of the organisation but are useful in the current network as a provider of capacity. They may evolve into the two other types of plants with the appropriate support. The true network players are essential to the competitive evolution of the organisation. The integrated plants are often plants with a long history and which are, almost by default, in the position of a network player. But they may underperform compared to what the company expects from them. If true, they represent a lost opportunity for the organisation. Vereecke found that many 'mother' plants are actually of this nature, which suggests that the original plant from which the company's activity started may remain in a position of 'satisfactory underperformance'.

MANAGEMENT OF THE NETWORK

Managing international networks has been the topic of many studies in international management (see for example Doz, 1986 and Bartlett and Ghoshal, 1989). The biggest challenge in international manufacturing is the coordination and sharing of learning.

In an integrated international network of manufacturing there are many forms of flows: there is the flow of goods, human resources, administrative information and actionable process knowledge, and capital. The coordination of these flows requires special attention.

Martinez and Jarillo (1989) make a distinction between formal and informal coordination mechanisms. Coordination is not an exclusive activity of international corporations. In any multi-site corporation, and even in larger one site corporations, the coordination issue is of utmost importance. But coordination becomes more complicated in the case of a multi-country network.

The five formal coordination mechanisms described by Martinez and Jarillo are: grouping of organizational units; the centralisation of decision making; formalization and standardisation of procedures; planning; and output and behaviour control. In addition, the three informal coordination mechanisms they describe are: lateral or cross departmental informal relations through direct contacts (throught temporary teams and taskforces, committees and integrative departments); informal communication; and socialisation (i.e., building an organisational culture of shared values and objectives by education, transfer of personnel, and appropriate performance appraisal and reward systems).

Technological and process knowledge is the most difficult flow to manage in a network of plants. The international transfer of explicit knowledge is a challenge, but can be planned and implemented through information systems, procedures, and rules about updating of standard operating procedures. But the transfer of tacit knowledge requires the interaction between human beings through face-to-face contacts (Nonaka and Takeuchi, 1994). That means a lot of travelling!

Sophisticated communication technology could overcome the need for geographical closeness (De Meyer, 1991). It is true that sophisticated videoconferencing systems and computer conferencing can, to some extent, overcome the need for face-to-face interactions. But two lessons emerge from research:

- In order to transfer tacit knowledge effectively through electronic media one needs to trust the person at the other side. Creating that trust still needs personal interaction. Once that trust is created, it enables people to collaborate over large geographical distances. But that trust will gradually disappear once both parties are separated. Misunderstandings could arise, time zone differences could lead to lack of availability precisely when a quick response is needed from the other side, etc. This may gradually lead to a breakdown of trust.
- Electronic media does not remove the decay of trust but decreases the rate of decay.

The transfer of tacit knowledge could be enhanced considerably by using 'ambassadors', i.e. representatives from one site who go and work for a considerable time (one to two years) at the other site, and who know that one of their major tasks is to build up a network of contacts. Once they return back home, the network, so created, could be used to keep sharing tacit knowledge.

MANAGING THE DYNAMICS OF THE NETWORK

The roles of factories will evolve over time. First, the architecture of the network in Figure 1 evolves over time because the strategy of the organisation evolves. Secondly, Vereecke pointed out that the isolated or blueprint factory is vulnerable over time, and thus must evolve towards more integration, or will risk closure.

How does one manage this transition? It is clear that the strategy of the organisation should be the driving element: network architecture and plants' roles have to follow the intended strategic evolution of the firm. But how does one manage that evolution over time? An interesting parallel can be found in a company which has to integrate the factories of an acquired organisation. Two case studies illustrate this point.

First case is the acquisition of a plant in Berlin by Samsung (De Meyer and Pycke, 1996), the second one is the acquisition by Toyo Ink of a plant in France (De Meyer and Probert, 1998). These two acquisitions were quite different from each other. In one case there was a massive upfront investment to turn around the plant in a few months, in the other case, considerable time was allowed for observing the plant and its capabilities before a few limited and gradual changes were implemented. The choice of the speed of integration was determined by how the acquiring company evaluated the competencies of the acquired company. In the first case, there was an attitude that, apart from the buildings and the technical competence of the workers, there was little of value to the acquirer. In the second case, the acquiring company considered the investment in France as an opportunity to learn about international manufacturing and about Europe's markets. But, as different as they were, they had a few common characteristics:

- In both cases there were attempts to instill in the acquired company the values of the acquirer. But, in both cases, there was considerable opposition and success came out of the newly created culture.
- In both cases there was a similar sequence in the programs that were implemented. The first set of programs were focused on 'cleaning up' (from literally improving

the factory floor to cleaning up obsolete or unadapted procedures and traditions). The second wave of programs phased in company specific quality management procedures and approach. And the third wave of programs were aimed at turning the newly acquired factory into a 'true network player'.
- Success in both cases can be attributed to congruency of goals between the acquiring company, and the management and employees of the acquired company.

These two case studies illustrate the management of the dynamics in an international network of plants. Moving a plant from one role to another requires four conditions to be met:

- Management and employees of the plant must understand and accept the need for change. This is very much in line with the procedural justice or perceived fairness of all successful change programmes as described in Kim and Mauborgne (1997). The key issue here is communication.
- Moving to a new position requires new values, metrics, management systems, etc. It may be better to first 'clean up' the existing procedures and then implement new ones, as opposed to adding a new layer onto the existing procedures. Clean up, create new procedures, and then let the factories become network players.
- Speed can be important, but is not essential.
- Goal congruency between the company and the plant management is important.

Key Concepts: Globalisation; Integrated Plants; Isolated Plants; Mother Plant; Offshore Plants; Process Focused Plants; Product Focused Plants; Production Input Factors; Server Plants; Source Plants.

Related Articles: Manufacturing Strategy; Product Design for Global Markets.

REFERENCES

Bartlett, C. and S. Ghoshal (1989). *Managing Across Borders: The Transnational Solution.* Hutchinson Business Books.
Brandeau, M.L. and S.S. Chui (1989). "An overview of representative problems on location research." *Management Science,* 35, 645–74.
Collins, R. and R.W. Schmenner (1995). "Taking manufacturing advantage of Europe's single market." *European Management Journal,* 13, 257–68.
De Meyer, A. (1991). "Tech Talk: How managers are stimulating global R&D communication." *Sloan Management Review,* 49–58.
De Meyer, A. and B. Pycke (1996). *Samsung (Berlin).* INSEAD EAC case study 03/97-4672.
De Meyer, A. and A. Vereecke (1996). "International Operations." *International Encyclopedia of Business and Management.* (Ed M Werner) Routledge.
De Meyer, A. and J. Probert (1998). *Francolor Pigments: A Toyo Ink Acquisition.* INSEAD EAC case study 05/98-4756.
Doz, Y. (1986). *Strategic Management in Multinational Companies.* Pergamon Press, Oxford.
Ferdows, K. (1997). "Making the most of foreign factories." *Harvard Business Review,* 73–88.
Hayes, R.H. and R.W. Schmenner (1978). "How should you organise manufacturing?" *Harvard Business Review,* 105–19.
Kim, W.C. and R. Mauborgne (1997). "Fair Process: Managing in the Knowledge Economy." *Harvard Business Review,* Jul/Aug.

Martinez, J.I. and J.C. Jarillo (1989). "The evolution of research on coordination mechanisms in multinational corporations." *Journal of International Business Studies*, 20, 489–514.

Nonaka, I. and H. Takeuchi (1995). *The Knowledge Creating Company*. Oxford University Press, New York.

Schmenner, R.W. (1982). *Making Business Location Decisions*. Prentice Hall.

Thill, G. and D. Leonard-Barton (1994). *Hewlett-Packard: Singapore (A), (B) and (C)*. Harvard Business School Case Study.

Vereecke, A. (1997). *The Determinants of the Creation of an Optimal International Network of Production Plants: An Empirical Study*. (Unpublished doctoral dissertation) The Vlerick School of Management, University of Ghent, Belgium.

XII. OPPORTUNITIES IN TACKLING ENVIRONMENTAL PROBLEMS

30. ENVIRONMENTAL ISSUES AND COMPETITIVE MANUFACTURING

ROBERT D. KLASSEN

University of Western Ontario, London, Ontario, Canada

ABSTRACT

The management of the natural environment is becoming increasingly important within manufacturing as customers, suppliers and the public demand that manufacturers minimize any negative environmental effects of their products and operations. Managers play a critical role in determining the environmental impact of manufacturing operations through choices of raw materials used, energy consumed, pollutants emitted and wastes generated. Over the past three decades, conceptual thinking on environmental issues have slowly expanded from a narrow focus on pollution control to include a large set of management decisions, programs and technologies. Pressures to apply the concept of sustainable development to manufacturing underscore the need to think strategically about environmental issues. In this context, sustainable development translates into the integration of environmental management into manufacturing design and technology decisions. However, at this point, the implications of environmental management for manufacturing performance appear mixed. Limited empirical research and anecdotal evidence point to benefits such as reduced waste, and new markets.

The management of environmental issues draws on elements of strategy, capabilities and performance. At the strategic level, product design, process technologies and managerial systems are major determinants in the

environmental performance of manufacturing firms. For example, if complex multi-layer plastics are used for packaging, closed-loop recycling of
materials becomes very difficult, if not impossible. Choices of particular
process technologies, such as the kraft bleaching process in the pulp and
paper industry, demand long-term commitments to particular raw materials
and waste streams that often are regulated, either now or in the future. The
kraft process requires chlorine or chorine dioxide for bleaching, which leads
to the emission of chlorinated toxins.

The practices of the following companies are discussed here: **3M; ABB
Group; Baxter International; BMW; British Airways; Daimler-Benz;
Eli-Lilly and Co.; Hermann Miller; Kodak; Northern Telecom;
Sony;**

THE IMPORTANCE OF ENVIRONMENTAL ISSUES

Management of the natural environment has become increasingly important to manufacturing firms as regulatory requirements tighten, customer demands change, employee concerns multiply, and public scrutiny increases. In its broadest sense, management of environmental issues encompasses a diverse set of management decisions,
programs, tools, and technologies that respond to, or integrate environmental issues
into all aspects of business competitiveness. While much research is underway across
a variety of other disciplines, including public policy, "green" marketing, corporate
social performance, economics and environmental engineering, this article is limited
to the area of production and manufacturing management. Because of the rapid pace
of change in this area, managerial practice often has been forced to lead or at least
match research.

At first glance, environmental issues may appear to concern only a small set of heavily polluting industries. The popular press is filled with examples of companies that
demonstrate poor environmental management—oil spills, explosions, leaking storage
tanks or contaminated ground water—and to a lesser extent, examples of firms that exhibit exemplary performance. In virtually all cases, operations managers played a critical
role in developing management systems and implementing decisions related to these
outcomes. However, at this time, manufacturers outside historically polluting industries
also must concern themselves with environmental issues. Product design and process
technology typically determine the types of pollutants emitted, the solid and hazardous
wastes generated, resources harvested and energy consumed. In addition, upstream and
downstream supplier partnerships, transportation and logistics, and customer relationships further magnify or attenuate environmental risks related to production.

At a more general level, public and private concerns over the environment recently
have been embodied in the concept of sustainable development, with its proposition
that economic growth can occur while simultaneously protecting the environment
(World Commission on Environment and Development). Other organizations, including World Business Council for Sustainable Development (WBCSD), United Nations
Environmental Programme (UNEP) and Coalition for Environmentally Responsible

Economies (CERES), also advocate principles that promote sustainable development. In response, there has been significant international movement toward standardized certification of environmental management systems, an example being ISO 14000 (comparable to ISO 9000 for quality certification). Increasingly, international agreements also have implications for operations systems and their target markets. One recent example is the Montreal Protocol, which eliminates the use of chloro-fluorocarbons (CFCs) by the year 2000.

HISTORY

Historically firms have viewed environmental management as a narrowly defined corporate legal function, responsible for minimizing the impact of environmental legislation on the firm and for ensuring regulatory compliance. Not surprisingly, environmental regulation, which for the most part only began to seriously develop in the late 1960's and 1970's, was the primary driver for environmental improvement. As a result, investment in pollution abatement expenditures in many developed nations showed a corresponding increase as manufacturing firms attempted to control their pollutant emission into air and water and onto land (Lanjouw and Mody, 1996). This limited view has evolved and expanded recently because of intense pressures from many different stakeholder directions: governments, typically through regulation, customers, public interest groups, plant neighbors and employees.

Paralleling this evolution, early research and conceptual thinking on environmental issues focused only on the concept of pollution control (Bragdon and Marlin, 1972). More recent research has broadened this basic view to integrate the consideration of environmental impacts into and across functional areas with implications for strategy and functional area decisions.

Research and practice can be structured along at least two dimensions: degree of specificity; and intended level of impact for operations (Figure 1). The first dimension, degree of specificity, captures the relative level of focus, which ranges from narrowly defined tools to broad, directional philosophies. In contrast, the second dimension considers the broadest level at which implications are felt or changes occur in operations. While other dimensions could be proposed, these two provide a conceptual map for identifying the relative position of various management initiatives in operations on environmental issues. For illustrative purposes, several management initiatives are plotted in Figure 1, with the caveat that specific initiatives by any individual firm may vary somewhat from the figure.

WHY ENVIRONMENTAL ISSUES ARE IMPORTANT?

Many initiatives to better manage environmental issues within operations are directed at improving productivity, raising resource efficiency and fostering innovation (Porter and Van der Linde, 1995). Moreover, adept management of environmental issues can establish barriers to entry for competitors as new regulation is introduced. For example, regulatory agencies continue to tighten acceptable emission and exposure standards, citing technology that is used by industry leaders, thereby forcing other firms to invest in

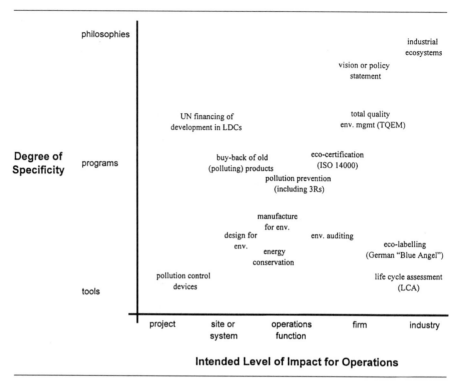

Figure 1. Management initiatives on environmental issues.

new technology. Manufacturing firms often must evaluate difficult trade-offs between acquiring or developing new technology, buying pollution credits, or developing new in-house management and control systems. Unfortunately, the risks associated with each of these alternatives often are difficult to predict and quantify.

Research is continuing on the difficult theoretical task of integrating manufacturing strategy concepts, including structural and infrastructural investments, with the current policies, programs and tools for managing environmental issues noted in Figure 1. As a first step, the complex interactions between multiple stakeholders, including customers, government agencies and environmental interest groups, must be explicitly recognized (Klassen, 1993). The most obvious linkages are to elements of operations strategies that leverage product design (Zhang et al., 1997), process technologies (Gupta, 1995; Sarkis, 1995), and product recovery (Brennan et al., 1994; Thierry et al., 1995). Newman and Hanna (1996) report empirical evidence that expanding manufacturing's strategic role in the firm tends, although not necessarily, to favor more proactive management of environmental issues.

Empirical studies characterizing approaches to the management of environmental issues align along two general perspectives: strategic choice; and developmental progression. The first perspective proposes that manufacturing firms make distinct choices from among a set of alternative positions, typically on a spectrum from reactive to proactive (Newman and Hanna, 1996; Smith and Melnyk, 1996). In contrast, the developmental progression captures the notion that technical understanding, skills and cross-functional integration build over time, thereby allowing firms to move from a naïve, crisis-oriented beginner to an enlightened, proactive innovator (Petulla, 1987; Hunt and Auster, 1990). While both perspectives recognize that capabilities change over time, the strategic perspective suggests that operations *might* use increasingly sophisticated environmental capabilities to defend their current levels of environmental performance. In contrast, the second perspective implies a subtle linkage between values/motivation and capabilities, where increased capabilities are associated with additional efforts to improve environmental leadership and performance.

THE ROLE OF TECHNOLOGY

Environmental technologies can be broadly defined to include any design, equipment, method or systems that conserve resources, minimize environmental burden, or otherwise protect the natural environment. Technologies are often further broadly classified as (1) *pollution abatement*, alternatively labeled end-of-pipe or pollution control, and (2) *pollution prevention*. Other related terms are clean technologies, design for environment, 3R (reduce, reuse, and recycle) technologies or green manufacturing, to list just a few terms (OECD, 1995). In general, much of the environmental management literature, usually in the context of proactive environmental management, espouses the use of pollution prevention as the most effective means by which to improve environmental performance. In contrast, pollution abatement merely tries to contain or control wastes and pollutants.

Much has been written about barriers to greater development and implementation of pollution prevention (for more complete reviews, see Ashford, 1993; and OECD, 1995). These barriers are broadly summarized as:

- *public policies* which favor command-and-control regulations, are inflexible, and enact incremental reductions;
- *financial constraints* in firms which focus on short term profitability and limit R&D expenditures to develop new technologies;
- *workforce related concerns* which resist new technological change that may displace jobs or require significant retraining;
- *technological capabilities* and risk which may hurt current manufacturing processes optimized for quality, cost and efficiency; and
- *managerial attitudes and structures* which cause inertia, poor communication, education and redirection of resources elsewhere.

Of these, the two barriers that receive the greatest amount of attention are public policy and technology. Of necessity, environmental regulation is a critical concern for most operations managers. Collectively, and often incrementally, regulations develop at the local or national level to form very different legislative systems, with potentially very different impacts on technological choices by manufacturers.

For example, U.S. regulations tend to limit the amount of pollutants released into the environment based on the availability of existing pollution control technologies, commonly termed a command-and-control regulatory system. Efforts to take risks and to innovate with environmental technologies generally are not rewarded. In contrast, while Germany certainly has not ignored the emission of pollutants, much greater emphasis has been placed on regulating the total product system, through environmental consumer labeling (i.e. "Blue Angel" eco-label) and mandated requirements to take back and recycle high percentages of packaging.

In response, firms in the U.S. have focused on end-of-pipe technologies installed at the end of the manufacturing process (Lanjouw and Mody, 1996), whereas German firms have been pushed to more actively pursue process redesign, material recycling, and product disassembly (Cairncross, 1992). Broader recognition of this barrier and additional research into the managerial implications of public policies is gradually encouraging the adoption of more flexible regulation, such as tradable emission permits and multi-media (air, water and land) regulations (Ashford, 1993).

EFFECT ON PERFORMANCE

Two main theoretical arguments link the management of environmental issues to manufacturing and firm performance. First, if traditional economic analysis is made, where social (public) costs of pollution are simply transferred back to the (private) firm, manufacturers that improve their environmental performance are at an economic and competitive disadvantage. Alternatively, if the costs of improved environmental performance are overshadowed by other benefits, including increased productivity, reduced liabilities, access to new markets and better public reputation, firms could be in a better competitive position. However, critics contend that new environmental regulations and specific actions to improve environmental performance often impose significant financial burdens and risk, and only a small fraction are likely to offer any competitive benefit. Furthermore, these "easy wins"—low risk, easily implemented solutions with positive financial returns—may have already been implemented in most companies (Walley and Whitehead, 1994).

Much research in fields such as finance, corporate social performance, economics, accounting and environmental management has been devoted to testing the relationship between environmental and economic performance (see Klassen and McLaughlin, 1996). Environmental performance, such as emission levels or reputation, is typically regressed against some form of past, present or future economic performance. Unfortunately, the results generally have been ambiguous, possibly because any direct effects were masked at such an aggregate level. Yet, based on firm- and industry-level studies, a growing number of researchers strongly advocate that an integrative approach coupled with pollution prevention can be competitively advantageous (Bragdon and Marlin, 1972; Cairncross, 1972; Schmidheiny, 1992; Porter and Van der Linde, 1995).

More specifically, case examples cite both positive and negative outcomes from investments in environmental technologies in manufacturing firms; several examples of positive outcomes related to pollution prevention are noted in the next section. More generally, manufacturing and environmental performance have been found to be related to the proportion of investment targeted at pollution prevention versus relative pollution control (Klassen and Whybark, 1999). Earlier studies based on the analysis of confidential U.S. Bureau of Census plant-level data indicate that an increase in pollution control costs translated into worse total factor productivity costs (Gray Shadbegian, 1993).

Managers would be better served by identifying circumstances under which particular kinds of environmental investments deliver returns to shareholders (Reinhardt, 1999). More research is needed to shed greater light on the performance implications of pollution prevention technologies in order to offer operations managers clearer guidance for action.

EXAMPLES

Many examples of strong or weak management of environmental issues are part of the public record. Rather than emphasize poor actors, a few notable positive examples are highlighted here. As a general indication of the perceived importance of environmental performance, an increasing number of manufacturing firms are now publishing annual environmental reports, similar to financial reports. Examples of multinational companies, covering a range of industries, include: ABB Group, British Airways, Daimler-Benz, Eastman Kodak, Eli Lilly & Co., Northern Telecom and Sony. Baxter International has published a particularly notable report on the estimation of costs and benefits of environmental management.

One firm with a notable track record is 3M, with its Pollution Prevention Pays program (now called 3P Plus) which claims to have saved over $790 million through 4,500 projects over a 20-year period. As implied by its name, the program is very pragmatic. Employees attempt to explicitly link environmental improvements with product or technological improvements that provide business advantages. Managers reported three primary lessons: strong commitment is essential from senior management, including having a senior manager responsible for environment; environmental issues need to be integrated into business plans; and employees must be involved in achieving environmental goals which can be best supported through formal recognition (for further discussion of this case, and other examples of benefits arising from improved environmental practice, see Schmidheiny (1992) and Shrivastava (1995)).

Examples of manufacturers taking a systems-oriented perspective on environmental management include BMW and Hermann Miller, with their emphasis on taking responsibility for their products, automobiles and office furniture, respectively, at the end of their useful life (Cairncross, 1992; Smith and Melnyk, 1996). Historically, while much of the metal and many high value components have been recovered from old automobiles, virtually none of the plastic material, accounting for 10% of the weight of an automobile, was recovered. By building on existing recycling capabilities, BMW is focusing on developing new product designs and production technologies that will meet their recycling target of 80% for plastics. A plant has been constructed to pilot vehicle disassembly, an increasing number of parts are being refurbished, and designs have been modified to reduce the use of complex plastics, to label the composition of parts and to use more recycled materials (Gupta, 1995; Thierry et al., 1995).

Thus manufacturers seeking practical approaches to integrate environmental issues into operations can turn to real examples of firms that have, in fact, introduced environmentally responsive management systems, process technology and product design. While some of these initiatives carry significant costs and risks, these firms have generally discovered that insightful and proactive management of environmental issues represent more than addressing regulatory constraints, but also pursuing new market opportunities and cost savings.

Key Concepts: 3 "R" Technologies; Clean Technologies; Design for the Environment (DFE); End-of-Pipe Technologies; ISO 9000; ISO 14000; Manufacturing Strategy; Montreal Protocol; Pollution Abatement or Pollution Control; Pollution Prevention; Total Quality Management.

Related Articles: Core Manufacturing Competencies; Manufacturing Strategy.

REFERENCES

Ashford, N.A. (1993). "Understanding Technology Responses of Industrial Firms to Environmental Problems: Implications for Government Policy." In K. Fischer and J. Schot (Eds.), *Environmental Strategies for Industry*. Washington, DC: Island Press, 277–307.

Bragdon, J. and J. Marlin (1972). "Is Pollution Profitable?" *Risk Management*, (April), 9–18.

Brennan, L., S.M. Gupta, and K. Taleb (1994). "Operations Planning Issues in an Assembly/Disassembly Environment." *International Journal of Production and Operations Management*, 14 (9), 57–67.

Cairncross, F. (1992). *Costing the Earth*. Boston, MA: Harvard and Business School Press.

Gray, W.B. and R.J. Shadbegian (1993). *Environmental Regulation and Manufacturing Productivity at the Plant Level* Report No. CES 93-6. Washington, DC: Center for Economics Studies, U.S. Bureau of the Census.

Gupta, M.C. (1995). "Environmental Management and Its Impact on the Operations Function." *International Journal of Operations and Production Management*, 15 (8), 34–51.

Hunt, C.B. and E.R. Auster (1990). "Proactive Environmental Management: Avoiding the Toxic Trap." *Sloan Management Review*, 31 (2) 7–18.

Klassen, R.D. (1993). "Integration of Environmental Issues into Manufacturing." *Production and Inventory Management Journal*, 34 (1), 82–88.

Klassen, R.D. and C.P. McLaughlin (1996). "The Impact of Environmental Management on Firm Performance." *Management Science*, 42 (8), 1199–1214.

Klassen, R.D. and D.C. Whybark (1999). "The impact of selection of environmental technologies on manufacturing performance." *Academy of Management Journal*, 41 (6), 599–615.

Lanjouw, J.O. and A.Mody (1996). "Innovation and the International Diffusion of Environmentally Responsive Technology." *Research Policy*, 25, 549–571.

Newman, W.R. and M.D. Hanna (1996). "An Empirical Exploration of the Relationship Between Manufacturing Strategy and Environmental Management." *International Journal of Operations and Production Management*, 16 (4), 69–87.

OECD (1995). *Technologies for Cleaner Production and Products*. Paris, France: OECD.

Petulla, J.M. (1987). "Environmental Management in Industry." *Journal of Professional Issues in Engineering*, 113 (2), 167–183.

Porter, M.E. and C. van der Linde (1995). "Green and Competitive: Ending the Stalemate." *Harvard Business Review*, 73 (5), 120–134.

Reinhardt, F.L. (1999). "Bringing the environment down to earth." *Harvard Business Review*, 77 (4), 149–157.

Sarkis, J. (1995). " Manufacturing Strategy and Environmental Consciousness." *Technovation*, 15 (2), 79–97.

Schmidheiny, S. (1992). *Charging Course: A Global Business Perspective on Development and the Environment.* Cambridge, MA: MIT Press.

Shrivastava, S. (1995). "Environmental Technologies and Competitive Advantage." *Strategic Management Journal*, 16 (3), 183–200.

Smith, R.T. and S.A. Melnyk (1996). *Green Manufacturing: Integrating the Concerns of Environmental Responsibility with Manufacturing Design and Execution.* Dearborn, MI: Society for Manufacturing Engineering.

Thierry, M., M. Salomon, J. Van Nunen, and L. Van Wassenhove (1995). "Strategic Issues in Product Recovery Management." *California Management Review*, 37 (2), 114–135.

Walley, N. and B.Whitehead (1994). "Its Not Easy Being Green." *Harvard Business Review*, 72 (3), 46–52.

World Commission on Environment and Development (1987). *Our Common Future.* New York: Oxford University Press.

Zhang, H.C., T.C. Kuo, H. Lu, and S.H. Huang (1997). "Environmentally conscious design and manufacturing: A state-of-the-art survey." *Journal of Manufacturing System*, 16 (5), 352–371.

XIII. THE REVOLUTION IN COSTING AND PERFORMANCE MEASUREMENT

31. ACTIVITY-BASED COSTING

MONTE R. SWAIN
STANLEY E. FAWCETT
Brigham Young University, Provo, Utah, USA

ABSTRACT

Activity-Based Costing (ABC) has the potential to enhance strategic management by reducing cost distortions. Occasionally, product costs are used solely to establish prices. In other cases, however, open competitive markets effectively establish prices. In the latter more common context, the strategic use of product costs is twofold: first, to determine if the organization is able to compete effectively in the market, and second, to identify opportunities for target costing. Uncertainty or error in understanding the cost of providing a product to the customer may negatively affect profitability, organizations may miss market entrance and exit opportunities or miss opportunities for cost improvements. Hence, accurate understanding of product costs within a complex multi-production organization can provide a valuable competitive edge in a market of shrinking profit margins.

Increasing the accuracy of cost data via ABC also allows an organization to implement its market strategy more effectively. Another strategic use of the ABC model is its ability to trace activity costs to a number of different cost objects. This allows a company to determine the value to the organization of various cost objects. For example, it is not difficult to trace revenues generated by a geographic region, customer, or product group. By accurately tracing activities (and attendant costs) associated with cost objects, profits generated by various cost objects can be determined. More importantly, with this information, management develops insight into various cost objects, such as employees, customers, regions, or product groups, leading to better management of activity costs.

WHAT IS ABC?

Activity-based costing was created in response to the problem Johnson & Kaplan noticed. Johnson and Kaplan (1987:1) argued that:

today's management accounting information, driven by the procedures and cycle of the organization's financial reporting system, is too late, too aggregated, and too distorted to be relevant for managers' planning and control decisions.

The traditional cost model

The classic product costing model, in harmony with Generally Accepted Accounting Principles (GAAP) of the United States, maintains that:

cost of products = direct materials + direct labor + manufacturing overhead

Typically, the most difficult challenge facing cost accountants involves attributing the manufacturing overhead costs to the product. The traditional approach of assigning overhead costs to products uses a two-stage process. The first step allocates costs of manufacturing support centers to production centers, then combines all allocated and overhead costs in production centers to compute an overhead burden rate based on some input activity (such as direct labor hours or machine hours). The traditional model (Unit-based costing) may be characterized as:

Costs in Support Centers → Costs in Production Center → Costs assigned to Products

The ABC model

The traditional two-stage allocation of overhead costs seems to indicate that products consume or incur costs. However, according to activity-based costing (ABC) theory, costs are consumed by activities and activities are necessary to manufacture products. In ABC cost attribution process is also accomplished in two stages. Costs incurred are first attributed to activities; then each activity is studied to determine its relationship to products (or other relevant cost objects such as customers, projects, geographic regions, office locations, etc.) Product costs are then determined on the basis of these activities. The flow of costs in the ABC model is characterized as

Costs of Resources → Cost Pools based on Activities → Cost assigned to Objects

Obviously, direct materials, direct labor, and manufacturing overhead do not make up the total set of costs in an organization. Yet, the costs allocated to products in the traditional GAAP-based model are strictly limited to these production-related costs (i.e., product costs). Other costs, not directly connected to any aspect of the production process, include selling and general administration costs. These costs, typically not included in the product cost category, fall under the heading "period costs." However, ABC does not distinguish between product and period costs. In an effort to provide better management of period costs, some selling and general administrative costs are in many cases attributed to products and other cost objects in ABC systems. Because of such handling of costs, output of the ABC system cannot be used for external financial statements. Thus, firms using ABC systems need multiple costs systems. (Kaplan, 1988).

The ABC system has a broader focus then the traditional system. Rather than focusing solely on measuring costs of products, ABC systems may identify a variety of cost objects for measurement, such as products, customers, sales representatives, operations, factory or office locations, geographic regions, etc.

Essentially, ABC proponents hold that the classic product cost view (cost of products = direct materials + direct labor + manufacturing overhead) is the wrong perspective (Turney, 1992). Further, ABC theorists view production costs to be the result of a hierarchy of activities: facility-support activities, product-sustaining activities, batch activities, and unit-level activities (Cooper, 1990). Identifying these activities and their relation to products produced by the organization is essential to the development of *multiple* activity drivers that can be used to create a comprehensive view of cost relationships affecting the product and other cost objects within the organization.

History

The structure of traditional product cost systems was established during the nineteenth century by the textile mills, railroads, and steel mills of the United States and Great Britain (Johnson and Kaplan, 1987). It has changed little since then. However, the structure of actual product costs has changed dramatically over the last 80 years. Improvements in production processes through improved use of technology and less labor have resulted in important cost efficiencies and enhanced product quality. Consequently, direct labor costs continue to become a smaller percentage of total product costs, while overhead costs are becoming increasingly larger (Boer, 1994). In addition, in response to growing diversity in customer demands, as well as advances in production technology, many manufacturing organizations face increasing diversity in both their product mix and their production processes. Within these organizations, legacy (i.e., traditionally-focused and well-established) manufacturing accounting systems often struggle with the task of assigning a diversity of production costs on the basis of direct labor inputs, leading to grossly misstated cost assignments. Cost accountants and executives started becoming particularly sensitive to the difficulties caused by inadequate cost systems as a result of the seminal publication *Relevance Lost* by Thomas Johnson and Robert Kaplan (1987).

Cooper's hierarchical model of activities

As a result of dissatisfaction with legacy systems, ABC applications gained popularity beginning in the late 1980s as product costing and product cost management

Exhibit 1. Cooper's hierarchical model of activities (adapted from Cooper and Kaplan [1991] Figure 5.3).

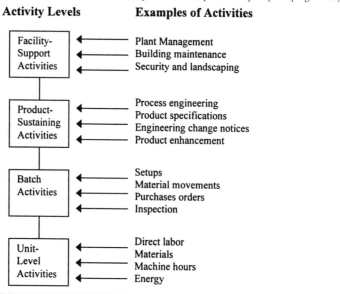

alternatives to traditional costs systems (referred to as unit-based costing, or UBC, systems). Other individuals also made significant contributions to develop and disseminate the ABC model of manufacturing, service, and retail organizations. Professor Robin Cooper published a series of articles in 1988 and 1989 that initially codified the ABC model, identified the need for ABC information, and described issues involved in developing accurate activity drivers for the ABC system (Cooper, 1988a; Cooper, 1988b; Cooper, 1989a; Cooper, 1989b). Subsequently, Cooper further refined the concept of activities with his hierarchical model of activities (Cooper, 1990). Cooper's hierarchy is shown in Exhibit 1.

One value of Cooper's hierarchy of activities is the insight it provides into selecting activity drivers useful for attributing activity costs to products. For example, unit-level activities have costs that vary in proportion to the number of units produced. Costs related to unit-level activities can be directly attributed to product units. On the other hand, costs of batch activities can only be attributed to batches of products, and product-sustaining activity costs are attributed to individual product lines. Perhaps most important, Cooper's hierarchy illustrates that it is difficult to intelligently attribute costs of facility-support activities to individual products, batches, or product lines.

Turney's cross-view of the ABC model

Peter Turney also expanded ABC theory with his cross-view of the ABC model that complements ABC with Activity-Based Management (ABM) as shown in Exhibit 2 (Turney, 1992).

Exhibit 2. Turney's cross view of the ABC/ABM model (adapted from Turney [1992] Figure 4.2).

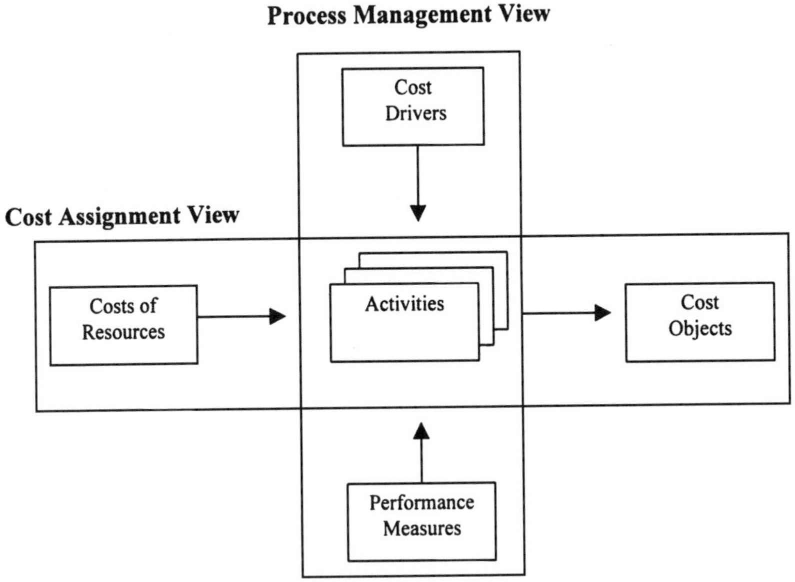

Process Management View

ABM expands the basic ABC theory by separating the concept of product costing using activity-based relationships (the cost assignment view) from the task of directly managing activities to achieve cost efficiencies and to enhance quality in business processes (the process management view). In Turney's cross-view of ABC/ABM, opportunities for cost efficiencies are created by separately tracking multiple cost drivers for each critical activity. ABM provides a framework of cost measures as well as nonfinancial performance measures, that are used to support planning, control, and evaluation of critical activities identified in the original ABC-based analysis. Continuous improvement of the organization is achieved using performance measures for critical activities in the organization. Managing cost drivers and performance measures for a particular activity is a management task. Unless management uses the insights gleaned to change to more efficient and effective production processes, the ABC effort will be underused as an approach to *strategic* cost management within the organization.

Technology support for ABC

The *idea* of defining activities and using multiple drivers to attribute costs was introduced long before Johnson and Kaplan published *Relevance Lost* in 1987. As early as 1921, academics and professionals published articles arguing that complex manufacturing costs were misrepresented by simplistic product costing systems (Harrison, 1921). ABC-type cost models have been promoted at least since the 1960s (see, for example, Drucker, 1963).

Given the criticism of product cost systems, why has it taken so long for ABC, and other improved cost model, to emerge? The answer is twofold. First, in recent years, most markets have experienced increased competition as a result of the expanding global market, increasing consumer sophistication, and government trends to deregulate key industries. Increased competition, with attending tight profit margins, results in a *need* for tracking product and process costs better than ever before. Furthermore, as a result of the relatively recent proliferation of inexpensive computing power and the development of advanced database technology, firms have the *opportunity* to track cost data more cost-effectively than ever before. The most profound difference between ABC and traditional UBC models is the amount of data generated when an organization moves from single to multiple drivers of product costs. As Johnson and Kaplan were preparing their seminal text *Relevance Lost* (1987), the emergence of the Information Age was being heralded by cheaper, faster, and more sophisticated computer technology. As a result, the ability to access, store, and process large amounts of data quickly and inexpensively is significant to the implementation of sophisticated ABC models in manufacturing organizations. Essentially, ABC models finally took a foothold in manufacturing during the 1980s and continue to grow in popularity and application today.

COST DISTORTIONS

The traditional process of using overhead burden rates to allocate the costs of support departments to units produced can create serious cost distortions and cross-subsidization between groups of products. Essentially, product cost cross-subsidization is the result of using UBC that allocates average overhead costs across all types of products within the company. Hence, high-volume products may be forced to carry part of the costs of producing specialty products. In the effort to establish a new product, such subsidization might be a sensible strategy. However, in situations in which subsidy is unknown to the management, poor decisions may result. This undesirable cross-subsidization is the major result of six different information distortions created by the traditional UBC system (Cooper and Kaplan, 1991). The six distortions are explained below.

Allocating Unrelated Costs: The first potential cost distortion is the allocation of costs to unrelated products. For example, consider the situation in which costs related to creating engineering change notices (ECN) are included in the total pool of overhead costs, the overhead burden rate is on the basis of direct labor hours. As each product moves through the manufacturing process and uses direct labor hours, the overhead allocated to the product includes a measure of ECN costs. However, in a diversified product mix, some well-established products do not generate ECN activities. Nonetheless, under UBC, these standard products will carry part of the burden of this engineering activity, and new or specialized products will not carry their "fair share" of overhead costs.

Omitting Relevant Costs: The second source of cost distortion is the disregard of certain costs incurred by a particular product. This is likely the result of accounting systems established to satisfy external auditors. Financial accounting regulators allow only the inclusion of product costs (i.e., direct labor, direct materials, and manufacturing overhead) in the valuation of inventory and costs of goods sold. However, the movement of a product through the various phases of development, manufacturing, selling, and warranting incurs many more costs than those recognized by the traditional product cost model. Once manufactured and placed in inventory for sale, a product continues to require handling and movement, and security costs. Marketing expenses are those realized as the market for a particular product is developed and supported. Selling and delivery costs may vary with the product type. Finally, warranty costs are generated as products fail to perform for customers. These costs are as real as the traditional product cost components. Thus, for a company to maintain a strategic view of its market position, these costs must also be understood and included in the market price.

Omitting Cost Objects: The third cost distortion can be introduced by costing only a subset of the relevant cost objects of the organization. For example, when the output of an organization includes both tangible (i.e., manufactured) and intangible (i.e., service and support) products, traditional cost systems may assign costs only to tangible outputs. Obviously, this causes reported costs of tangible objects to be too high and does not support the effective management of costs of intangible outputs.

Erroneous Burden Rates: The fourth type of distortion is related to the first stage of the traditional UBC allocation model. In the first stage of developing the burden rate, overhead costs associated with to support departments are allocated to production centers. If, for example, costs of a machine maintenance group are allocated to production centers based on the number of square feet in each production area, the overhead costs borne by each production center may not truly reflect the maintenance costs actually incurred by each center. This data is then subsequently used to develop burden rates within the production center. Since distorted burden rates are used, faulty product costs will result.

Erroneous Allocation Bases: The fifth source of cost distortion is perpetuated within the second stage of the UBC model. The second stage uses the burden rate developed before a given production period in order to allocate overhead costs during the given production period to units produced. If the allocation basis is improperly selected, then overhead costs will be improperly assigned to individual products. Suppose, for example, that the costs to set up (e.g., clean and calibrate) a machine for a production run is the same regardless of the number of units scheduled for the production run. It would not be accurate, then, to allocate this overhead cost with all the other overhead costs in the production center based on any of the unit-based cost drivers (e.g., number of units, direct labor hours, machine hours).

Allocating Common Costs: The sixth cost distortion results from trying to allocate costs that simply cannot be sensibly assigned to specific products. These unallocatable costs, the common costs of the production facility, are typically related to activities defined by Cooper's (1990) hierarchical model as facility-support activities. Any attempt to allocate or attribute these costs to specific products cannot be done objectively. Consider, for example, the salary and office costs of the cost accountant assigned to track all production cost information. Clearly this individual is essential to the manufacturing activity. If the company chose to change from a manufacturing to a merchandising organization, this particular job description would be terminated. Therefore, the expense of having this accountant is manufacturing overhead by definition. However, the cost of the accountant may not be traceable to any particular product group. Attempts to allocate these costs will be arbitrary and may mislead the decision-maker trying to analyze individual product costs.

ALTERNATIVE VIEWS OF COST OBJECTS

Rather than strictly focusing on products, consider specific customers as the cost objects. In the past few years, many companies have initiated customer rationalization programs in order to provide world-class service to their best customers. In a separate study we conducted, these firms had adopted a policy by which customers were rated based on their importance. The few "A" customers (most important) received the greatest attention, with considerable resources being dedicated to customize products, orders, and deliveries normally not provided to lower-rated customers. By contrast, "C" customers were often poorly served and at times ignored. In a few cases, the firms indicated that they were no longer interested in the business of the "C" customers, since meeting their needs dissipated resources that were needed to keep the critical "A" customers satisfied. Firms discovered that some key accounts with important customers were actually losing money only after more rigorous and accurate ABC systems were implemented with customers as the cost object. Exhibit 3 illustrates the ABC attribution approach that led to this discovery. Much to the surprise

Exhibit 3 Costing the service requirements of selected customers

Traditional UBC View		ABC View	
Salaries	$120,000	Receive Material	$86,600
Supplies	30,000		
Depreciation	20,000	Move Material	84,600
Overtime	15,000		
Space	30,000	Expedite Material	58,800
Other	5,000		
Total	$230,000		$230,000

The H.S. Sponge account represents 25% of units sold, yet consumes 25% of the receiving activities, 30% of the movement activities, and 40% of the expediting activities. How much does it cost to meet H.S. Sponge's service requirements?

Traditional UBC View			ABC View		
Overhead	% of Units	Customer Cost	Activity Costs	of Units	Customer Cost
			$86,600	× 25%	= $21,650
			$84,600	× 30%	= $25,380
$230,000	× 25%	= $57,500	$58,800	× 40%	= $23,520
					$70,550

of some companies, the new approach to costing revealed that many "C" customers whom the companies had previously disparaged as unimportant and too small were actually their most profitable accounts. When the costs of "extra effort" activities were averaged across all products and customers the companies found they were under-pricing these special customers. Similar to the product cross-subsidization problem, these ABC analyses supported very important (though potentially painful) decisions to increase prices or discontinue some key account customers. This example illustrates well the strategic effect of ABC data on the management decision process.

THE IMPLEMENTATION OF ABC

The first stage of ABC

Pragmatically, how does one specifically trace costs to cost objects in the ABC model? This is a two-stage process. The first step is to identify and cost the activities within the organization. Rather than thinking of the organization as a pool of departments (e.g., production, quality control, personnel, accounts payable, etc.), the organization is viewed as a pool of activities. Many of these activities will cut across departments, the departments often participating in many different activities. Since the ABC imple-mentation team needs to orient itself by activities, rather than by departments, a good starting point is to understand that there are three essential cycles in a business: the Acquisition/Payment Cycle, the Conversion Cycle, and the Sales/Collection Cycle (Denna et al., 1993). All activities in the company relate to the process of acquiring and paying for resources, converting resources into salable goods or services, or selling goods/services and collecting revenues. This view of the organization can be quite valuable in the processes of identifying and mapping critical activities for an ABC implementation.

Attaching costs of resources to activities can be a rather imprecise process involving many estimates. In terms of materials resources (e.g., supplies), it is not as difficult for the organization to use the current accounting system to directly trace the cost of supplies required by a particular activity. On the other hand, labor and overhead costs are more challenging to trace to activities. It is possible to require employees to fill out detailed time records in order to track labor resources flowing to particular activities. However, such expensive and obtrusive record keeping is unlikely for most organizations. A more viable alternative than maintaining precise time sheets is to simply interview department managers for estimates of effort expended on specific activities. However, interviewing is not the only way to gather essential insight about activities. Other methods include simple observation of direct production and essential support activities, computerized or manual timekeeping systems that track work done and time spent on each activity, questionnaires, and storyboards (in order to discover consensus on specific activities among employees directly involved). These methods can be used alone or in conjunction with each other.

The second stage of ABC

Once the organization has traced all resource costs to the appropriate activities and established activity cost pools, the second ABC stage is to determine who or what consumes each activity and the basis for the activity use. The output of this analysis is activity drivers that provide the link between activities and cost objects.

In analyzing for activity drivers, it is important to remember that products are not the only cost objects that consume activity costs. As described above, customers, market segments, and distribution channels all consume activities and can be identified as cost objects. The main idea of Cooper's (1990) hierarchical activity model is that costs and activities should be managed at the level at which they occur. Forcing all costs to behave as production unit costs is an artificial simplification of the complex reality of modern business. Additionally, activities themselves can also be cost objects. For example, many companies may find that their payroll activity requires computer maintenance activity.

In identifying activity drivers, Cooper's (1990) hierarchical model (Exhibit 2) provides valuable insight. For example, the "Quality Design" activity in many manufacturing organizations is required to support individual product types. Within Cooper's hierarchy, it would be classified a product-support activity and its costs may be assigned to overall product lines possibly based on number of engineering change notices or on the number of components within the product. In another example, investment and maintenance costs may be assigned to a "Production Machinery" activity. This activity is consumed by the production of various products or product components, and may be assigned to these cost objects on a per-unit basis. The "Purchase Order" activity, on the other hand, if required each time a new batch is initiated, should be assigned on a batch basis. The process of determining activity drivers is the same process the ABC implementation team follows in determining activity pools. Interviews, observations, database analysis, and sampling procedures may be used in this identification process. In fact, as part of the process of successfully developing activity pools, activity drivers become transparent as the consumers of each activity are identified.

TECHNOLOGY FOR IMPLEMENTATION

Most organizations need to reengineer their whole approach to information system design in order to take advantage of ABC. Current computerized accounting information systems are generally limited to financial measurements. This limitation affects the ABC model and its need to track nonfinancial cost drivers and performance measures. In accounting systems, preclassification in terms of debits, credits, assets, liabilities, and equity in very common. However, the ABC model is based on a set of activities, not on a chart of accounts. As a result, useful integration of financial and nonfinancial data necessary for the effective implementation of ABC does not exists in many organizations.

There are several proposed technology solutions to implementing ABC. Essentially, a disaggregated relational database could be developed that maintains detailed data for use in any model of decision support including ABC. Only a few years ago such enterprise-wide information warehouses were not possible. Today, with decreasing technology costs and modern advances in information theory, organizations can create a robust information environment conducive to ABC. Perhaps the best solution currently is an approach referred to as *events-driven system design* (Swain and Denna, 1998). Essentially, events-driven systems are designed around *all* the business events (or activities) of interest rather than shoehorning some events into an information artifact like the general ledger. For example, consider the "purchase inventory" activity. Why artificially separate pieces of this event into the accounts labelled "Inventory," "Purchases," and "Accounts Payable," and then fail to track other essential information like the time of delivery, who made the delivery, who placed the order, and the quality of the inventory? Events-driven systems integrate all these data into a single database (usually a relational database). This database becomes the organization's information warehouse—an asset that is becoming increasingly affordable as the costs of information technology continue to come down. Software companies like Lawson Software and SAP® provide off-the-shelf technology that incorporates many of these concepts.

WHEN TO USE ABC?

Continuing to use UBC systems that do not accurately portray complex production cost flows can lead to information distortions. However, implementing an ABC system is an expensive and time-consuming process. Organizations need to determine carefully the need for sophisticated cost information before beginning an ABC implementation project. Some production settings may be adequately served by a traditional UBC system.

Market Warning Signs: Competitive markets can be helpful in analyzing the need for ABC. Organizations can watch for specific warning signs in the market which may indicate that an ABC system may be justified (Cooper, 1987).

Confusing Profit Margins: First, management should evaluate honestly whether profit margins across the product mix make sense. One way to compare product groups within the overall product mix is to differentiate between high-volume standard products and low-volume customized products. Low-volume customized products indicate a desire to "delight the customer." These customized products are obviously more difficult to produce. Products are often customized to satisfy small pockets of customers.

However, management may be confused—though not unhappy—at the apparent ability of these customized products to show healthy profit margins with little or no markup on price. Most customized products logically either require significant price premiums or, in the effort to build customer loyalty, display acceptably tight or negative margins. If management cannot explicitly explain why the unexpectedly large profit margins exist, it is possible that reported margins are masking a problematic product mix relationship—low-volume customized products are being unknowingly subsidized by high-volume standard products.

Unexplained Market Barriers: Another indicator of the potential that high-volume products are subsidizing low-volume products is the inexplicable dominance of a market segment. Management may find itself in a market for one of its low-volume custom products with little or no competition. Again, management may be happy, but puzzled. Why do competitors not attempt to penetrate this market in order to share in the seemingly significant profits? A good ABC system may reveal what competition already knows—the company is pursuing phantom profits that do not actually exist!

Similarly, management may be frustrated at their inability to penetrate hold the market for one of their high-volume standard products. The company may have developed a very efficient production system and should be able to compete. However, others may be bringing the product to the market at what seems to be an impossibly low price. Again, this may denote the inadvertent inclusion of the cost of producing low-volume products within the standard product's cost base. With ABC, the cross-subsidization may be removed from the high volume product to reveal previously unknown profit potential.

Bidding Mysteries: Bidding mysteries can also indicate a need for ABC. First, management should carefully evaluate their successes and failures in bidding for contract work. Regular loss of contracts for which bids are prepared with very tight profit margins, particularly for high-volume standard products, may indicate cross-subsidization. Additionally, if the company surprisingly wins contracts it didn't expect to win, this may indicate that cross-subsidization is taking place. Similar need for ABC may be seen in analyses that recommend outsourcing work which the company should be able to do itself profitably.

LESSONS FROM IMPLEMENTATION

Despite the introduction of ABC systems in the late 1980s, accountants and managers in many organizations today consider the tremendous investments in money and time required to bring online a new cost system to be a major impediment to implementation. Complicating the cost/benefit analysis of an implementation decision is the mixed success some organizations have had in adopting ABC as a product cost system (1995c). In a three-part series of articles, Player and Keys (1995a, 1995b, 1995c) discuss the difficulties involved in successfully launching an ABC system and provide important caveats to managers of implementation teams. These lessons are valuable to those who are considering ABC investment.

Perpetrators effort

Player and Keys (1995a) note that companies often fail to appreciate the effort required to initiate successfully an ABC system. Without the involvement of top management in the organization, ABC implementation is likely to stall. Everyone involved must clearly understand exactly why the organization needs a new cost system and what can be expected from making this investment. The lack of involvement across the organization is typically revealed by underfunded projects staffed by employees who are not adequately trained in the essential principles and technology of ABC. As a result of misunderstanding the commitment required, there are many industry examples of failed ABC projects. Further, ABC is not simply a financial or accounting project. The focus on activities requires that managers and employees from functional areas such as operations, engineering, and marketing must be involved. Otherwise, the new system will fail to deal effectively with operational or strategic issues. Finally, ABC is not an isolated, one-time project. To operate effectively as a strategic cost management system, ABC must be integrated with other strategic initiates that the organization is pursuing. For example, the cross-view of the ABC/ABM model as proposed by Turney (1992) in Exhibit 2 should be used to develop performance measurements and ideas for continuous improvement for the organization's just-in-time (JIT) or total quality management (TQM) initiatives.

The ABC pilot

"It is extremely difficult to implement a comprehensive [ABC] system without doing a pilot project first" (Player and Keys, 1995b, 20). By first launching an ABC pilot, the implementation team is able to generate greater support throughout the organization when the mainstream system is brought online. Further, the pilot program provides opportunities to train personnel in interviewing and observation methods necessary to identify critical activities and develop relevant drivers that link activity costs to cost objects. The pilot can also alert management to potential software problems or to difficulties in obtaining the necessary data required to support the ABC model. Finally, the organization can use the pilot to understand the critical balance between too much and too little detail in its proposed ABC system. As a result of a poorly designed ABC system, management within the organization can be frustrated by "information overload." The ABC design team must, therefore, be mindful of the effect of information detail provided to management. Too much information results in "information overload" and decision-makers, who are either confused by the information provided by the ABC system or who simply refuse to use it! On the other hand, the ABC pilot should incorporate enough new data to ensure that it supports insights into processes and costs that are significantly different from those provided by the legacy system.

Moving from the pilot to mainstream

Player and Keys (1995c) are careful to point out in their studies of ABC implementation projects that behavioral issues tend to be much more crucial than technical issues in successfully establishing an ABC system. Implementing ABC is a large undertaking. The implementation team should expect that both individuals and departments will naturally resist change until it is clearly understood how the information provided by the new system will impact jobs and relations within and outside of the organization. To move successfully from the pilot phase of the ABC project to full mainstream implementation, the organization must have in place formal procedures involving how and when reports will be developed and disseminated. Further, it must be clear to everyone how the organization plans to act on the ABC information and how those actions will affect individuals and profits. Finally, the decision to invest in the ABC system must be continually evaluated in terms of costs and benefits in order to justify the constant investments required to update the ABC system to reflect changing business processes within a dynamic organization.

CONCLUSION

Dissatisfaction with traditional cost systems, coupled with the advent of inexpensive and sophisticated computer technology, has led to the emergence of ABC as a viable, strategic cost-management system. ABC can enhance an organization's product and market strategy by reducing cost distortions and by allowing flexibility in defining cost objects. Market events can provide valuable signals to help the organization evaluate the potential benefits of an ABC system. The process of designing and implementing an ABC system is both costly and time-consuming, and some organizations have struggled in their launch of ABC systems. Most of the pitfalls that can undermine an ABC implementation project relate to behavioral, rather than technical issues of managing the change.

Key Concepts: ABC Classification; Activity-Based Management; Enterprise Resource Planning; Information Warehouse; SAP; Target Costing; Unit-Based Costing (UBC).

Related Articles: The Evolution of Enterprise Resource Planning; Just-in-Time Manufacturing; Target Costing.

REFERENCES

Boer, G. (1994). "Five modern management accounting myths." *Management Accounting* (USA, January) 22–27.

Cooper, R. (1987). "Does your company need a new cost system?" *Journal of Cost Management* (Spring) 45–49.

——— (1988a). "The rise of activity-based costing—part one: What is an activity-based cost system?" *Journal of Cost Management* (Summer) 45–54.

——— (1988b). "The rise of activity-based costing—part two: When do I need an activity-based cost system?" *Journal of Cost Management* (Fall) 41–48.

——— (1989a). "The rise of activity-based costing—part three: How many cost drivers do you need, and how do you select them?" *Journal of Cost Management* (Winter) 34–46.

——— (1989b). "The rise of activity-based costing—part four: What do activity-based cost systems look like?" *Journal of Cost Management* (Spring) 38–49.

——— (1990). "Cost classifications in unit-based and activity-based manufacturing cost systems." *Journal of Cost Management for the Manufacturing Industry* (Fall) 4–14.

Cooper, R. and R.S. Kaplan (1991). *The Design of Cost Management Systems*, Prentice Hall, Englewood Cliffs.

Drucker, P.F. (1963). "Managing for business effectiveness." *Harvard Business Review* (May–June), 59–66.

Denna, E.L., J.O. Cherrington, D.P. Andros, and A.S. Hollander (1993). *Event-Driven Business Solutions.* Business One Irwin, Homewood, IL.

Harrison, G.C. (1921). "What is wrong with cost accounting?" *N.A.C.A. Bulletin.*

Johnson, H.T. and R.S. Kaplan (1987). *Relevance Lost! The Rise and Fall of Management Accounting*, Harvard Business School Press, Boston, MA.

Kaplan R.S. (1988). "One cost system isn't enough." *Harvard Business Review* (January–February), 61–66.

Player, R.S. and D.E. Keys (1995a). "Lessons from the ABM battlefield: Getting off to the right start." *Journal of Cost Management* (Spring), 26–38.

——— (1995b). "Lessons from the ABM battlefield: Developing the pilot." *Journal of Cost Management* (Summer), 20–35.

——— (1995c). "Lessons from the ABM battlefield: Moving from pilot to mainstream." *Journal of Cost Management* (Fall) 31–41.

Turney, P.B.B. (1992). *Common Cents: The ABC Performance Breakthrough*, Cost Technology, Portland, OR.

Swain, M.R. and E.L. Denna (1998). "Making new wine: Useful management accounting systems," *International Journal of Applied Quality Management.* 1, 25–44.

32. TARGET COSTING

RAM RAMANAN

University of Notre Dame, Notre Dame, IN, USA

ABSTRACT

Target Costing is an approach to "cost management" or "cost planning" over the life of a new product. It is a valuable competitive tool in the global manufacturing environment which has become very competitive. In a competitive market, sales price of products has to be evaluated carefully by management in several key industries, taking into account potential losses in market share to competitors offering lower prices. Therefore, in order to remain competitive, there is a great need for controlling of costs. Further, in a rapidly changing business environment, it becomes difficult to use just one factor, such as quality, to sustain a competitive advantage. The prevailing environment is unforgiving of poor quality, mistakes or delay, and is very demanding since consumers today are better informed about products.

Target costing, as a management initiative, responds to this environment by anticipating costs before they are incurred. It focuses externally on customer requirements and competitive threat, and attempts to systematically link an organization to its suppliers, dealers, customers, and recyclers in a cohesive and integrated profit and cost planning system. Using this approach, firms attempt to continually improve product and process designs at optimal costs. Read here about the use of target costing in the following mostly Japanese companies: **Daihatsu Motors; Komatsu; Matshushita Electric Works; Mercedes-Benz; Olympus Optical Company; Panasonic; Sharp; Toyota; Yokohama Corporation.**

WHAT IS TARGET COSTING?

Target costing is the concept of *price-based costing* instead of *cost-based pricing*. A *target price* is the estimated price for a product or service that potential customers will be willing to pay. A target cost is the estimated long-run cost of a product or service that allows the firm to achieve a targeted profit. *Target cost* is derived by subtracting the target profit from the target price.

Target costing is widely used. For example, Mercedes and Toyota in the automobile industry, Panasonic and Sharp in the electronic industry, and Apple and Toshiba in the personal computer industry use target costing (Maher, Stickney and Weil, 1997). This approach is quite different from standard costing. Target costing begins with identifying customers' needs and estimating an acceptable sales price for the product. Working backwards, the full cost of the product is established so as to earn an estimated profit. Target costs serve as goals for research and development, design, production personnel, and other departments in the value chain. If the target cost cannot be achieved, management must take a close look at the viability of making the proposed product.

HISTORY

In the late 1960's, as personal income in Japan increased rapidly, people diversified their tastes. As consumer needs became diversified, product diversification came about. As a result of these changes, Japanese companies had to develop and produce numerous products with quite different characteristics. An improvement in production methods from mass production was needed to meet these diverse social needs. In particular, the rapid progress of industrial robots and factory automation had made it possible to produce multiple products at low cost. Sakurai (1989) defines factory automation to include *computer-aided design, computer-aided manufacturing, flexible manufacturing systems*, and *office automation*. In this article, Sakurai provides an excellent description of the developments in Japan and the applicability of target costing to various settings.

In addition, product life cycles became shorter as a result of more diversified needs by customers. Therefore, the development, planning, and design phases of a product became critical to cost management. Target costing emerged as a management tool to plan needed features or functions of a product at a cost that would permit the firm to earn a targeted profit. This method has become prevalent in the United States only since the late 80's. The loss of market share to Japanese companies has been a major motivation for the U.S. companies adopting target costing.

IMPLEMENTATION OF TARGET COSTING

Target costing relies on product design for cost reduction to meet a specified profit, that is, it is applied primarily to new product development efforts. The relationship of target costing activities to the product development cycle, as developed in Ansari (1997) et al., is shown in Exhibit 1. Typically, target costing occurs in two stages that correspond roughly to the first and second halves of product development cycle. First is the establishment phase that involves establishing a target cost during the product

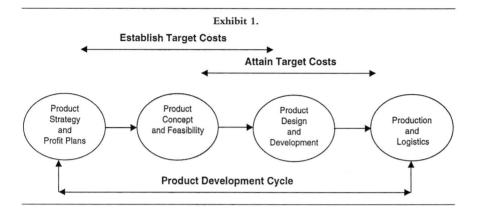

Exhibit 1.

Establish Target Costs

Attain Target Costs

Product Strategy and Profit Plans → Product Concept and Feasibility → Product Design and Development → Production and Logistics

Product Development Cycle

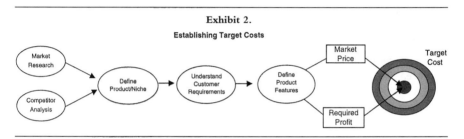

Exhibit 2.

Establishing Target Costs

planning and development stages. The attainment phase involves achieving a target cost and takes place during the design development and production stages of the cycle. The following seven major activities are usually carried out by companies during the establishment phase (see Exhibit 2):

1. Market research is carried out to gain information about unmet needs and wants of customers. This research defines the market and product niche to be exploited.
2. Competitor analysis is done to determine how customers evaluate competitor products, and how they might react to a company's new product introductions.
3. A customer or product niche is then defined by analyzing market and competitor information to decide what particular customer requirements to target.
4. Customer requirements are fine-tuned by introducing an initial product concept and asking customers for their reaction. The products are then designed to meet customer requirements.
5. Product features are defined by setting specific requirements for the features the product will have and the level of performance for those features.
6. A market price is established that is acceptable to customers without affecting the firm's competitive position.
7. The firm then decides on the required profit to be made on the product. This will leave the target, or allowable, cost.

The efforts within the firm to actually attain target costs during the attainment phase can be described by the following three steps:

1. The cost gap is computed. This is the difference between the current estimate of the cost of producing something based on current cost factors or models, and the allowable cost that has been established.
2. The next step is to design costs out of a product. This means reduction of cost through product design, and is the most critical step in attaining target costs.
3. The design is released and the company gets on the continuous improvement path. This is an attempt to reducing costs beyond that which is possible through design alone, and is achieved by improving production yields and eliminating wastes along the way.

Narrowing the cost gap can be an extremely difficult process though. The initial concept design is the beginning of the cost planning process. Subsequently, the product and the processes are concurrently designed. Tools used at this stage are Computer Aided Design and Computer Aided Manufacturing. Cost parameters may be built into these computer models so cost impact of changes in design can be simulated concurrently.

Next comes cost analysis. This involves identifying cost targets for major sub-components and parts of a product as well as cost reduction goals for selected components. The activities include the development of a list of the components, the function(s) of these components and their current estimated costs. The next step is to do a functional cost breakdown. This involves identifying the cost of each of the functions of the product.

The manufacturer's view of the product as a sum of functions needs to be reconciled with the customer's view of the product as a set of important features. A tool called Quality Function Deployment (QFD, borrowed from the quality literature) is used to systematically arrange information about features, functions, and competitive evaluation. This tool highlights the relationships between customer requirements and design parameters. Moreover, a concerted effort is made to include information about how customers evaluate competitor offering on these same features. QFD provides information that allows a manager to convert product feature rankings into functional or component rankings. This analysis provides useful insights that facilitate systematic trade-offs in the design of the various components using value engineering.

VALUE ENGINEERING

Value engineering, which is at the core of target costing, is an organized effort directed at analyzing the functions of the various components of the product for the purpose of achieving these functions at the lowest overall cost without reductions in performance, reliability, maintainability, quality, safety, recyclability, and usability. It analyzes product and manufacturing process design and reduces cost by simplifying both.

The first step in value engineering is identifying components for cost reduction. A Value Index is typically computed in order to select components for cost reduction or for design enhancements. The value index is the ratio of the degree of importance to the customer (expressed as a percentage) and the cost of each component (also expressed as a percentage of the total product cost). Components with a value index of less than one are targeted for cost reduction while those with a value index of greater than one are targeted for enhancements.

To illustrate this, consider two components A and B. Component A has a relative importance of 35% to customers and costs 46% of the total product cost. The value index is 0.76 and this component is a prime candidate for cost reduction efforts, since it is costing too much relative to its importance to customers. Component B has a relative importance of 10% to customers and it costs 4% of total product costs. The value index of 2.5 suggests that this component could be enhanced since current spending is not sufficient relative to its importance to customers. When the value index is 1.0 or thereabouts, no action for either cost reduction or enhancement is implied.

The generation of cost reduction ideas is usually through a brainstorming exercise about what can be reduced, eliminated, combined, substituted, rearranged, or enhanced to provide the same level of functionality from a component at less cost. These ideas are evaluated on technical feasibility and acceptability to customers, prior to implementation.

Estimating achievable cost is an activity that takes place at every iteration of the design and production cycle. Estimates of achievable cost should become more and more accurate as the process moves on and the company gains experience with the product. This not only includes manufacturing costs but also distribution, marketing, and support costs.

EFFECT ON PERFORMANCE

Target costing has definite effects on the performance of employees from different functional areas. Target costing requires all the key players, such as design engineers, process planning engineers, marketing personnel, and management accountants to have early involvement with a product. They are required to provide good estimates for each design iteration, and they must look ahead and provide information about products from incomplete design data. They must help each other on the team. Thus, it calls for an expansion of the traditional role of employees used to narrow functional silos.

Target costing requires different behaviors from all members of an organization. Team members must learn to live with ambiguity because design is an incomplete process, which means hard numbers are not always available. Members of the organization must work closely with other disciplines, as team playing is extremely important. This also calls for more effective communication among department personnel.

A customer focus is achieved because the target costing process must always focus on how it creates value for a customer, as customer requirements drive these activities. The evaluation of employees geared to target costing goals will insure the involvement of all workers, keep their focus on cross functional teamwork, encourage open sharing of information, and value customer inputs.

Target costing is not without its limitations. Overemphasis on design may lead to longer product development times, which could cause a costly delay in getting a product to the market. Usually, targets for cost, quality, and time must be made. Pressure to attain demanding targets can cause employee burnout and frustration. If targets are not attained despite much overtime and hard effort, employees' future levels of interest and effort may be reduced. This is why employees must participate in target setting. A small amount of slack should also be used to allow an organization to devote extra effort during crisis periods. Also, to reduce stress, the focus should be on continuous improvement and not radical changes.

Another potential problem of target costing is that too much attention to customers can lead to adding features without regard to extra costs involved, and proliferation of product models that can cause market confusion. This can be controlled if engineers are very aware of the costs of new features.

If the proper steps to avoid these potential problems are implemented, then target costing can be a great source of competitive advantage. It needs to have the support of all levels of the organization, otherwise it will breakdown somewhere along the line. Ansari et al. (1997) point out that target costing focuses on the product as it moves through time, across units, across organizations, and across activities. Further, they argue that since target costing is accomplished by cross-functional teams, who have a product and process focus, target costing cannot function in an organization that is not ready to adopt a process orientation.

WHEN IS TARGET COSTING APPROPRIATE?

Target costing is most appropriate in the beginning stages of product development. It needs to start as early as the new product brainstorming stage. If a product is already on the market, target costing to change design features and processes is probably not worth the extra time and effort. It is appropriate in an organization where cross-departmental communication is easily facilitated. It is also better used in an environment where there is a strong understanding of cost drivers and work flows, so that, production process costs can be efficiently estimated. Finally, it is applicable mainly for *assembly-oriented* industries that produce a variety of products in medium to small volumes than for *process-oriented* industries.

EXAMPLES OF USE

Ansari, et al. (1997) provide a comprehensive example of target costing with the help of a hypothetical company, Kitchenhelp Inc. They illustrate all the key activities for product concept that combines a coffee grinder and a drip system into a single coffee-maker. These authors also include an excellent discussion of the various facets of target costing as a management tool.

Shank (1995) presents an illustration of target costing for a forest products company, which adopted price-based cost planning initiatives including re-engineering for their continued viability. Cooper (1994) provides another example of target costing practices in the automobile industry in Japan, as it relates to Yokohama Corporation which is a supplier to Tokyo Motors. Cooper and Slagmulder (1997) describe the experiences of several Japanese corporations with regard to target costing and value engineering. In their book, the authors also provide an extensive reference to the literature including an annotated bibliography of selected articles.

Brausch (1994) describes the implementation of target costing by an American textile manufacturer. This practitioner article emphasizes the importance of using a cross-functional, team-based approach to target costing. Cooper and Chew (1996) emphasize the need for feedforward cost management in the new global economy with intense competition on price, quality and functionality. The article cites the experiences of two Japanese firms, Olympus Optical Company and Komatsu, to illustrate effective application of target costing. In a similar vein, Fisher (1995) documents the use of target costing by Matshushita Electric Works and Toyota Motors, Tanaka (1993) describes Toyota Motors' unique differential approach to cost planning, and Hiromota (1988) describes target costing practices at Daihatsu Motors.

The Kato (1993) explores the contributions of target costing to cost management in Japanese firms. Information systems necessary to support this process are described, including support systems for market research, target profit computation, R&D, and value engineering. The article also warns against potential disadvantages of target costing, such as the extreme demands it places on the employees. Other potential downsides of target costing, such as longer development cycles, employee burnout, market confusion, and organizational conflicts are very well presented in Kato, Boer and Chow (1995).

Key Concepts: Cost Drivers; Cost Management; Cost Planning; Cost-based Pricing; Price-based Costing; Quality Function Deployment (QFD); Standard Costing; Target Price; Value Engineering; Value Index.

Related Articles: Activity-Based Costing; Concurrent Engineering; Product Development and Concurrent Engineering.

REFERENCES

Ansari, S., J. Bell, T. Klammer, and C. Lawrence (1997). *Target Costing*. Irwin.

Brausch, J.M. (1994). "Beyond ABC: Target Costing for Profit Enhancement." *Management Accounting*, November, 45–49.

Cooper, R. (1994). "Yokohama Corporation Ltd. (A)." *Harvard Business School Case* 5-195-071.

Cooper, R. and W.B. Chew (1996). "Control Tomorrow's Costs through Today's Designs." *Harvard Business Review*, January–February, 88–97.

Cooper, R and R. Slagmulder (1997). *Target Costing and Value Engineering*, Institute of Management Accountants, Productivity Press.

Fisher, J. (1995). "Implementing Target Costing." *Journal of Cost Management*, Summer, 50–59.

Hiromoto, T. (1988). "Another Hidden Edge-Japanese Management Accounting." *Harvard Business Review*, July–August, 22–26.

Kato, Y. (1993). "Target Costing Support Systems: Lessons from Leading Japanese Companies." *Management Accounting Research*, 4 (4) 33–47.

Kato, Y., G. Boer, and C.W. Chow (1995). "Target Costing: An Integrative Management Process." *Journal of Cost Management*, Spring, 39–51.

Sakurai, M. (1989). "Target Costing and How to Use It." *Cost Management*, Summer.

Shank, J. (1995). "Strategic Cost Management: A Target Costing Field Study." The Amos Tuck School of Business Administration, Dartmouth College.

Tanaka, T. (1993). "Target Costing at Toyota." *Journal of Cost Management*, Spring, 4–11.

33. BALANCED SCORECARDS

RAMACHANDRAN RAMANAN

University of Notre Dame, Indiana, U.S.A.

ABSTRACT

An overemphasis on achieving and maintaining short-term financial results can cause companies to overinvest in short-term fixes and to under invest in long-term value creation, particularly in the intangible and intellectual assets that generate future growth. To illustrate, Larry Brady, President of F.C.: "*As a highly diversified company, the return-on-capital-employed (ROCE) measure was especially important to us. At year-end, we rewarded division managers who delivered predictable financial performance. We had run the company tightly for the past 20 years and had been successful. But it was becoming less clear where future growth would come from and where the company should look for breakthroughs into new areas. We had become a high return-on-investment company but had less potential for further growth. It was also not at all clear from our financial reports what progress we were making in implementing long-term initiatives.*" Similarly, many senior executives realize that, in several cases, employees are not able to link the applicability of corporate mission and vision to their day-to-day activities.

Breakthroughs in performance require change, and change is required in the measurement and management systems as well. Balanced scorecard offers this framework. As argued by Kaplan and Norton in their book, balanced scorecard can help organizations in three identifiable ways. First, the scorecard describes the organization's vision of the future to the entire organization and promotes shared understanding. Second, the scorecard creates a holistic model of the strategy that allows all employees to see how they contribute to organizational success. Without such a linkage, individuals and departments can optimize their local performance but not contribute to achieving strategic objectives. Third, the scorecard also focuses attention on change efforts. If the right objectives and measures are identified, successful implementation will likely occur. If not, investments and initiatives will be wasted.

The book and articles by Kaplan and Norton describe how the balanced scorecard has been developed and used in many companies. Mostly, it has been used at the top management level where it supports the company's strategic management system. The authors observe that the balanced scorecard concept has also been helpful to both top and middle management to shape and clarify organizational goals and strategy. Moreover, it has been useful at the worker level, when the complex trade-offs implied by the balanced scorecard are translated into simple performance measures. For example, in the Analog Devices Case, the "Corporate Scorecard" contained, in addition to several traditional financial measures, performance measures relating to customer delivery times, quality and cycle times of manufacturing processes, and effectiveness of new product developments.

It is worth noting that this system provides an organization-wide framework to integrate the benefits of several improvement initiatives, such as total quality management, just-in-time production and distribution systems, time-based competition, lean production/lean enterprise, building customer-focused organizations, activity-based cost management, employee empowerment, and reengineering. Kaplan and Norton claim that the programs like those listed above, if fragmented and not linked to the organization's strategy, could and will lead to disappointing results. Read here about the use of balanced scorecards in the following organizations: **Ben & Jerry Ice Cream; Metro Bank; and National Insurance.**

WHAT IS A BALANCED SCORECARD?

The *balanced scorecard* is a performance measurement system that uses a set of performance targets and results to show an organization's performance in meeting its objectives relating to its various stakeholders. It recognizes that organizations are responsible to different stakeholder groups, such as employees, suppliers, customers, business partners, the community and shareholders. This method of measuring performance focuses attention on achieving strategic organizational objectives relating to the above stakeholders. Further, the balanced scorecard provides a system to channel the energies, abilities, and specific knowledge held by people throughout the organization toward achieving these strategic objectives.

Sometimes different stakeholders have different wants. For example, employees depend on an organization for their employment. Customers expect a quality product with the shortest possible delivery time and competitive price. Shareholders depend on an organization to maintain and grow their investment. The organization must balance these competing wants. For many years, organizations focused only on financial results, which reflected mainly the shareholders' interests. In recent years, organizations have shifted attention to customer issues, such as quality and service. They also pay attention to the needs of employees and the community. For example, Ben & Jerry's Ice Cream measures its social performance along with financial performance and has a social audit next to its financial audit in its annual report.

The concept of a balanced scorecard is to measure how well the organization is doing in view of competing stakeholder wants. A balanced scorecard could be used as a strategic management system. Such a scorecard requires the balancing of the efforts of the organization among the *financial, customer, internal business process,* and *learning and growth* (innovation) objectives. For each of the objectives, the balanced scorecard provides a framework for the firm to develop measures of outcomes relating to the objective, to set target outcomes for a period and to identify key initiatives that would help in achieving the desired outcomes. A comparison of actual to the target at the end of the period can then be used for feedback, corrective actions and performance evaluation.

HISTORY

The development of the balanced scorecard method can be traced back to 1990 when the Nolan Norton Institute, the research arm of the accounting firm KPMG, sponsored a one-year study, "Measuring Performance in the Organization of the Future." This study was motivated by a belief that existing performance measurement approaches relying on financial accounting measures, were becoming obsolete. David Norton, CEO of Nolan Norton, was the study leader and Robert Kaplan, Professor of Accounting, Harvard University was the academic consultant. Senior managers from 12 corporations participated in this year-long study. All of these managers shared a belief that exclusive reliance on summary performance measures was hindering organizations' abilities to create future economic value.

Participants of this group shared their experiences in bimonthly meetings and several ideas emerged in the group discussions over the year. The group examined various innovative performance measurement systems, such as shareholder value, productivity and quality improvements and compensation plans, were studied. The participants started focusing on multidimensional scorecards as offering the most promise for their needs. This led to the design of a rough "Corporate Scorecard." and evolved into what is now entitled "Balanced Scorecard." Kaplan and Norton (1996) state that, "...the name reflected the balance provided between short- and long-term objectives, between financial and non-financial measures, between lagging and leading indicators, and between external and internal performance perspectives." The scorecard measures organizational performance across four balanced perspectives: financial, customers, internal business processes, and learning and growth (innovation).

Seminal work on this topic is done by Kaplan and Norton, the recognized architects of balanced scorecard. They have published several articles since 1992 and a book in 1996. In their recently published book, the authors demonstrate how senior executives in industries such as banking, oil, insurance, and retailing are using the balanced scorecard both to guide current performance and target future performance. They show how to use measures in financial performance, customer knowledge, internal business processes and learning and growth to align individual, organizational and cross-departmental initiatives. They describe how these measures can be used to identify entirely new processes for meeting customer and shareholder objectives. The book also provides the steps that managers in any company can use to build their own balanced scorecard.

Interested readers could also gain useful insights from the various articles by these authors. For example, Kaplan and Norton (1992), summarizes the findings of their study group in 1990. In a subsequent article in 1993, the authors describe the importance of choosing measures based on strategic success and reflects their consulting experience with some corporations. More recently, Kaplan and Norton (1996), suggest how the use of balanced scorecard could be improved by extending it from a measurement system to a core management system. Based on their experiences of consulting partnerships with various organizations implementing this method, Kaplan and Norton (1996) posit that the balanced scorecards provides managers with the instrumentation they need to navigate to future competitive success. This approach to performance evaluation translates an organization's mission and strategy into a comprehensive set of performance measures that provides the framework for a strategic measurement and management system.

IMPLEMENTATION

Kaplan and Norton (1993) provide a typical profile of the process of building a balanced scorecard. They point out that each organization is unique and so may follow its own path for building this system. They propose the following eight-step process that firms may adapt in their systematic efforts to develop a balanced scorecard:

(1) *Preparation:* The organization must first define the business unit for which the top-level scorecard is to be developed. Typically a business unit would have its own customers, distribution channels, production facilities (if appropriate) and financial and performance measures.

(2) *Interviews, First Round:* The balanced scorecard facilitator (either outside consultant or executive from the firm) conducts interviews of about 90 minutes each with senior managers to obtain their inputs on the company's objectives. Interviews with principal shareholders and key customers may also be conducted to gather inputs on their expectations.

(3) *Executive Workshops, First Round:* The top management team is brought together with the facilitator to debate the proposed mission and strategy statements until consensus is reached. Subsequently the group brainstorms about the development of the measures. At this meeting, it is not necessary to reach consensus on the proposed measures as long as the group develops a menu of measures for each objective.

(4) *Interviews, Second Round:* The facilitator consolidates the executive discussions and seeks further input from senior managers about the measures and the process of implementation.

(5) *Executive Workshop, Second Round:* At this time more middle managers are involved in the group meeting in addition to the top management team. Participants work in groups, debate and develop the measures and start to develop an implementation plan. They could also be asked to propose preliminary targets for each objective and measure.

(6) *Executive Workshop, Third Round:* The top management team is brought together to reach consensus on the balanced scorecard. This is the time for the group to develop an implementation plan, a communication plan, and an information system to support the scorecard.

(7) *Implementation:* It may be beneficial to form a team for implementation. This team may also encourage the development of second-level metrics for decentralized units.

(8) *Periodic Review:* Each quarter or month, top management may undertake reviews of the balanced scorecard information and discussions with managers of decentralized divisions and departments. The authors recommend that senior managers revisit the balanced scorecard metric annually as part of their strategic planning, goal setting and resource allocation processes.

GUIDELINES FOR IMPLEMENTATION

Some general guidelines are offered below with regard to selecting measures for each of the four perspectives. First, the *financial objectives* have first to be related to the product life cycle (harvest, sustain or growth). Three typical measures encountered are revenue growth and mix, earnings improvement, asset utilization. Second, it is important that the *customer-related objectives* be identified with more specificity. For example, objectives relating to market share should not go unqualified; rather "market share of targeted customers" would be more appropriate to get across what actual measure the managers are seeking to define. Other common measures include percentage growth of business with existing customers to capture customer loyalty and the solicitation cost per new customer acquired to capture market development effort.

Third, while traditional approaches may seek to improve existing internal business processes, balanced scorecard encourages identification of *new processes* and seeks to meet needs of future. In this context, product and service performance attributes that are likely to create value for the customer become critical in addition to measures such as response time, cost and quality. Finally, in the context of *learning and growth,* balanced scorecard provides a framework to reveal gaps between existing capabilities of people, systems, and procedures and those needed to meet the strategic goals of the organization.

In general, the authors recommend between 10 and 25 metrics in total as a benchmark and indicate that a typical firm takes about 16 weeks to develop the balanced scorecard system. These are carefully selected as strategic measures that focus on factors that will lead to competitive breakthroughs. Keeping the list short and manageable is key for top management to monitor progress without too much clutter.

Several case study examples are used to illustrate the concepts by Kaplan and Norton in their book and articles. *Metro Bank,* for example, developed a scorecard with measures that focused **directly** on achieving their strategy. In this case, the use of both "lead" and "lag" indicators was especially important, with "lead" performance driver measures signaling strategy to managers (e.g. hours spent with customer enforcing the importance of a new relationship-based sales approach) and "lag" core outcome measures providing feedback on the process. For *National Insurance* initially, only "lagging" core outcome measures could be identified. The importance of "lead" performance driver measures was obvious due to the severe "lag" of data in insurance industry. As a result of balanced scorecard thinking, the firm developed a survey to assess policy-holder satisfaction and the results became leading indicators with regard to their effort in the area of customer acquisition and retention.

It is important to stress corporate-wide knowledge of balanced scorecard's implementation and use. Firms could make effective use of a "corporate billboard" displaying the balanced scorecarded in high-traffic areas such as meeting rooms, main thoroughfares, etc. and encouraging all the employees to periodically monitor performance in their functions.

CONCLUSION

For the balanced scorecard approach to be effective, it must reflect the strategic vision of the senior executive group and have the full support of this group. Benchmarking with other company measures is traditionally *not* good since competitive environments and strategies typically vary from firm to firm. Further, if implementation takes too long, a loss of momentum will occur in the organizational shift. Accordingly, even if the measures are not perfect to begin with, the implementation should go on since balanced scorecard is meant to be a continually reviewed process with many opportunities for evolution.

Key Concepts: Activity-Based Cost Management; Corporate Billboard; Financial Accounting Measures; Non-Financial Performance Measures; Performance Drivers; Stakeholder Groups.

Related Articles: Activity-Based Costing; Performance Excellence: Malcolm Baldrige National Quality Award Criteria.

REFERENCES

Kaplan, R.S. and D.P. Norton (1992). "The balanced scorecard—measures that drive performance." *Harvard Business Review*, January–February.
Kaplan, R.S. and D.P. Norton (1993). "Putting the balanced scorecard to work." *Harvard Business Review*, September–October.
Kaplan, R.S. and D.P. Norton (1196). *The Balanced Scorecard*, HBS Press 1996
Kaplan, R.S. and D.P. Norton (1996). "Using the balanced scorecard as a strategic management system." *Harvard Business Review*, January–February.
http://www.inphase.com/bbs.htm

34. PERFORMANCE EXCELLENCE: THE MALCOLM BALDRIDGE NATIONAL QUALITY AWARD CRITERIA

JAMES R. EVANS

University of Cincinnati, Cincinnati, Ohio, USA

ABSTRACT

The Malcolm Baldrige National Quality Award (MBNQA) recognizes U.S. companies that excel in quality management practices and performance excellence. The Baldrige Award does not exist simply to recognize product excellence, nor does it exist for the purpose of "winning." Its principal focus is on promoting high-performance management practices throughout the economy that lead to customer satisfaction and business results. The award promotes the awareness of quality as an increasingly important element in competitiveness, an understanding of the requirements for performance excellence, and sharing of information on successful quality management practices and the benefits derived from implementation of these strategies. Up to two companies can receive a Baldrige Award in each of the categories of manufacturing, small business, and service. In 1995, pilot programs in education and health care were instituted. See Table 2 for a list of award winners. This article describes the following award winners: **ADAC Laboratories; Trident Precision Machining; and others.**

HISTORY

Recognizing that U.S. productivity was declining, President Reagan signed legislation mandating a national study/conference on productivity in October 1982. The American Productivity and Quality Center (formerly the American Productivity

Center) sponsored seven computer networking conferences in 1983 to prepare for an upcoming White House Conference on Productivity. The final report on these conferences recommended that "a National Quality Award, similar to the Deming Prize in Japan, be awarded annually to those firms that successfully challenge and meet the award requirements. These requirements and the accompanying examination process should be very similar to the Deming Prize system to be effective." The Baldrige Award was signed into law (Public Law 100-107) on August 20, 1987. The award is named after President Reagan's Secretary of Commerce who was killed in an accident shortly before the Senate acted on the legislation. Malcolm Baldrige was highly regarded by world leaders, having played a major role in carrying out the administration's trade policy, resolving technology transfer differences with China and India, and holding the first Cabinet-level talks with the Soviet Union in seven years that paved the way for increased access for U.S. firms in the Soviet market. The Award is a public-private partnership, funded primarily through a private foundation.

PURPOSE OF THE AWARD

The purposes of the Award are to:

- Help stimulate American companies to improve quality and productivity for the pride of recognition while obtaining a competitive edge through increased profits
- Recognize the achievements of those companies that improve the quality of their goods and services and provide an example to others
- Establish guidelines and criteria that can be used by business, industrial, governmental, and other enterprises in evaluating their own quality improvement efforts
- Provide specific guidance for other American enterprises that wish to learn how to manage for high quality by making available detailed information on how winning enterprises were able to change their cultures and achieve eminence.

The award examination is based upon a rigorous set of criteria designed to encourage companies to enhance their competitiveness through efforts toward dual, results-oriented goals:

1. Delivery of ever-improving value to customers, resulting in improved marketplace success
2. Improvement of overall company performance and capabilities.

The primary emphasis of the Baldrige program is not on winning awards; but rather on educating American business.

AWARD CRITERIA

The award criteria are built upon a set of core values and concepts that are described below.

Customer-Driven Quality: Quality is judged by customers. All product and service features and characteristics that contribute value to customers and lead to customer satisfaction and preference must be a key focus of a company's management system. Customer-driven quality is directed toward customer retention, market share gain, and growth. It demands constant sensitivity to emerging customer and market requirements, and measurement of the factors that drive customer satisfaction and retention. It also demands awareness of developments in technology and of competitors' offerings, and rapid and flexible response to customer and market requirements. In addition, the company's success in recovering from defects and errors ("making things right for the customer") is crucial to building customer relationships and to customer retention.

Leadership: A company's senior leaders set direction and create customer orientation, clear and visible values, and high expectations. Reinforcement of the values and expectations requires personal commitment and involvement. The leaders need to guide the creation of strategies, systems, and methods for achieving excellence and building capabilities. Through their personal involvement in planning, communications, review of company performance, and employee recognition, the senior leaders serve as role models, reinforcing the values and building leadership and initiative throughout the company.

Continuous Improvement and Learning: Achieving the highest levels of performance requires a well-executed approach to continuous improvement. The term "continuous improvement" refers to both incremental and "breakthrough" improvement. Improvements may be of several types: (1) enhancing value to customers through new and improved products and services; (2) reducing errors, defects, and waste; (3) improving responsiveness and cycle time performance; (4) improving productivity and effectiveness in the use of all resources; and (5) improving the company's performance in fulfilling its public responsibilities and serving as a corporate citizenship role model. Continuous improvement must contain cycles of planning, execution, and evaluation with a basis for assessing progress and for deriving information for future cycles of improvement.

Employee Participation and Development: Companies need to invest in the development of the work force through education, training, and opportunities for continuing growth. Increasingly, training, development, and work organizations need to be tailored to a diverse work force and to more flexible, high performance work practices. Major challenges in the area of work force development include: (1) integration of human resource management selection, performance, recognition, training, and career advancement; and (2) aligning human resource management with business plans and strategic change processes.

Fast Response: Success in competitive markets increasingly demands ever-shorter cycles for new or improved product and service introduction. Also, faster and more flexible response to customers is now a more critical requirement. Response time improvements often drive simultaneous improvements in organization, quality, and productivity.

Design Quality and Prevention: Business management should place strong emphasis on design quality and waste prevention achieved through building quality into products and services and efficiency into production and delivery processes. Increasingly, design quality also includes the ability to incorporate information gathered from diverse sources and data bases, that combine factors such as customer preference, competitive offerings, marketplace changes, and external research findings and developments. Consistent with the theme of design quality and prevention, improvement needs to emphasize interventions "upstream" at early stages in processes.

Long-Range View of the Future: Pursuit of market leadership requires a strong future orientation and a willingness to make long-term commitments to all stakeholders-customers, employees, suppliers, stockholders, the public, and the community. Planning needs to anticipate many types of changes including those that may affect customers' expectations of products and services, technological developments, changing customer segments, evolving regulatory requirements, community/societal expectations, and thrusts by competitors.

Management by Fact: Modern business management systems depend upon measurement, data, information, and analysis. Measurements must derive from the company's strategy and encompass all key processes and the outputs and results of those processes. Analysis refers to extracting larger meaning from data to support evaluation and decision making at all levels within the company. Facts, data, and analysis support a variety of company purposes, such as planning, reviewing company performance, improving operations, and comparing company performance with competitors' or with "best practices" benchmarks. A system of measures or indicators tied to customer and/or company performance requirements represents a clear basis for aligning all activities with the company's goals.

Partnership Development: Companies should seek to build internal and external partnerships to better accomplish their overall goals. Internal partnerships might include those that promote labor-management cooperation, such as agreements with unions. External partnerships might be with customers, suppliers, and education organizations for a variety of purposes, including education and training. Internal and external partnerships should seek to develop longer-term objectives, thereby creating a basis for mutual investments.

Corporate Responsibility and Citizenship: A company's management should stress corporate responsibility and encourage corporate citizenship. Corporate responsibility refers to basic expectations of company—business ethics and protection of public health, safety, and the environment. Inclusion of public responsibility areas within a performance system means meeting all local, state, and federal laws and regulatory requirements. It also means treating these and related requirements as areas for continuous improvement "beyond mere compliance." This requires that appropriate measures be created and used in managing performance. Corporate citizenship refers to leadership and support of publicly important purposes, including areas of corporate responsibility.

Results Orientation: A company's performance system needs to focus on results. Results should be guided by and balanced by the interests of all stakeholders—customers, employees, stockholders, suppliers and partners, the public, and the community. To meet the sometimes conflicting and changing aims that balance implies, company strategy

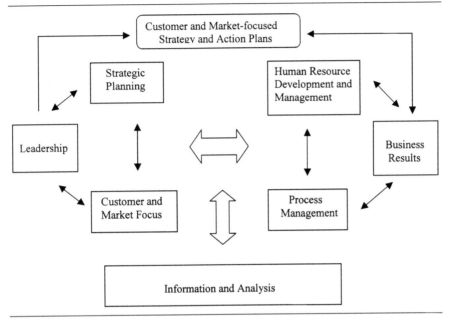

Figure 1. Malcolm Baldrige national quality award framework.

needs to explicitly address all stakeholder requirements to ensure that actions and plans meet the differing needs and avoid adverse impact on any stakeholders. The use of a balanced composite of performance measures offers an effective means to communicate requirements, to monitor actual performance, and to marshal support for improving results.

The Criteria consist of a hierarchical set of *categories, items,* and *areas to address.* The seven categories form the organization of the Criteria as illustrated in Figure 1. Leadership, Strategic Planning, and Customer and Market Focus represent the "leadership triad," and suggest the importance of integrating these three functions. Human Resource Development and Management and Process Management represent how the work in an organization is accomplished and lead to business results. These functions are linked to the leadership triad. Finally, Information and Analysis supports the entire framework by providing the foundation for performance assessment and management by fact.

Each category consists of several items that focus on the requirements that businesses should understand to function successfully in a competitive environment. Each item, in turn, consists of a small number of "areas to address," which seek specific information on *approaches* used to ensure and improve competitive performance, the *deployment* of these approaches, or *results* obtained from such deployment.

"Areas to address" request information and begin with the word "how"; that is, they fundamentally address key management processes that should form an effective system

Table 1 Key management processes reflected in the 1997 Baldrige criteria

Leadership
1. How senior leadership creates and sustains values, sets company directions, and develops, sustains, and improves an effective leadership system.
2. How the company addresses impacts of its products, services, facilities and operations on the society.
3. How the company practices good community citizenship.

Strategic Planning
4. The process of developing strategy, taking into account customers and markets, competitors, risks, company capabilities, and supplier/partner capabilities.
5. The process of translating strategy into action plans and tracking performance.

Customer and Market Focus
6. The process of determining longer-term requirements, expectations, and preferences of key customer groups and/or market segments, and evaluating and improving listening and learning strategies.
7. Approaches to providing access and information to enable customers to seek assistance, conduct business, and voice complaints to strengthen relationships and improve products and services.
8. The process of determining customer satisfaction and satisfaction relative to competitors.

Information and Analysis
9. The process of selecting, managing, and using information and data to support key company processes and improve company performance.
10. The process of selecting, managing, and using comparative information to improve overall performance and competitive position.
11. The process of integrating and analyzing customer, operational, competitive, and financial/market performance data.
12. The process of reviewing company performance and capabilities to assess progress and determine improvement priorities.

Human Resource Development and Management
13. The process of designing work and jobs to encourage all employees to contribute effectively to achieving the company's performance and learning objectives.
14. The design of compensation and recognition systems to reinforce work systems and performance.
15. The process of designing, delivering, evaluating, and improving education and training to build knowledge and capabilities and contribute to improved employee performance and development.
16. The design of a safe and healthful work environment.
17. The support of employee well-being, satisfaction, and motivation via services, facilities, and opportunities.
18. The process of determining employee well-being, satisfaction, and motivation, and using the results to identify improvement activities.

Process Management
19. The process of designing products, services, and production delivery processes to incorporate changing customer requirements, meet quality and operational performance requirements, and ensure trouble-free introduction and delivery of products and services.
20. The processes of managing, evaluating, and improving key product and service production/delivery processes to maintain process integrity, meet operational and customer requirements, and achieve better performance.
21. The process of evaluating and improving product and service production/delivery processes.
22. Approaches for managing and improving supplier and partnering processes, relationships, and performance.
23. The process of evaluating a balanced scorecard of business results, including customer satisfaction, financial/market, human resource, supplier/partner, and other efficiency and effectiveness measures.

for any quality-minded organization. The key business processes that are embedded within the 1997 Baldrige Criteria are listed in Table 1. Note that not all Criteria items and areas to address are reflected in Table 1, because some deal explicitly with key information or results, and do not specifically address management processes.

KEY CHARACTERISTICS OF THE CRITERIA

(1) *The criteria are directed toward business results:* The criteria focus principally on seven key areas of business performance: customer satisfaction and retention; market share and new market development; product and service quality; financial indicators, productivity, operational effectiveness, and responsiveness; human resource performance and development; supplier performance and development; and public responsibility and corporate citizenship. Improvements in these seven areas contribute significantly to overall company performance—including financial performance. Emphasis on results balances strategies in such a way that they avoid inappropriate tradeoffs among important stakeholders or objectives.

(2) *The criteria are nonprescriptive:* The criteria are a set of basic interrelated, results-oriented requirements, but they do not prescribe specific quality tools, techniques, technologies, systems, or starting points. Companies are encouraged to develop and demonstrate creative, adaptive, and flexible approaches to meeting basic requirements.

(3) *The criteria are comprehensive:* The criteria address all internal and external requirements of the company, including those related to fulfilling public responsibilities. Accordingly, all processes of all company work units are tied to these requirements.

(4) *The criteria include interrelated learning cycles:* Learning takes place via feedback among the process and results elements. A learning cycle has four stages:
  - planning, selection of indicators, and deployment of requirements
  - execution of plans
  - assessment of progress
  - revision of plans based upon assessment findings.

(5) *The criteria emphasize alignment:* The criteria call for improvement at all levels and in all parts of the company. Such improvement is achieved via interconnecting and mutually reinforcing measures and indicators, derived from overall company requirements. These measures and indicators tie directly to customer value and operational performance and provide a communications tool and a basis for deploying consistent customer and operational performance requirements to all work units.

(6) *The criteria are part of a diagnostic system:* Using the criteria provides a profile of strengths and areas for improvement that directs attention to processes and actions that contribute to business success.

THE BALDRIGE AWARD SCORING SYSTEM

Each examination item is assigned a point value that can be earned during the evaluation process. It is based upon three evaluation dimensions: approach, deployment, and results.

Approach refers to the methods the company uses to achieve the requirements addressed in each category. The factors used to evaluate approaches include:

- The appropriateness of the methods, tools, and techniques to the requirements
- The effectiveness of methods, tools, and techniques

- The degree to which the approach is systematic, integrated, and consistently applied
- The degree to which the approach embodies effective evaluation/improvement cycles
- The degree to which the approach is based upon quantitative information that is objective and reliable
- Evidence of unique and innovative approaches, including the significant and effective new adaptations of tools and techniques used in other applications or types of businesses.

Deployment refers to the extent to which the approaches are applied to all relevant areas and activities addressed and implied in each category. The factors used to evaluate deployment include:

- Appropriate and effective use of the approach in key processes
- Appropriate and effective application of the approach in the development and delivery of products and services
- Appropriate and effective use of the approach in all interactions with customers, suppliers of goods and services, and the public.

Results refers to the outcomes and effects in achieving the purposes addressed and implied in the criteria. The factors used to evaluate results include:

- Current performance levels
- Performance levels relative to appropriate comparisons and benchmarks
- Rate of performance improvement
- Breadth and importance of performance improvements
- Demonstration of sustained improvement or sustained high-level performance.

Examiners evaluate the applicant's response to each examination item, listing major strengths and areas for improvement relative to the criteria. Strengths demonstrate an effective and positive response to the criteria. Areas for improvement do not prescribe specific practices or examiners' opinions on what the company should be doing, but rather how management can better address the criteria. Based on these comments, a percentage score is given to each item. Each examination item is evaluated on approach/deployment or results.

Like quality itself, the specific award criteria are continually improved each year. The initial set of criteria in 1988 had 62 items with 278 areas to address. By 1991, the criteria had only 32 items and 99 areas to address. The 1995 criteria were streamlined significantly to 24 items and 54 areas to address. More significantly, the word quality has been judiciously dropped throughout the document. For example, until 1994, Category 3 had been titled "Strategic Quality Planning." The change to "Strategic Planning" signifies that quality should be a part of business planning, not a separate issue. Throughout the document, the term performance has been substituted for quality as a deliberate attempt to recognize that the principles of TQM are the foundation for a company's management system, not just the quality system. In 1997, the criteria were further streamlined and reorganized, with the goal being to develop the shortest list of key requirements necessary to compete in today's marketplace, improve the linkage between process and results, and make the criteria more generic and user-friendly. Only a few minor changes were made to the 1998 Criteria. As Curt Reimann, director of the Baldrige Award Program noted, "The things you do to win a Baldrige Award are exactly the things you'd do to win in the marketplace. Our strategy is to have the Baldrige Award criteria be a useful daily tool that simulates real competition."

A single free copy of the criteria can be obtained from the National Institute of Standards and Technology. Contact the Malcolm Baldrige National Quality Award, National Institute of Standards and Technology, Route 270 & Quince Orchard Road, Administration Building, Room A537, Gaithersburg, MD 20899 (301) 975-2036, FAX (301) 948-3716, or visit the web site at http://www.quality.nist.gov/

THE BALDRIGE AWARD EVALUATION PROCESS

The Baldrige evaluation process is rigorous. In the first stage, each application is thoroughly reviewed by up to 15 examiners chosen from among leading quality professionals in business, academia, health care, and government (all of whom are volunteers). The scores are reviewed by a panel of nine judges without knowledge of the specific companies. The higher scoring applications enter a consensus stage in which a selected group of examiners discuss variations in individual scores and arrive at consensus scores for each item. The panel of judges then reviews the scores and selects the highest scoring applicants for site visits. At this point, six or seven examiners visit the company for up to a week to verify information contained in the written application and resolve issues that are unclear. The judges use the site visit reports to recommend award recipients. Final contenders each receive more than 400 hours of evaluation.

All applicants receive a feedback report that critically evaluates the company's strengths and areas for improvement relative to the award criteria. The feedback report, frequently 30 or more pages in length, contains the evaluation team's response to the written application. It includes a distribution of numerical scores of all applicants and a scoring summary of the individual applicant. This feedback is one of the most valuable aspects of the Baldrige Award program.

RESULTS OF THE AWARD PROGRAM

The Baldrige Award has generated an incredible amount of interest in quality, both within the United States and internationally. Winning companies have made thousands of presentations describing their quality management approaches and practices. Many states have instituted awards similar to the Baldrige and based on the Baldrige Criteria. The award criteria have been used to train hundreds of thousands of people and as a self-assessment tool within organizations.

In assessing the applications submitted for the Baldrige Award, several key strengths and common weaknesses are evident. Most companies that apply have strong senior management leadership and are driven by the needs of customers and the marketplace. These firms have aggressive goals and high expectations. Companies have strong information systems that provide an excellent basis for assessing the state of quality and link external customer satisfaction measurements with internal measurements such as process quality and employee satisfaction. Most companies have invested heavily in human resource development, and employee involvement is continuing and expanding. Common weaknesses among firms that do not score well include:

- Weak information systems
- Delegation of quality responsibility to lower levels of the company
- A partial quality system, for example, strong in manufacturing but weak in support services
- An unclear definition of what quality means in the organization
- A lack of alignment among diverse functions within the firm; that is, not all processes are driven by common goals or use the same approaches
- Failure to use all listening posts to gather information critical to decision making.

Among the more disappointing results found by Baldrige Award administrators is that relatively few organizations have integrated systems. Many lack a quality vision or do not effectively translate this vision into a business strategy. Many companies still emphasize the negative side of quality—defect reduction, rather than customer focus. Finally, the gaps between the best and the average companies is quite large, indicating that much opportunity and hard work remains.

Baldrige Award-winning companies have shown excellent financial results. The Commerce Department reported that a person who would have invested $1000 in common stock in each publicly traded winning company on the first business day in April of the year they won the Baldrige Award (or the date when they began public trading, if it is later); or for subsidiaries, $1,000 times the percent of the whole company's employee base the subunit represented at the time of its application, would have outperformed the S&P 500 by approximately 3 to 1, achieving a 324.9% return. The group of five whole company winners outperformed the S&P 500 by greater than 3.5 to 1, achieving a 380.2% return.

A hypothetical sum was also invested in each 1990–1995 publicly-traded, site visited company. The 48 publicly-traded site visited applicants, as a group, outperformed the S&P 500 by 2 to 1, achieving a 167.5% return compared to a 83.3% return for the S&P 500. The group of 35 site visited applicants without the winners outperformed the S&P 500 by 68%, achieving a 138.3% return compared to a 82.0% return for the S&P 500. (Names of Baldrige applicants are kept confidential.)

AWARD WINNERS

A list of winners through 1997 is given in Table 2. In 1997, Solectron became the first company to receive the Award a second time. The 1996 winners included one manufacturing company, one service company, and two small businesses. ADAC Laboratories produces high-technology healthcare products for diagnostic imaging and

Table 2 Baldrige Award Winners through 1997

ADAC Laboratories (1996)
Ames Rubber Corporation (1993)
Armstrong World Industries (1995)
AT&T Consumer Communications Services (1994)
AT&T Network Systems Group, Transmission Systems Business Unit
  (now part of Lucent Technologies) (1992)
AT&T Universal Card Services (1992)
Cadillac Motor Car Company (1990)
Corning Telecommunications Products Division (1995)
Custom Research Inc. (1996)
Dana Commercial Credit Corporation (1996)
Eastman Chemical Company (1993)
Federal Express Corporation (1990)
Globe Metallurgical Inc. (1988)
Granite Rock Company (1992)
GTE Directories Corporation (1994)
IBM Rochester (1990)
Marlow Industries (1991)
Merrill Lynch Credit Corporation (1997)
Milliken & Company (1989)
Motorola Inc. (1988)
Solectron Corporation (1991, 1997)
Texas Instruments Defense Systems & Electronics Group (1992)
The Ritz-Carlton Hotel Company (1992)
3M Dental Products Division (1997)
Trident Precision Manufacturing Inc. (1996)
Wainwright Industries (1994)
Wallace Co. Inc. (1990)
Westinghouse Electric Corporation Commercial Nuclear Fuel Division (1988)
Xerox Business Services (1997)
Xerox Corporation Business Products & Systems (1989)
Zytec Corporation (1991)

healthcare information systems. Their management system based on Baldrige helped transform the company into a world leader after successfully coming out of a turnaround in the mid-1980s. Between 1990 and 1995, overall efficiency improvements resulted in an increase in revenue per employee from $200,000 to $330,000, 65% better than its competitors. ADAC consistently brings products to market faster than its larger competitors. A 1995 independent survey rated the first-month reliability of ADAC cameras as best in the industry. If a system breaks down, technicians will have it back up within an average of 17 hours after receiving a customer's call, less than a third of the time it took in 1990. On eight measures of customer service satisfaction, ADAC was the sole leader in five categories and tied for the top spot in the remaining ones.

Dana Commercial Credit Corporation provides leasing and financing services to a broad range of commercial customers. A collection of quality-linked "scoring processes" assesses how the company is progressing in its pursuit of continuous improvement goals set for all key areas of the business. Since 1994, customer-satisfaction scores for the Capital Markets Group have exceeded four on a five-point scale, and in 1995, topped the industry average by almost two points. The Dealer Products Group has scored between eight and nine on a 10-point scale, or nearly three points higher than the industry average. Rates of return on equity and assets have increased more than 45% since 1991.

Trident Precision Manufacturing, Inc. is a 167-person manufacturer of precision sheet metal components, electro-mechanical assemblies, and custom products. Quality is its basic business plan, focused on five key business drivers: customer satisfaction, employee satisfaction, shareholder value, operational performance, and supplier partnerships. Employee turnover has declined from 41% in 1988 to 5% in 1994 and 1995. Defect rates have fallen to a point where the company offers a full guarantee against defects for its custom products, and delivery performance has improved from 87% to 99.94 percent from 1990 to 1995.

Custom Research Inc. is a national marketing research firm, employing only about 100 people. With an intense focus on customer satisfaction, a team-oriented workforce, and information technology, CRI pursues individualized service and satisfied customers through a Surprise and Delight strategy. In 1996, most CRI employees received over 134 hours of training. CRI "meets or exceeds" clients expectations on 97% of its projects, and more remarkably, "exceeds expectations" on 70% of projects.

LESSONS

The Baldrige Award criteria form a blueprint for quality improvement in any organization. Hundreds of thousands of applications are distributed each year, yet only a handful of completed applications are received. Many companies are using the award criteria to evaluate their own quality programs, set up and implement TQM programs, communicate better with suppliers and partners, and for education and training. Even the U.S. Postal Service has decided to use the Baldrige Criteria as a basis to reestablish a quality system by identifying the areas that need the most improvement and providing a baseline to track progress. Using the award criteria as a self-assessment tool provides an objective framework, sets a high standard, and compares units that have different systems or organizations. It is also being used as a basis for giving awards within companies and at the local, state, and federal levels.

Many different philosophies and quality improvement programs exist. Organizations just getting started in quality improvement often have problems defining the quality system and setting objectives. The Baldrige Award addresses the full range of quality issues and can help those setting up new systems to get a complete picture of TQM.

The Baldrige Award criteria assist companies with internal communications, communications with suppliers, and communications with other companies seeking to share information. The criteria provide a focus on what to communicate and a framework for comparing strategies, methods, progress, and benchmarks. Finally, the Baldrige Award examination is being used for training and education, particularly for management, because it summarizes major issues that managers must understand. It draws the distinction between excellence and mediocrity.

Key Concepts: Benchmarking; Return on Equity; Strategic Quality Planning.

Related Articles: Balanced Scorecard; Lean Manufacturing Implementation; Total Productive Maintenance; Total Quality Management.

REFERENCES

Bemowski, K. (1996). "Baldrige Award Celebrates Its 10th Birthday With a New Look." *Quality Progress*, 29, 49–54.
DeCarlo, N.J. and W.K. Sterret (1990). "History of the Malcolm Baldrige National Quality Award." *Quality Progress*, 23, 21–27.
Evans, J.R. (1996). "Something old, something new: a process view of the Baldrige criteria." *International Journal of Quality Science*, 1, 62–68.
Evans, J.R. and M.W. Ford (1997). "Value-Driven Quality." *Quality Management Journal*, 4 (4), 19–31.
Evans, J.R. and W.M. Lindsay (1998). *The Management and Control of Quality.* (4th Edition), South-Western, Cincinnati.
Neves, J.S. and B. Nakhai (1994). "The Evolution of the Baldrige Award." *Quality Progress*, 27, 65–70.
U.S. Department of Commerce (1997). *Malcolm Baldrige National Quality Award Criteria For Performance Excellence.*
U.S. Department of Commerce (undated). *Malcolm Baldrige National Quality Award, Profiles of Winners.*

INDEX OF COMPANIES

GENERAL INDEX